STUDENT SOLUTIONS MANUAL

Thomas Engel • Philip Reid
University of Washington

PHYSICAL CHEMISTRY
SECOND EDITION

Thomas Engel • Philip Reid

Prentice Hall
New York   Boston   San Francisco.
London   Toronto   Sydney   Tokyo   Singapore   Madrid
Mexico City   Munich   Paris   Cape Town   Hong Kong   Montreal

Acquisitions Editor: Dan Kaveney
Editor in Chief, Chemistry and Geosciences: Nicole Folchetti
Marketing Manager: Erin Gardner
Assistant Editors: Jessica Neumann and Carol DuPont
Managing Editor, Chemistry and Geosciences: Gina M. Cheselka
Project Manager: Traci Douglas
Operations Specialist: Maura Zaldivar
Supplement Cover Manager: Paul Gourhan
Supplement Cover Designer: Tina Krivoshein
Cover Credit: Corbis/Superstock

© 2010 Pearson Education, Inc.

Pearson Prentice Hall

Pearson Education, Inc.

Upper Saddle River, NJ 07458

Pearson Prentice Hall™ is a trademark of Pearson Education, Inc.

The author and publisher of this book have used their best efforts in preparing this book. These efforts include the development, research, and testing of the theories and programs to determine their effectiveness. The author and publisher make no warranty of any kind, expressed or implied, with regard to these programs or the documentation contained in this book. The author and publisher shall not be liable in any event for incidental or consequential damages in connection with, or arising out of, the furnishing, performance, or use of these programs.

Printed in the United States of America

10   9   8   7   6   5   4   3   2   1

ISBN-13:   978-0-321-61626-5
ISBN-10:   0-321-61626-X

Prentice Hall
is an imprint of

www.pearsonhighered.com

Contents

*Chapters 14, 25, 26, and 28 do not contain solutions in the Solutions Manual

Chapter 1: Fundamental Concepts of Thermodynamics

P1.1) Approximately how many oxygen molecules arrive each second at the mitochondrion of an active person? The following data are available: Oxygen consumption is about 40. mL of O_2 per minute per kilogram of body weight, measured at $T = 300.$ K and $P = 1.00$ atm. In an adult with a body weight of 64 kilograms there are about 1.0×10^{12} cells. Each cell contains about 800. mitochondria.

We first calculate the number of moles of O_2 consumed per unit time and convert this quantity into molecules per unit time using the Avogadro number.

$$n = \frac{PV}{RT} = \frac{1 \, atm \times 4.0 \times 10^{-2} \, L}{8.206 \times 10^{-2} L \, atm \, mol^{-1} \, K^{-1} \times 300. \, K}$$

$$= 1.62 \times 10^{-3} \, mol \text{ in one minute per kg}$$

$$n' = n \times \frac{1 \, min}{60 \, s} \times 64 \, kg = 1.62 \times 10^{-3} \, mol \, s^{-1}$$

$$N = N_A \times n' = 1.73 \times 10^{-3} \, mol \, s^{-1} \times 6.022 \times 10^{23} \, mol^{-1} = 1.3 \times 10^{6} \, \text{molecules} \, s^{-1}$$

P1.2) A compressed cylinder of gas contains 2.25×10^3 g of N_2 gas at a pressure of 4.25×10^7 Pa and a temperature of 19.4°C. What volume of gas has been released into the atmosphere if the final pressure in the cylinder is 1.80×10^5 Pa? Assume ideal behavior and that the gas temperature is unchanged.

Let n_i and n_f be the initial and final number of mols of N_2 in the cylinder.

$$\frac{n_i RT}{P_i} = \frac{n_f RT}{P_f}$$

$$n_f = n_i \frac{P_f}{P_i} = \frac{2.25 \times 10^3 g}{28.01 g \, mol^{-1}} \times \frac{1.80 \times 10^5 Pa}{4.25 \times 10^7 Pa} = 0.340 \, mol$$

$$n_i = \frac{2.25 \times 10^3 g}{28.01 g \, mol^{-1}} = 80.3 \, mol$$

The volume of gas released into the atmosphere is given by

$$V = \frac{\left(n_f - n_i\right) RT}{P} = \frac{(80.3 - 0.340) mol \times 8.206 \times 10^{-2} \, L \, atm \, mol^{-1} K^{-1} \times (273.15 + 19.4) \, K}{1 \, atm}$$

$$= 1.92 \times 10^3 \, L$$

P1.5) A gas sample is known to be a mixture of ethane and butane. A bulb having a 215.0-cm^3 capacity is filled with the gas to a pressure of 108.5×10^3 Pa at 19.2°C. If the weight of the gas in the bulb is 0.3554 g, what is the mole percent of butane in the mixture?

n_1 = moles of ethane n_2 = moles of butane

$$n_1 + n_2 = \frac{PV}{RT} = \frac{108.5 \times 10^3 \ Pa \times 0.2150 \times 10^{-3} \ m^3}{8.314 \ J \ mol^{-1} \ K^{-1} \times (273.15 + 19.2) \ K} = 9.60 \times 10^{-3} \ mol$$

The total mass is

$$n_1 M_1 + n_2 M_2 = 0.3554 \ g$$

Dividing this equation by $n_1 + n_2$

$$\frac{n_1 M_1}{n_1 + n_2} + \frac{n_2 M_2}{n_1 + n_2} = \frac{0.3554 \ g}{9.60 \times 10^{-3} mol} = 37.0 \ g \ mol^{-1}$$

$$x_1 M_1 + x_2 M_2 = (1 - x_2) M_1 + x_2 M_2 = 37.0 \ g \ mol^{-1}$$

$$x_2 = \frac{37.0 \ g \ mol^{-1} - M_1}{M_2 - M_1} = \frac{37.0 \ g \ mol^{-1} - 30.069 \ g \ mol^{-1}}{58.123 \ g \ mol^{-1} - 30.069 \ g \ mol^{-1}} = 0.248$$

mole % = 24.8%

P1.6) One liter of fully oxygenated blood can carry 0.20 liters of O_2 measured at T = 273 K and P = 1.00 atm. Calculate the number of moles of O_2 carried per liter of blood. Hemoglobin, the oxygen transport protein in blood has four oxygen binding sites. How many hemoglobin molecules are required to transport the O_2 in 1.0 L of fully oxygenated blood?

$$n_{O_2} = \frac{PV}{RT} = \frac{1 \ atm \times 0.20 \ L}{8.206 \times 10^{-2} \ L \ atm \ mol^{-1} \ K^{-1} \times 273 \ K}$$

$$= 8.93 \times 10^{-3} mol$$

$$N_{hemoglobin} = \frac{n_{O_2} \times N_A}{4} = \frac{8.93 \times 10^{-3} mol \times 6.022 \times 10^{23}}{4}$$

$$= 1.34 \times 10^{21} \ molecules$$

P1.12) A rigid vessel of volume 0.455 m^3 containing H_2 at 26.45°C and a pressure of 755×10^3 Pa is connected to a second rigid vessel of volume 0.875 m^3 containing Ar at 38.9°C at a pressure of 255×10^3 Pa. A valve separating the two vessels is opened and both are cooled to a temperature of 14.5°C. What is the final pressure in the vessels?

$$n_{H_2} = \frac{PV}{RT} = \frac{755 \times 10^3 \ Pa \times 0.455 \ m^3}{8.314 \ J \ mol^{-1} K^{-1} \times (273.15 + 26.4) K} = 138 \ mol$$

$$n_{Ar} = \frac{PV}{RT} = \frac{255 \times 10^3 \text{ Pa} \times 0.875 \text{ m}^3}{8.314 \text{ J mol}^{-1}\text{K}^{-1} \times (273.15 + 38.9)\text{K}} = 86.0 \text{ mol}$$

$$P = \frac{nRT}{V} = \frac{(138 + 86.0)\text{mol} \times 8.314 \text{ J mol}^{-1}\text{K}^{-1} \times (273.15 + 14.5)\text{K}}{(0.455 + 0.875) \text{ m}^3} = 4.03 \times 10^5 \text{ Pa}$$

P1.15) Devise a temperature scale, abbreviated G, for which the magnitude of the ideal gas constant is $3.14 \text{ J G}^{-1} \text{ mol}^{-1}$.

Let T and T' represent the Kelvin and G scales, and R and R' represent the gas constant in each of these scales. Then

$$PV = nRT = nR'T'$$

$$T' = \frac{R}{R'}T = \frac{8.314}{3.14}T = 2.65T$$

The temperature on the G scale is the value in K multiplied by 2.65.

P1.16) Aerobic cells metabolize glucose in the respiratory system. This reaction proceeds according to the overall reaction

$$6O_2(g) + C_6H_{12}O_6(s) \rightarrow 6CO_2(g) + 6H_2O(l)$$

Calculate the volume of oxygen required at STP to metabolize 0.010 kg of glucose ($C_6H_{12}O_6$). STP refers to standard temperature and pressure, that is, T = 273 K and P = 1.00 atm. Assume oxygen behaves ideally at STP.

From the stoichiometric equation, we see that 6 mols of O^2 are required for each mol of glucose. Therefore

$$V_{O_2} = \frac{n_{O_2}RT}{P} = \frac{\dfrac{10. \text{ g}}{180.18 \text{ g mol}^{-1}} \times 8.206 \times 10^{-2} \text{L atm mol}^{-1} \text{ K}^{-1} \times 273 \text{ K}}{1 \text{ atm}}$$

$$= 7.46 \text{ L}$$

P1.18) A mixture of 2.50×10^{-3} g of O_2, 3.51×10^{-3} mol of N_2, and 4.67×10^{20} molecules of CO are placed into a vessel of volume 4.65 L at 15.4°C.

a. Calculate the total pressure in the vessel.

b. Calculate the mole fractions and partial pressures of each gas.

a) $n_{O_2} = \dfrac{2.50 \times 10^{-3} \text{ g}}{32.0 \text{ g mol}^{-1}} = 7.81 \times 10^{-5} \text{ mol}$;

$n_{total} = n_{O_2} + n_{N_2} + n_{CO} = 7.81 \times 10^{-5} \text{ mol} + 3.51 \times 10^{-3} \text{ mol} + 7.75 \times 10^{-4} \text{ mol} = 4.36 \times 10^{-3} \text{ mol}$

$P_{total} = \dfrac{nRT}{V} = \dfrac{4.36 \times 10^{-3} \text{ mol} \times 8.314 \times 10^{-2} \text{ L bar mol}^{-1}\text{K}^{-1} \times (273.15 + 15.4) \text{ K}}{4.65 \text{ L}} = 2.25 \times 10^{-2} \text{ bar}$

$n_{CO} = \dfrac{4.67 \times 10^{20} \text{ molecules}}{6.022 \times 10^{23} \text{ molecules mol}^{-1}} = 7.75 \times 10^{-4} \text{ mol}$

b) $x_{O_2} = \dfrac{7.81 \times 10^{-5} \text{ mol}}{4.36 \times 10^{-3} \text{ mol}} = 0.0179$; $x_{N_2} = \dfrac{3.51 \times 10^{-3} \text{ mol}}{4.36 \times 10^{-3} \text{ mol}} = 0.803$; $x_{CO} = \dfrac{7.75 \times 10^{-4} \text{ mol}}{4.36 \times 10^{-3} \text{ mol}} = 0.178$

$P_{O_2} = x_{O_2} P_{total} = 0.0179 \times 2.25 \times 10^{-2} \text{ bar} = 4.03 \times 10^{-4} \text{ bar}$

$P_{N_2} = x_{N_2} P_{total} = 0.803 \times 2.25 \times 10^{-2} \text{ bar} = 1.81 \times 10^{-2} \text{ bar}$

$P_{CO} = x_{CO} P_{total} = 0.177 \times 2.25 \times 10^{-2} \text{ bar} = 4.00 \times 10^{-3} \text{ bar}$

P1.19) Calculate the pressure exerted by benzene for a molar volume of 1.76 L at 685 K using the Redlich-Kwong equation of state:

$$P = \frac{RT}{V_m - b} - \frac{a}{\sqrt{T}} \frac{1}{V_m(V_m + b)} = \frac{nRT}{V - nb} - \frac{n^2 a}{\sqrt{T}} \frac{1}{V(V + nb)}$$

The Redlich-Kwong parameters a and b for benzene are 452.0 bar dm^6 mol^{-2} K$^{1/2}$ and 0.08271 dm^3 mol^{-1}, respectively. Is the attractive or repulsive portion of the potential dominant under these conditions?

$P = \dfrac{RT}{V_m - b} - \dfrac{a}{\sqrt{T}} \dfrac{1}{V_m(V_m + b)}$

$= \dfrac{8.314 \times 10^{-2} \text{ bar dm}^3\text{mol}^{-1}\text{K}^{-1} \times 685 \text{ K}}{1.76 \text{ dm}^3\text{mol}^{-1} - 0.08271 \text{ dm}^3\text{mol}^{-1}}$

$- \dfrac{452.0 \text{ bar dm}^6\text{mol}^{-2}\text{K}^{\frac{1}{2}}}{\sqrt{685 \text{ K}}} \times \dfrac{1}{1.76 \text{ dm}^3\text{mol}^{-1} \times (1.76 \text{ dm}^3\text{mol}^{-1} + 0.08271 \text{ dm}^3\text{mol}^{-1})}$

$P = 28.6 \text{ bar}$

$P_{ideal} = \dfrac{RT}{V} = \dfrac{8.3145 \times 10^{-2} \text{ L bar mol}^{-1}\text{K}^{-1} \times 685 \text{ K}}{1.76 \text{ L}} = 32.4 \text{ bar}$

Because $P < P_{ideal}$, the attractive part of the potential dominates.

P1.21) An initial step in the biosynthesis of glucose $C_6H_{12}O_6$ is the carboxylation of pyruvic acid $CH_3COCOOH$ to form oxaloacetic acid $HOOCCOCH_2COOH$

$$CH_3COCOOH(s) + CO_2(g) \rightarrow HOOCCOCH_2COOH(s)$$

If you knew nothing else about the intervening reactions involved in glucose biosynthesis other than no further carboxylations occur, what volume of CO_2 is required to produce 0.50 g of glucose? Assume $P = 1$ atm and $T = 310.$ K.

From the stoichiometric equation,

$$n_{CO_2} = n_{glucose} = \frac{m_{glucose}}{M_{glucose}} = \frac{0.50 \text{ g}}{180.18 \text{ g mol}^{-1}} = 2.78 \times 10^{-3} \text{ mol}$$

$$V_{CO_2} = \frac{n_{CO_2} RT}{P}$$

$$= \frac{2.78 \times 10^{-3} \text{ mol} \times 8.206 \times 10^{-2} \text{ L atm mol}^{-1} \text{ K}^{-1} \times 310. \text{ K}}{1 \text{ atm}}$$

$$= 0.071 \text{ L}$$

P1.23) Assume that air has a mean molar mass of 28.9 g mol^{-1} and that the atmosphere has a uniform temperature of 25.0°C. Calculate the barometric pressure in Pa at Boulder, for which $z = 5430$ ft. Use the information contained in Problem P1.20.

$$P = P^0 e^{-\frac{M_i g z}{RT}} = 10^5 \text{ Pa exp}\left(-\frac{28.9 \times 10^{-3} \text{ kg} \times 9.81 \text{ m s}^{-2} \times 5430. \text{ ft} \times 0.3048 \text{ m ft}^{-1}}{8.314 \text{ J mol}^{-1} \text{ K}^{-1} \times 300 \text{ K}} \right)$$

$$= 8.29 \times 10^4 \text{ Pa}$$

P1.30) Carbon monoxide competes with oxygen for binding sites on the transport protein hemoglobin. CO can be poisonous if inhaled in large quantities. A safe level of CO in air is 50. parts per million (ppm). When the CO level increases to 800. ppm, dizziness, nausea, and unconsciousness occur, followed by death. Assuming the partial pressure of oxygen in air at sea level is 0.20 atm, what proportion of CO to O_2 is fatal?

$$x_{O_2} = \frac{0.20 \text{ atm}}{1 \text{ atm}} = 2.0 \times 10^5 \text{ ppm}$$

$$\frac{x_{CO}}{x_{O_2}} = \frac{800. \text{ ppm}}{2.0 \times 10^5 \text{ ppm}} = 4.0 \times 10^{-3}$$

P1.36) A glass bulb of volume 0.225 L contains 1.251 g of gas at 759.0 Torr and 121.0°C. What is the molar mass of the gas?

$$n = \frac{m}{M} = \frac{PV}{RT}; M = m\frac{RT}{PV}$$

$$M = 1.251 \text{ g} \times \frac{8.206\times10^{-2} \text{ L atm mol}^{-1}\text{K}^{-1}\times(273.15 + 121.0)\text{K}}{\frac{759}{760} \text{ atm}\times 0.225 \text{ L}} = 180. \text{ amu}$$

Chapter 2: Heat, Work, Internal Energy, Enthalpy, and the First Law of Thermodynamics

P2.1) 4.25 moles of an ideal gas with $C_{V,m}=3/2R$ initially at a temperature $T_i = 325$ K and $P_i = 1.00$ bar is enclosed in an adiabatic piston and cylinder assembly. The gas is compressed by placing a 575 kg mass on the piston of diameter 20.0 cm. Calculate the work done in this process and the distance that the piston travels. Assume that the mass of the piston is negligible.

We first calculate the external pressure and the initial volume.

$$P_{external} = \frac{F}{A} = 10^5 \text{ Pa} + \frac{mg}{\pi r^2} = 1.00 \times 10^5 \text{ Pa} + \frac{575 \text{ kg} \times 9.81 \text{ ms}^{-2}}{\pi \times (0.100 \text{ m})^2} = 2.80 \times 10^5 \text{ Pa}$$

$$V_i = \frac{nRT}{P_i} = \frac{4.25 \text{ mol} \times 8.314 \text{ J mol}^{-1}\text{K}^{-1} \times 325 \text{ K}}{10^5 \text{ Pa}} = 0.115 \text{ m}^3 = 115 \text{ L}$$

Following Example Problem 2.6,

$$T_f = T_i \left(\frac{C_{V,m} + \dfrac{RP_{external}}{P_i}}{C_{V,m} + \dfrac{RP_{external}}{P_f}} \right) = 298 \text{ K} \times \left(\frac{12.47 \text{ J mol}^{-1}\text{K}^{-1} + \dfrac{8.314 \text{ J mol}^{-1}\text{K}^{-1} \times 2.80 \times 10^5 \text{ Pa}}{1.00 \times 10^5 \text{ Pa}}}{12.47 \text{ J mol}^{-1}\text{K}^{-1} + \dfrac{8.314 \text{ J mol}^{-1}\text{K}^{-1} \times 2.80 \times 10^5 \text{ Pa}}{2.80 \times 10^5 \text{ Pa}}} \right)$$

$$= 558 \text{ K}$$

$$V_f = \frac{nRT}{P_f} = \frac{4.25 \text{ mol} \times 8.314 \text{ J mol}^{-1}\text{K}^{-1} \times 558 \text{ K}}{2.80 \times 10^5 \text{ Pa}} = 7.06 \times 10^{-2} \text{ m}^3$$

$$w = -P_{external}(V_f - V_i) = -2.80 \times 10^5 \text{ Pa} \times (7.05 \times 10^{-2} \text{ m}^3 - 11.5 \times 10^{-2} \text{ m}^3) = 12.4 \times 10^3 \text{ J}$$

$$h = -\frac{V_f - V_i}{\pi r^2} = \frac{4.43 \times 10^{-2} \text{ m}^3}{3.14 \times 10^{-2} \text{ m}^2} = 1.41 \text{ m}$$

P2.3) 1.65 moles of an ideal gas, for which $C_{V,m} = 3/2R$, is subjected to two successive changes in state: (1) From 39.0°C and 100. × 10^3 Pa, the gas is expanded isothermally against a constant pressure of 16.5 × 10^3 Pa to twice the initial volume. (2) At the end of the previous process, the gas is cooled at constant volume from 39.0°C to –25.0°C. Calculate q, w, ΔU, and ΔH for each of the stages. Also calculate q, w, ΔU, and ΔH for the complete process.

a) $V_i = \dfrac{nRT}{P_i} = \dfrac{1.65 \text{ mol} \times 8.314 \text{ J mol}^{-1}\text{K}^{-1} \times 312 \text{ K}}{100. \times 10^3 \text{ Pa}} = 4.28 \times 10^{-2} \text{ m}^3$

$V_f = 2V_i = 8.56 \times 10^{-2} \text{ m}^3$

$w = -P_{ext}\left(V_f - V_i\right) = -16.5 \times 10^3 \text{ Pa} \times \left(8.56 \times 10^{-2} \text{ m}^3 - 4.28 \times 10^{-2} \text{ m}^3\right) = -707 \text{ J}$

ΔU and $\Delta H = 0$ because $\Delta T = 0$

$q = -w = 707 \text{ J}$

b) $\Delta U = nC_{V,m}\left(T_f - T_i\right) = 1.65 \text{ mol} \times 1.5 \times 8.314 \text{ J mol}^{-1}\text{K}^{-1} \times \left(248 \text{ K} - 312 \text{ K}\right)$

$= -1.32 \times 10^3 \text{ J}$

$w = 0$ because $\Delta V = 0$

$q = \Delta U = -1.32 \times 10^3 \text{ J}$

$\Delta H = nC_{P,m}\left(T_f - T_i\right) = n\left(C_{V,m} + R\right)\left(T_f - T_i\right)$

$\quad = 1.65 \text{ mol} \times 2.5 \times 8.314 \text{ J mol}^{-1}\text{K}^{-1} \times \left(248 \text{ K} - 312 \text{ K}\right)$

$\quad = -2.19 \times 10^3 \text{ J}$

$\Delta U_{total} = 0 - 1.32 \times 10^3 \text{ J} = -1.32 \times 10^3 \text{ J}$

$w_{total} = 0 - 707 \text{ J} = -707 \text{ J}$

$q_{total} = 707 \text{ J} - 1.32 \times 10^3 \text{ J} = -610. \text{ J}$

$\Delta H_{total} = 0 - 2.19 \times 10^3 \text{J} = -2.19 \times 10^3 \text{ J}$

P2.5) Count Rumford observed that using cannon-boring machinery, a single horse could heat 11.6 kg of ice water ($T = 273$ K) to $T = 355$ K in 2.5 hours. Assuming the same rate of work, how high could a horse raise a 150. kg weight in one minute? Assume the heat capacity of water is 4.18 J K^{-1} g^{-1}.

$Rate = \dfrac{C_p m_{water} \Delta T}{time_1} = \dfrac{4.18 \text{ J K}^{-1} \text{ g}^{-1} \times \left(355 - 273\right)\text{K}}{2.5 \text{ hr} \times 3600 \text{ s hr}^{-1}} = 442 \text{ J s}^{-1}$

$h = \dfrac{Rate \times time_2}{m_{weight} g} = \dfrac{442 \text{ J s}^{-1} \times 60 \text{ s}}{150. \text{ kg} \times 9.81 \text{ m s}^{-2}} = 18 \text{ m}$

P2.6) 2.25 moles of an ideal gas at 35.6°C expands isothermally from an initial volume of 26.0 dm^3 to a final volume of 70.0 dm^3. Calculate w for this process (a) for expansion against a constant external pressure of 1.00 $\times 10^5$ Pa and (b) for a reversible expansion.

a) $w = -P_{external}\Delta V = -1.00 \times 10^5 \text{ Pa} \times \left(70.0 - 26.0\right) \times 10^{-3} \text{ m}^3 = -4.40 \times 10^3 \text{ J}$

b)
$$w_{reversible} = -nRT \ln\frac{V_f}{V_i} = -2.25 \text{ mol}\times 8.314 \text{ J mol}^{-1}\text{K}^{-1}\times(273.15+35.6) \text{ K}\times\ln\frac{70.0 \text{ dm}^3}{26.0 \text{ dm}^3}$$

$$= -5.72\times10^3 \text{ J}$$

P2.7) Calculate *q*, *w*, ΔU, and ΔH if 1.65 mol of an ideal gas with $C_{V,m}$ = 3/2R undergoes a reversible adiabatic expansion from an initial volume V_i = 7.75 m³ to a final volume V_f = 20.5 m³. The initial temperature is 300. K.

$q = 0$ because the process is adiabatic.

$$\frac{T_f}{T_i} = \left(\frac{V_f}{V_i}\right)^{1-\gamma}$$

$$T_f = \left(\frac{20.5 \text{ L}}{7.75 \text{ L}}\right)^{1-\frac{5}{3}}\times T_i = 157 \text{ K}$$

$$\Delta U = w = nC_{V,n}\Delta T = 1.65 \text{ mol}\times\frac{3\times8.314 \text{ J mol}^{-1}\text{K}^{-1}}{2}\times(157 \text{ K}-300. \text{ K}) = -2.95\times10^3 \text{ J}$$

$$\Delta H = \Delta U + nR\Delta T = -2.95\times10^3 \text{ J} + 1.65 \text{ mol}\times8.314 \text{ J mol}^{-1}\text{K}^{-1}\times(157 \text{ K}-300. \text{ K})$$

$$\Delta H = -4.91\times10^3 \text{ J}$$

P2.10) A muscle fiber contracts by 2.0 cm and in doing so lifts a weight. Calculate the work performed by the fiber and the weight lifted. Assume the muscle fiber obeys Hooke's law *F* = -*k x* with a force constant, *k*, of 800. N m⁻¹.

$$w = \frac{1}{2}kx^2 = \frac{1}{2}\times800. \text{ N m}^{-1}\times(2.0\times10^{-2} \text{ m})^2 = 0.16 \text{ J}$$

P2.12) In the adiabatic expansion of 2.25 mol of an ideal gas from an initial temperature of 32.0°C, the work done on the surroundings is 1450. J. If $C_{V,m}$ = 3/2R, calculate *q*, *w*, ΔU, and ΔH.

$q = 0$ because the process is adiabatic

$\Delta U = w = -1450.\ J$

$\Delta U = nC_{V,m}\left(T_f - T_i\right)$

$T_f = \dfrac{\Delta U + nC_{V,m}T_i}{nC_{V,m}}$

$= \dfrac{-1450.\ J + 2.25 \times 1.5 \times 8.314\ J\ mol^{-1}K^{-1} \times 305\ K}{2.25 \times 1.5 \times 8.314\ J\ mol^{-1}K^{-1}}$

$= 253\ K$

$\Delta H = nC_{P,m}\left(T_f - T_i\right) = n\left(C_{V,m} + R\right)\left(T_f - T_i\right)$

$= 2.25 \times 2.5 \times 8.314\ J\ mol^{-1}K^{-1}\left(253\ K - 305\ K\right)$

$= -2.42 \times 10^3\ J$

P2.14) 2.75 moles of an ideal gas is expanded from 375 K and an initial pressure of 4.75 bar to a final pressure of 1.00 bar, and $C_{P,m} = 5/2R$. Calculate w for the following two cases:

a. The expansion is isothermal and reversible.

b. The expansion is adiabatic and reversible.

Without resorting to equations, explain why the result to part (b) is greater than or less than the result to part (a).

a)

$w = -nRT\ln\dfrac{V_f}{V_i} = -nRT\ln\dfrac{P_i}{P_f}$

$= -2.75\ mol \times 8.314\ J\ mol^{-1}K^{-1} \times 450\ K \times \ln\dfrac{4.75\ bar}{1.00\ bar} = -13.3 \times 10^3\ J$

b) Because $q = 0$, $w = \Delta U$. In order to calculate ΔU, we first calculate T_f.

$\dfrac{T_f}{T_i} = \left(\dfrac{V_f}{V_i}\right)^{1-\gamma} = \left(\dfrac{T_f}{T_i}\right)^{1-\gamma}\left(\dfrac{P_i}{P_f}\right)^{1-\gamma} ; \left(\dfrac{T_f}{T_i}\right)^{\gamma} = \left(\dfrac{P_i}{P_f}\right)^{1-\gamma} ; \dfrac{T_f}{T_i} = \left(\dfrac{P_i}{P_f}\right)^{\frac{1-\gamma}{\gamma}}$

$T_f = T_i \times \left(\dfrac{4.75\ bar}{1.00\ bar}\right)^{\frac{1-\frac{5}{3}}{\frac{5}{3}}} = 201\ K$

$w = \Delta U = nC_{V,m}\Delta T = 2.75\ mol \times \dfrac{3 \times 8.314\ J\ mol^{-1}K^{-1}}{2} \times \left(201\ K - 375\ K\right) = -5.96 \times 10^3\ J$

Less work is done on the surroundings in part b) because in the adiabatic expansion, the temperature falls and therefore the final volume is less than that in part a).

P2.16) One mole of an ideal gas with $C_{V,m} = 3/2R$ initially at 325 K and 1.50×10^5 Pa undergoes a reversible adiabatic compression. At the end of the process, the pressure is 2.50×10^6 Pa. Calculate the final temperature of the gas. Calculate q, w, ΔU, and ΔH for this process.

$q = 0$ because the process is adiabatic.

$$T_f = T_i \left(\frac{P_i}{P_f} \right)^{\frac{1-C_{P,m}/C_{V,m}}{C_{P,m}/C_{V,m}}} = 325 \text{ K} \times \left(\frac{1.50 \times 10^5 \text{ Pa}}{2.50 \times 10^5 \text{ Pa}} \right)^{\frac{1-5/3}{5/3}} = 1.00 \times 10^3 \text{ K}$$

$$w = \Delta U = nC_{V,m}\Delta T = 1 \text{ mol} \times \frac{3 \times 8.314 \text{ J mol}^{-1}\text{K}^{-1}}{2} \times (1000 \text{ K} - 325 \text{ K}) = 8.44 \times 10^3 \text{ J}$$

$$\Delta H = \Delta U + \Delta(PV) = \Delta U + R\Delta T = 5.62 \times 10^3 \text{ J} + 8.314 \text{ J mol}^{-1}\text{K}^{-1} \times (1000 \text{ K} - 325 \text{ K})$$

$$\Delta H = 14.1 \times 10^3 \text{ J}$$

P2.17) A vessel containing 2.25 mol of an ideal gas with $P_i = 1.00$ bar and $C_{P,m} = 5/2R$ is in thermal contact with a water bath. Treat the vessel, gas, and water bath as being in thermal equilibrium, initially at 312 K, and as separated by adiabatic walls from the rest of the universe. The vessel, gas, and water bath have an average heat capacity of $C_P = 6250.$ J K^{-1}. The gas is compressed reversibly to $P_f = 10.5$ bar. What is the temperature of the system after thermal equilibrium has been established?

Assume initially that the temperature rise is so small that the reversible compression can be thought of as an isothermal reversible process. If the answer substantiates this assumption, it is valid.

$$w = -nRT_1 \ln \frac{V_f}{V_i} = -nRT_1 \ln \frac{P_i}{P_f}$$

$$= -2.25 \text{ mol} \times 8.314 \text{ J mol}^{-1}\text{K}^{-1} \times 312 \text{ K} \times \ln \frac{1.00 \text{ bar}}{10.5 \text{ bar}} = 13.7 \times 10^3 \text{ J}$$

$$\Delta U_{combined\ system} = C_P \Delta T$$

We use C_P in the above equation because the heat capacity is dominated by the water bath, and for a liquid $C_P \approx C_V$.

$$\Delta T = \frac{\Delta U_{combined\ system}}{C_P} = \frac{13.7 \times 10^3 \text{ J}}{6250 \text{ J K}^{-1}} = 0.976 \text{ K}$$

$$T_f \approx 313 \text{ K}$$

The result justifies the assumption.

P2.21) The heat capacity of solid lead oxide is given by

$$C_{P,m} = 44.35 + 1.47 \times 10^{-3} \frac{T}{K} \text{ in units of J K}^{-1}\text{mol}^{-1}$$

Calculate the change in enthalpy of 3.25 mol of PbO(s) if it is cooled from 750. to 300. K at constant pressure.

$$\Delta H = n \int_{T_i}^{T_f} C_{p,m} dT$$

$$= 3.25 \times \int_{750}^{300} \left(44.35 + 1.47 \times 10^{-3} \frac{T}{K} \right) d\left(\frac{T}{K} \right)$$

$$= 3.25 \times \left(\begin{array}{c} 44.35 \times (300 \text{ K} - 750 \text{ K}) \\ + \left[\frac{1.47 \times 10^{-3}}{2} \left(\frac{T}{K} \right)^2 \right]_{750 \text{ K}}^{300 \text{ K}} \end{array} \right)$$

$$= -64.9 \times 10^3 \text{ J} - 1.13 \times 10^3 \text{ J}$$

$$= -66.0 \times 10^3 \text{ J}$$

P2.25) A major league pitcher throws a baseball with a speed of 150. kilometers per hour. If the baseball weighs 220. grams and its heat capacity is 2.0 J g^{-1} K^{-1}, calculate the temperature rise of the ball when it is stopped by the catcher's mitt. Assume no heat is transferred to the catcher's mitt and that the catcher's arm does not recoil when he/she catches the ball.

$$v = 150. \times 10^3 \text{ m hr}^{-1} \times \frac{\text{hr}}{3600 \text{ s}} = 41.7 \text{m s}^{-1}$$

$$q_P = C_P \Delta T = \tfrac{1}{2} m v^2$$

$$\Delta T = \frac{\tfrac{1}{2} m v^2}{C_P m} = \frac{0.5 \times 0.220 \text{ kg} \times \left(41.7 \text{ m s}^{-1}\right)^2}{2000 \text{ J g}^{-1} \text{ K}^{-1}} = 0.43 \text{ K}$$

P2.26) A 1.65 mol sample of an ideal gas for which $C_{V,m}$=3/2R undergoes the following two-step process: (1) From an initial state of the gas described by $T = 14.5°C$ and $P = 2.00 \times 10^4$ Pa, the gas undergoes an isothermal expansion against a constant external pressure of 1.00×10^4 Pa until the volume has doubled. (2) Subsequently, the gas is cooled at constant volume. The temperature falls to –35.6°C. Calculate q, w, ΔU, and ΔH for each step and for the overall process.

a) For the first step, $\Delta U = \Delta H = 0$ because the process is isothermal.

$$V_i = \frac{nRT_i}{P_i}$$

$$= \frac{1.65 \text{ mol} \times 8.314 \text{ J mol}^{-1}\text{K}^{-1} \times (273.15 + 14.5)\text{K}}{2.00 \times 10^4 \text{Pa}} = 0.197 \text{m}^3$$

$$w = -q = -P_{external}\Delta V = -1.00 \times 10^4 \text{ Pa} \times 2 \times 0.197 \text{ m}^3$$
$$= -1.97 \times 10^3 \text{ J}$$

b) For the second step, $w = 0$ because $\Delta V = 0$.

$$q = \Delta U = nC_V \Delta T = 1.65 \text{ mol} \times \frac{3 \times 8.314 \text{ J mol}^{-1}\text{K}^{-1}}{2} \times (-35.6°C - 14.5°C) = -1.03 \times 10^3 \text{ J}$$

$$\Delta H = \Delta U + \Delta(PV) = \Delta U + nR\Delta T = -1.03 \times 10^3 \text{ J}$$
$$+ 1.65 \text{ mol} \times 8.314 \text{ J mol}^{-1}\text{K}^{-1} \times (-35.6°C - 14.5°C)$$

$$\Delta H = -1.72 \times 10^3 \text{ J}$$

For the overall process, $w = -1.97 \times 10^3$ J, $q = -1.03 \times 10^3$ J $+ 1.97 \times 10^3$ J

$= 941$ J, $\Delta U = -1.03 \times 10^3$ J, and $\Delta H = -1.72 \times 10^3$ J.

P2.28) 1.75 mole of an ideal gas with $C_{V,m} = 3/2R$ is expanded adiabatically against a constant external pressure of 1.00 bar. The initial temperature and pressure are $T_i = 290.$ K and $P_i = 19.5$ bar. The final pressure is $P_f = 1.00$ bar. Calculate q, w, ΔU, and ΔH for the process.

$$\Delta U = nC_{V,m}\left(T_f - T_i\right) = -P_{external}\left(V_f - V_i\right) = w$$

$q = 0$ because the process is adiabatic.

$$nC_{V,m}\left(T_f - T_i\right) = -nRP_{external}\left(\frac{T_f}{P_f} - \frac{T_i}{P_i}\right)$$

$$T_f\left(nC_{V,m} + \frac{nRP_{external}}{P_f}\right) = T_i\left(nC_{V,m} + \frac{nRP_{external}}{P_i}\right)$$

$$T_f = T_i\left(\frac{C_{V,m} + \dfrac{RP_{external}}{P_i}}{C_{V,m} + \dfrac{RP_{external}}{P_f}}\right) = 290.\ K \times \left(\frac{1.5 \times 8.314\ \mathrm{J\,mol^{-1}K^{-1}} + \dfrac{8.314\ \mathrm{J\,mol^{-1}K^{-1}} \times 1.00\ bar}{19.5\,bar}}{1.5 \times 8.314\ \mathrm{J\,mol^{-1}K^{-1}} + \dfrac{8.314\ \mathrm{J\,mol^{-1}K^{-1}} \times 1.00\ bar}{1.00\ bar}}\right)$$

$$T_f = 180.\ K$$

$$\Delta U = w = nC_{V,n}\Delta T = 1.75\ \mathrm{mol} \times \frac{3 \times 8.314\ \mathrm{J\,mol^{-1}K^{-1}}}{2} \times (180\ K - 290\ K) = -2.40 \times 10^3\ J$$

$$\Delta H = \Delta U + nR\Delta T = -2.40 \times 10^3\ J + 1.75\,\mathrm{mol} \times 8.314\ \mathrm{J\,mol^{-1}K^{-1}} \times (180\ K - 290\ K)$$

$$\Delta H = -4.00 \times 10^3\ J$$

P2.30) For 2.25 mol of an ideal gas, $P_{external} = P = 200. \times 10^3$ Pa. The temperature is changed from 122°C to 28.5°C, and $C_{V,m} = 3/2R$. Calculate q, w, ΔU, and ΔH.

$$\Delta U = nC_{V,m}\Delta T = 2.25\ \mathrm{mol} \times \frac{3}{2} \times 8.314\ \mathrm{J\,mol^{-1}K^{-1}} \times (302\ K - 395\ K) = -2.63 \times 10^3\ J$$

$$\Delta H = nC_{P,m}\Delta T = n\left(C_{V,m} + R\right)\Delta T$$

$$= 2.25\ \mathrm{mol} \times \frac{5}{2} \times 8.314\ \mathrm{J\,mol^{-1}K^{-1}} \times (302\ K - 395\ K)$$

$$= -4.37 \times 10^3\ J$$

$$= q_P$$

$$w = \Delta U - q_P = -2.63 \times 10^3\ J + 4.37 \times 10^3\ J = 1.75 \times 10^3\ J$$

P2.37) Calculate ΔH and ΔU for the transformation of 1 mol of an ideal gas from 35.0°C and 1.00 atm to 422°C and 17.0 atm if

$$C_{P,m} = 20.9 + 0.042\frac{T}{K}\ \text{in units of J K}^{-1}\text{mol}^{-1}.$$

$$\Delta H = n \int_{T_i}^{T_f} C_{P,m} dT$$

$$= \int_{308K}^{695K} \left(20.9 + 0.042 \frac{T'}{K} \right) dT'$$

$$= 20.9 \times (695 \text{ K} - 308 \text{ K}) \text{J} + \left[0.021 T^2 \right]_{308K}^{695 \text{ K}} \text{J}$$

$$= 8.088 \times 10^3 \text{J} + 8.153 \times 10^3 \text{J}$$

$$= 1.62 \times 10^4 \text{J}$$

$$\Delta U = \Delta H - \Delta (PV) = \Delta H - nR\Delta T$$

$$= 1.62 \times 10^4 \text{J} - 8.314 \text{ J K}^{-1}\text{mol}^{-1} \times (695 - 308) \text{ K}$$

$$= 13.0 \times 10^3 \text{J}$$

P2.38) 2.50 moles of an ideal gas for which $C_{V,m} = 20.8$ J K^{-1} mol^{-1} is heated from an initial temperature of 10.5°C to a final temperature of 305°C at constant volume. Calculate q, w, ΔU, and ΔH for this process.

$w = 0$ because $\Delta V = 0$.

$$\Delta U = q = nC_V \Delta T = 2.50 \text{ mol} \times 20.8 \text{ J mol}^{-1}\text{K}^{-1} \times 294.5 \text{ K} = 15.3 \times 10^3 \text{ J}$$

$$\Delta H = \Delta U + \Delta (PV) = \Delta U + nR\Delta T = 15.3 \times 10^3 \text{J} + 2.50 \text{ mol} \times 8.314 \text{ J mol}^{-1}\text{K}^{-1} \times 275 \text{ K}$$

$$= 21.4 \times 10^3 \text{ J}$$

P2.42) DNA can be modeled as an elastic rod which can be twisted or bent. Suppose a DNA molecule of length L is bent such that it lies on the arc of a circle of radius R_c. The reversible work involved in bending DNA without twisting is $w_{bend} = \dfrac{BL}{2R_c^2}$ where B is the bending force constant. The DNA in a nucleosome particle is about 680. Å in length. Nucleosomal DNA is bent around a protein complex called the histone octamer into a circle of radius 55 Å. Calculate the reversible work involved in bending the DNA around the histone octamer if the force constant $B = 2.00 \times 10^{-28}$ J m^{-1}.

$$w_{bend} = \frac{BL}{2R_c^2} = \frac{2.00 \times 10^{-28} \text{ J m} \times 680. \times 10^{-10} \text{ m}}{2 \times (55 \times 10^{-10} \text{ m})^2} = 2.2 \times 10^{-19} \text{ J}$$

Chapter 3: The Importance of State Functions: Internal Energy and

Enthalpy

P3.2) Use the result of Problem P3.26 to show that $(\partial C_V / \partial V)_T$ for the van der Waals gas is zero. We use the relationship

$$\left(\frac{\partial C_V}{\partial V} \right)_T = T \left(\frac{\partial^2 P}{\partial T^2} \right)_V$$

$$P = \frac{RT}{V_m - b} - \frac{a}{V_m^2}$$

$$\left(\frac{\partial P}{\partial T} \right)_V = \frac{R}{V_m - b}$$

$$\left(\frac{\partial^2 P}{\partial T^2} \right)_V = \left(\frac{\partial \frac{R}{V_m - b}}{\partial T} \right)_V = 0$$

therefore $\left(\frac{\partial C_V}{\partial V} \right)_T = T \left(\frac{\partial^2 P}{\partial T^2} \right)_V = T \times 0 = 0$

P3.7) Integrate the expression $\beta = 1/V (\partial V / \partial T)_P$ assuming that β is independent of pressure. By doing so, obtain an expression for V as a function of T and β at constant P.

$$\beta = \frac{1}{V} \left(\frac{\partial V}{\partial T} \right)_P$$

$$\frac{dV}{V} = \beta dT$$

$$\int \frac{dV}{V} = \int \beta dT \text{ or } \ln \frac{V_f}{V_i} = \beta \left(T_f - T_i \right)$$

if β can be assumed constant in the temperature interval of interest.

P3.12) Calculate $w, q, \Delta H$, and ΔU for the process in which 1.75 mol of water undergoes the transition $H_2O(l, 373 \text{ K}) \rightarrow H_2O(g, 525 \text{ K})$ at 1 bar of pressure. The volume of liquid water at 373 K is $1.89 \times 10^{-5} \text{ m}^3 \text{ mol}^{-1}$ and the volume of steam at 373 and 525 K is 3.03 and $3.74 \times 10^{-2} \text{ m}^3 \text{ mol}^{-1}$

respectively. For steam, $C_{P,m}$ can be considered constant over the temperature interval of interest at 33.58 J mol^{-1} K^{-1}.

$$q = \Delta H = n\Delta H_{vaporization} + nC_{P,m}^{steam}\Delta T$$

$$= 1.75 \text{ mol} \times 40656 \text{ J mol}^{-1} + 1.75 \text{ mol} \times 33.58 \text{ J mol}^{-1}\text{K}^{-1} \times (525 \text{ K} - 373 \text{ K}) = 8.01 \times 10^{4} \text{ J}$$

$$w = -P_{external}\Delta V - -10^{5} \text{ Pa} \times \left(1.75 \times 3.03 \times 10^{-2} \text{ m}^{3} - 1.75 \times 1.89 \times 10^{-5} \text{ m}^{3}\right)$$

$$= -6.54 \times 10^{3} \text{ J}$$

$$\Delta U = w + q = -6.54 \times 10^{3} \text{ J} + 8.01 \times 10^{4} \text{ J} = 7.35 \times 10^{4} \text{ J}$$

P3.16) The Joule coefficient is defined by $(\partial T/\partial V)_{U} = (1/C_{V})[P - T(\partial P/\partial T)_{V}]$. Calculate the Joule coefficient for an ideal gas and for a van der Waals gas.

For an ideal gas

$$\left(\frac{\partial T}{\partial V}\right)_{U} = \frac{1}{C_{V,m}}\left[P - T\left(\frac{\partial}{\partial T}\frac{nRT}{V}\right)_{V}\right] = \frac{1}{C_{V,m}}\left[P - \frac{nRT}{V}\right] = 0$$

For a van der Waals gas

$$\left(\frac{\partial T}{\partial V}\right)_{U} = \frac{1}{C_{V,m}}\left[P - T\left(\frac{\partial}{\partial T}\left[\frac{RT}{V_{m} - b} - \frac{a}{V_{m}^{2}}\right]\right)_{V}\right] = \frac{1}{C_{V,m}}\left[P - \frac{RT}{(V_{m} - b)}\right] = -\frac{1}{C_{V}}\frac{a}{V_{m}^{2}}$$

P3.18) Show that the expression $(\partial U/\partial V)_{T} = T(\partial P/\partial T)_{V} - P$ can be written in the form

$$\left(\frac{\partial U}{\partial V}\right)_{T} = T^{2}\left(\partial\left[\frac{P}{T}\right]\bigg/\partial T\right)_{V} = -\left(\partial\left[\frac{P}{T}\right]\bigg/\partial\left[\frac{1}{T}\right]\right)_{V}$$

$$\left(\frac{\partial U}{\partial V}\right)_{T} = T\left(\frac{\partial P}{\partial T}\right)_{V} - P$$

$$\left(\frac{\partial[P/T]}{\partial T}\right)_{V} = P\left(\frac{\partial[1/T]}{\partial T}\right)_{V} + \frac{1}{T}\left(\frac{\partial P}{\partial T}\right)_{V}$$

$$= -\frac{P}{T^{2}} + \frac{1}{T}\left(\frac{\partial P}{\partial T}\right)_{V}$$

$$\left(\frac{\partial P}{\partial T}\right)_{V} = T\left(\left(\frac{\partial[P/T]}{\partial T}\right)_{V} + \frac{P}{T^{2}}\right)$$

$$\left(\frac{\partial U}{\partial V}\right)_{T} = T^{2}\left(\left(\frac{\partial[P/T]}{\partial T}\right)_{V} + \frac{P}{T^{2}}\right) - P$$

$$= T^{2}\left(\frac{\partial[P/T]}{\partial T}\right)_{V} + P - P = T^{2}\left(\frac{\partial[P/T]}{\partial T}\right)_{V}$$

We now change the differentiation to the variable $1/T$.

$$\left(\frac{\partial[P/T]}{\partial T}\right)_V = \left(\frac{\partial[P/T]}{\partial[1/T]}\right)_V \left(\frac{\partial[1/T]}{\partial T}\right)_V = -\frac{1}{T^2}\left(\frac{\partial[P/T]}{\partial[1/T]}\right)_V$$

$$\left(\frac{\partial U}{\partial V}\right)_T = T^2 \left(\frac{\partial[P/T]}{\partial T}\right)_V = T^2 \left(-\frac{1}{T^2}\frac{\partial[P/T]}{\partial[1/T]}\right)_V = -\left(\frac{\partial[P/T]}{\partial[1/T]}\right)_V$$

P3.19) Derive an expression for the internal pressure of a gas that obeys the Bethelot equation of state, $P = \dfrac{RT}{V_m - b} - \dfrac{a}{TV_m^2}$

The internal pressure of a gas is given by

$$\left(\frac{\partial V}{\partial T}\right)_T = T\left(\frac{\partial P}{\partial T}\right)_V - P$$

Using the Bethelot equation of state

$$\left(\frac{\partial P}{\partial T}\right)_V = \frac{R}{V_m - b} + \frac{a}{T^2 V_m^2}$$

$$\left(\frac{\partial U}{\partial V}\right)_T = \frac{RT}{V_m - b} + \frac{a}{TV_m^2}\left(\frac{RT}{V_m - b} - \frac{a}{TV_m^2}\right) = \frac{2a}{TV_m^2}$$

P3.20) Because U is a state function, $(\partial/\partial V\,(\partial U/\partial T)_V)_T = (\partial/\partial T\,(\partial U/\partial V)_T)_V$. Using this relationship, show that $(\partial C_V/\partial V)_T = 0$ for an ideal gas.

For an ideal gas, by definition, $\left(\dfrac{\partial U}{\partial V}\right)_T = 0$. Because

the order of differentiation can be changed for a state function,

$$\left(\frac{\partial}{\partial V}\left(\frac{\partial U}{\partial T}\right)_V\right)_T = \left(\frac{\partial C_V}{\partial V}\right)_T = \left(\frac{\partial}{\partial T}\left(\frac{\partial U}{\partial V}\right)_T\right)_V = 0$$

P3.22) Use $(\partial U / \partial V)_T = (\beta T - \kappa P)/\kappa$ to calculate $(\partial U / \partial V)_T$ for an ideal gas.

$$\beta = \frac{1}{V}\left(\frac{\partial V}{\partial T}\right)_P = \frac{1}{V}\frac{nR}{P}; \quad \kappa = -\frac{1}{V}\left(\frac{\partial V}{\partial P}\right)_T = \frac{nRT}{VP^2} = \frac{1}{P}$$

$$\left(\frac{\partial U}{\partial V}\right)_T = \frac{\beta T - \kappa P}{\kappa} = \frac{\dfrac{1}{V}\dfrac{nRT}{P} - 1}{\dfrac{1}{P}} - P(1-1) = 0$$

P3.24) A differential $dz = f(x,y)dx + g(x,y)dy$ is exact if the integral $\int f(x,y)dx + \int g(x,y)dy$ is independent of the path. Demonstrate that the differential $dz = 2xy\,dx + x^2\,dy$ is exact by integrating dz along the paths $(1,2) \rightarrow (7,2) \rightarrow (7,9)$ and $(1,2) \rightarrow (2,2) \rightarrow (2,6) \rightarrow (7,6) \rightarrow (7,9)$. The first number in each set of parentheses is the x coordinate, and the second number is the y coordinate.

$$\int dz = \int 2xy\,dx + \int x^2\,dy$$

Path 1

$$\int dz = 2\int_1^7 2x\,dx + 49\int_2^9 dy = 96 + 343 = 439$$

Path 2

$$\int dz = 2\int_1^2 2x\,dx + \int_2^6 4\,dy + 12\int_2^7 x\,dx + 49\int_6^9 dy$$
$$= 6 + 16 + 270 + 147 = 439$$

P3.28) Use the relation $C_{P,m} - C_{V,m} = T\left(\dfrac{\partial V_m}{\partial T}\right)_P \left(\dfrac{\partial P}{\partial T}\right)_V$, the cyclic rule, and the van der Waals equation of state to derive an equation for $C_{P,m} - C_{V,m}$ in terms of V_m, T, and the gas constants R, a, and b.

We use the cyclic rule to evaluate $\left(\dfrac{\partial V_m}{\partial T}\right)_P$.

$$\left(\frac{\partial V_m}{\partial T}\right)_P \left(\frac{\partial T}{\partial P}\right)_{V_m} \left(\frac{\partial P}{\partial V_m}\right)_T = -1$$

$$\left(\frac{\partial V_m}{\partial T}\right)_P = -\left(\frac{\partial P}{\partial T}\right)_{V_m} \left(\frac{\partial V_m}{\partial P}\right)_T$$

$$C_{P,m} - C_{V,m} = T\left(\frac{\partial V_m}{\partial T}\right)_P \left(\frac{\partial P}{\partial T}\right)_{V_m} = -T\left[\left(\frac{\partial P}{\partial T}\right)_{V_m}\right]^2 \left(\frac{\partial V_m}{\partial P}\right)_T = -T\frac{\left[\left(\frac{\partial P}{\partial T}\right)_{V_m}\right]^2}{\left(\frac{\partial P}{\partial V_m}\right)_T}$$

$$P = \frac{RT}{V_m - b} - \frac{a}{V_m^2}$$

$$\left(\frac{\partial P}{\partial T}\right)_{V_m} = \frac{R}{V_m - b}$$

$$\left(\frac{\partial P}{\partial V_m}\right)_T = \frac{-RT}{(V_m - b)^2} + \frac{2a}{V_m^3} = \frac{-RTV_m^3 + 2a(V_m - b)}{V_m^3(V_m - b)^2}$$

$$C_{P,m} - C_{V,m} = -T\frac{\left(\frac{R}{V_m - b}\right)^2}{\frac{-RT}{(V_m - b)^2} + \frac{2a}{V_m^3}} = -T\frac{R}{-T + \frac{2a(V_m - b)^2}{RV_m^3}} = \frac{R}{1 - \frac{2a(V_m - b)^2}{RTV_m^3}}$$

In the ideal gas limit, $a = 0$, and $C_{P,m} - C_{V,m} = R$.

P3.31) This problem will give you practice in using the cyclic rule. Use the ideal gas law to obtain the three functions $P = f(V,T)$, $V = g(P,T)$, and $T = h(P,V)$. Show that the cyclic rule $(\partial P/\partial V)_T (\partial V/\partial T)_P (\partial T/\partial P)_V = -1$ is obeyed.

$$P = \frac{nRT}{V}; \quad V = \frac{nRT}{P}; \quad T = \frac{PV}{nR}$$

$$\left(\frac{\partial P}{\partial V}\right)_T = -\frac{nRT}{V^2}; \quad \left(\frac{\partial V}{\partial T}\right)_P = \frac{nR}{P}; \quad \left(\frac{\partial T}{\partial P}\right)_V = \frac{V}{nR}$$

$$\left(\frac{\partial P}{\partial V}\right)_T \left(\frac{\partial V}{\partial T}\right)_P \left(\frac{\partial T}{\partial P}\right)_V = \left(-\frac{nRT}{V^2}\right)\left(\frac{nR}{P}\right)\left(\frac{V}{nR}\right) = \frac{-nRT}{PV} = -1$$

P3.32) Regard the enthalpy as a function of T and P. Use the cyclic rule to obtain the expression

$$C_P = -\left(\frac{\partial H}{\partial P}\right)_T \bigg/ \left(\frac{\partial T}{\partial P}\right)_H$$

$$\left(\frac{\partial H}{\partial P}\right)_T \left(\frac{\partial P}{\partial T}\right)_H \left(\frac{\partial T}{\partial H}\right)_P = -1$$

$$C_P = \left(\frac{\partial H}{\partial T}\right)_P = -\left(\frac{\partial H}{\partial P}\right)_T \left(\frac{\partial P}{\partial T}\right)_H = -\frac{\left(\frac{\partial H}{\partial P}\right)_T}{\left(\frac{\partial T}{\partial P}\right)_H}$$

P3.36) Prove that $C_V = -\left(\frac{\partial U}{\partial V}\right)_T \left(\frac{\partial V}{\partial T}\right)_U$

$$\left(\frac{\partial U}{\partial T}\right)_V \left(\frac{\partial T}{\partial V}\right)_U \left(\frac{\partial V}{\partial U}\right)_T = -1 \text{ from the cyclic rule}$$

$$C_V = \left(\frac{\partial U}{\partial T}\right)_V = \frac{-1}{\left(\frac{\partial T}{\partial V}\right)_U \left(\frac{\partial V}{\partial U}\right)_T} = -\left(\frac{\partial V}{\partial T}\right)_U \left(\frac{\partial U}{\partial V}\right)_T$$

Chapter 4: Thermochemistry

P4.2) At 1000. K, $\Delta H_R^\circ = -123.77 \text{ kJ mol}^{-1}$ for the reaction $N_2(g) + 3H_2(g) \rightarrow 2NH_3(g)$, with $C_{P,m} =$ 3.502R, 3.466R, and 4.217R for $N_2(g)$, $H_2(g)$, and $NH_3(g)$, respectively. Calculate ΔH_f° of $NH_3(g)$ at 300. K from this information. Assume that the heat capacities are independent of temperature.

$$\Delta H_{reaction}^\circ (300.\text{ K}) = \Delta H_{reaction}^\circ (1000.\text{ K}) + \int_{1000.\text{ K}}^{300\text{K}} \Delta C_P(T)dT$$

For this problem, the heat capacities are assumed to be independent of T.

$$\Delta H_{reaction}^\circ (300.\text{ K}) = \Delta H_{reaction}^\circ (1000.\text{ K}) + \Delta C_P \Delta T$$

$$= -123.77 \text{ kJ mol}^{-1} + \left[2C_{P,m}(NH_3, g) - C_{P,m}(N_2, g) - 3C_{P,m}(H_2, g) \right][-700.\text{ K}]$$

$$= -123.77 \text{ kJ mol}^{-1} + 8.314 \text{ J mol}^{-1}\text{ K}^{-1} \times [2 \times 4.217 - 3.502 - 3 \times 3.466][-700.\text{ K}]$$

$$= -91.96 \text{ kJ mol}^{-1}$$

$$\Delta H_f^\circ (NH_3, g) = 1/2 \Delta H_{reaction}^\circ (300.\text{ K}) = -45.98 \text{ kJ mol}^{-1}$$

P4.3) A sample of $K(s)$ of mass 2.380 g undergoes combustion in a constant volume calorimeter. The calorimeter constant is 1849 J K^{-1}, and the measured temperature rise in the inner water bath containing 1450. g of water is 2.91 K. Calculate ΔU_f° and ΔH_f° for K_2O.

$$2K(s) + \tfrac{1}{2}O_2(g) \rightarrow K_2O(s)$$

$$\Delta U_f^\circ = -\frac{M_s}{m_s}\left(\frac{m_{H_2O}}{M_{H_2O}} C_{H_2O,m} \Delta T + C_{calorimeter} \Delta T \right)$$

$$= -\frac{39.098 \text{ g mol}^{-1}}{2.380 \text{ g}} \times \frac{2 \text{ mol K}}{1 \text{ mol reaction}} \times \left(\begin{array}{l} \dfrac{1.450 \times 10^3 \text{ g}}{18.02 \text{ g mol}^{-1}} \times 75.3 \text{ J mol}^{-1}\text{ K}^{-1} \times 2.91°C \\[2mm] +1.849 \times 10^3 \text{ J}°C^{-1} \times 2.91°C \end{array} \right)$$

$$= -360. \text{ kJ mol}^{-1}$$

$$\Delta H_f^\circ = \Delta U_f^\circ + \Delta nRT$$

$$= -360. \text{ kJ mol}^{-1} - \frac{1}{2} \times 8.314 \text{ J K}^{-1}\text{mol}^{-1} \times 298.15 \text{ K} = -361 \text{ kJ mol}^{-1}$$

P4.10) The data below are a DSC scan of a solution of a T4 lysozyme mutant. From the data determine T_m. Determine also the excess heat capacity ΔC_P at $T = 308$ K. Determine also the intrinsic δC_P^{int} and transition δC_P^{trs} excess heat capacities at $T = 308$ K. In your calculations use the extrapolated curves, shown as dotted lines in the DSC scan, where the y axis shows C_P.

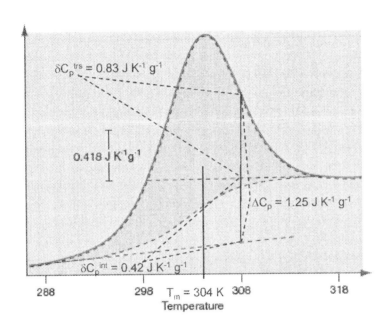

P4.11) At 298 K, $\Delta H_R^\circ = 131.28$ kJ mol^{-1} for the reaction C(*graphite*) + H$_2$O(g) $\rightarrow$ CO(g) + H$_2$(g), with $C_{P,m}$ = 8.53, 33.58, 29.12, and 28.82 J K^{-1} mol^{-1} for graphite, H$_2$O(g), CO(g), and H$_2$(g), respectively. Calculate ΔH_R° at 125°C from this information. Assume that the heat capacities are independent of temperature.

$$\Delta H_{reaction}^\circ \left(398\text{ K}\right) = \Delta H_{reaction}^\circ \left(298\text{ K}\right) + \int_{298\text{ K}}^{398\text{ K}} \Delta C_P\left(T\right) dT$$

For this problem, it is assumed that the heat capacities are independent of T.

$$\Delta H_{reaction}^\circ \left(398\text{ K}\right) = \Delta H_{reaction}^\circ \left(298\text{ K}\right)$$
$$+ \left[C_{P,m}\left(\text{H}_2,g\right) + C_{P,m}\left(\text{CO},g\right) - C_{P,m}\left(\text{C},graphite\right) - C_{P,m}\left(\text{H}_2\text{O},g\right)\right]\Delta T$$
$$= 131.28 \text{ kJ mol}^{-1} + \left[28.82 + 29.12 - 8.53 - 33.58\right] \text{J mol}^{-1}\text{ K}^{-1} \times 100.\text{ K}$$
$$= 132.86 \text{ kJ mol}^{-1}$$

P4.12) Consider the reaction TiO$_2$(s) + 2 C(*graphite*) + 2 Cl$_2$(g) $\rightarrow$ 2 CO(g) + TiCl$_4$(l) for which $\Delta H_{R,298K}^\circ$ = −80. kJ mol^{-1}. Given the following data at 25°C, (a) calculate ΔH_R° at 135.8°C, the boiling

23

point of TiCl$_4$, and (b) calculate ΔH_f° for TiCl$_4$ (l) at 25°C:

Substance	TiO$_2$(s)	Cl$_2$(g)	C(graphite)	CO(g)	TiCl$_4$(l)
ΔH_f° (kJ mol^{-1})	−945			−110.5	
$C_{P,m}$ (J K^{-1} mol^{-1})	55.06	33.91	8.53	29.12	145.2

Assume that the heat capacities are independent of temperature.

a) Calculate $\Delta H_{reaction}^{\circ}$ at 135.8°C, the boiling point of TiCl$_4$.

b) Calculate ΔH_f° for TiCl$_4$(l) at 25°C.

Assume that the heat capacities are independent of temperature.

a) $\quad \Delta H_{reaction}^{\circ}\left(409.0\ \mathrm{K}\right) = \Delta H_{reaction}^{\circ}\left(298\ \mathrm{K}\right) + \displaystyle\int_{298\ \mathrm{K}}^{409.0\ \mathrm{K}} \Delta C_{P,m}\, dT$

In this case, the heat capacities are assumed to be independent of T.

$\Delta H_{reaction}^{\circ}\left(409.0\ \mathrm{K}\right) = \Delta H_{reaction}^{\circ}\left(298\ \mathrm{K}\right) + \Delta C_{P,m}\left[409.0\ \mathrm{K} - 298\ \mathrm{K}\right]$

$\quad = -80.\ \mathrm{kJ\ mol}^{-1} + \left[C_{P,m}\left(\mathrm{TiCl}_4, l\right) + 2C_{P,m}\left(\mathrm{CO}, g\right) - C_{P,m}\left(\mathrm{TiO}_2, s\right)\right.$

$\qquad\qquad \left. - 2C_{P,m}\left(graphite, s\right) - 2C_{P,m}\left(\mathrm{Cl}_2, g\right)\right]\left[409.0\ \mathrm{K} - 298\ \mathrm{K}\right]$

$\quad = -80.\ \mathrm{kJ\ mol}^{-1} + \left[145.2 + 2\times 29.12 - 55.06 - 2\times 8.53 - 2\times 33.91\right]\left[409.0\ \mathrm{K} - 298\ \mathrm{K}\right]$

$\quad = -73\ \mathrm{kJ\ mol}^{-1}$

b) TiO$_2$(s) + 2C(graphite) + 2Cl$_2$(g) → 2CO(g) + TiCl$_4$(l)

$\Delta H_{reaction}^{\circ} = -80.\ \mathrm{kJ\ mol}^{-1} = 2\Delta H_f^{\circ}\left(\mathrm{CO}, g\right) + \Delta H_f^{\circ}\left(\mathrm{TiCl}_4, l\right) - \Delta H_f^{\circ}\left(\mathrm{TiO}_2, s\right)$

$\Delta H_f^{\circ}\left(\mathrm{TiCl}_4, l\right) = \Delta H_f^{\circ}\left(\mathrm{TiO}_2, s\right) - 2\Delta H_f^{\circ}\left(\mathrm{CO}, g\right) - 80.\ \mathrm{kJ\ mol}^{-1}$

$\quad = -945\ \mathrm{kJ\ mol}^{-1} + 2\times 110.5\ \mathrm{kJ\ mol}^{-1} - 80.\ \mathrm{kJ\ mol}^{-1}$

$\quad = -804\ \mathrm{kJ\ mol}^{-1}$

P4.13) Calculate ΔH_R° and ΔU_R° for the oxidation of benzene (g). Also calculate

$$\frac{\Delta H_R^{\circ} - \Delta U_R^{\circ}}{\Delta H_R^{\circ}}$$

15/2O$_2$(g) + C$_6$H$_6$(l) → 3H$_2$O(l) + 6CO$_2$(g)

From the data tables,

$$\Delta H^{\circ}_{combustion} = 3\Delta H^{\circ}_f\left(H_2O, l\right) + 6\Delta H^{\circ}_f\left(CO_2, g\right) - \Delta H^{\circ}_f\left(C_6H_6, l\right)$$

$$= -3\times285.8 \text{ kJ mol}^{-1} - 6\times393.5 \text{ kJ mol}^{-1} - 49.1 \text{ kJ mol}^{-1}$$

$$= 3268 \text{ kJ mol}^{-1}$$

$$\Delta U^{\circ}_{reaction} = \Delta H^{\circ}_{reaction} - \Delta n RT = -3268 \text{ kJ mol}^{-1} + 1.5\times8.314 \text{ J K}^{-1}\text{mol}^{-1}\times298.15 \text{ K}$$

$$= -3264 \text{ kJ mol}^{-1}$$

$$\frac{\Delta H^{\circ}_{reaction} - \Delta U^{\circ}_{reaction}}{\Delta H^{\circ}_{reaction}} = \frac{-3268 \text{ kJ mol}^{-1} + 3264 \text{ kJ mol}^{-1}}{-3268 \text{ kJ mol}^{-1}} = 0.0122$$

P4.23) Calculate ΔH°_R at 800. K for the reaction $4NH_3(g) + 6NO(g) \rightarrow 5N_2(g) + 6H_2O(g)$ using the temperature dependence of the heat capacities from the data tables. Compare your result with ΔH°_R at 298.15 K. Is the difference large or small? Why?

$$\Delta H^{\circ}_{reaction}\left(650K\right) = \Delta H^{\circ}_{reaction}\left(298.15K\right) + \int_{298.15}^{800} \Delta C_P\left(\frac{T}{K}\right) d\frac{T}{K}$$

$$\Delta C_P = 5C_{P,m}\left(N_2, g\right) + 6C_{P,m}\left(H_2O, g\right) - 4C_{P,m}\left(NH_3, g\right) - 6C_{P,m}\left(NO, g\right)$$

$$= \begin{bmatrix} \left(5\times30.81 + 6\times33.80 - 4\times29.29 - 6\times33.58\right) \\ -\left(5\times0.01187 + 6\times0.00795 + 4\times0.01103 - 6\times0.02593\right)\dfrac{T}{K} \\ +\left(5\times2.3968 + 6\times2.8228 - 4\times4.2446 - 6\times5.3326\right)\times10^{-5}\dfrac{T^2}{K^2} \\ -\left(5\times1.0176 + 6\times1.3115 - 4\times2.7706 - 6\times2.7744\right)\times10^{-8}\dfrac{T^3}{K^3} \end{bmatrix} \text{J K}^{-1}\text{ mol}^{-1}$$

$$= \left[38.21 + 0.00441\frac{T}{K} - 2.0053\times10^{-4}\frac{T^2}{K^2} + 1.4772\times10^{-7}\frac{T^3}{K^3}\right] \text{J K}^{-1}\text{ mol}^{-1}$$

$$\int_{298.15}^{800} \Delta C_P\left(\frac{T}{K}\right) d\frac{T}{K} = \left[\int_{298.15}^{800}\left(38.21 + 0.00441\frac{T}{K} - 2.0053\times10^{-4}\frac{T^2}{K^2} + 1.4772\times10^{-8}\frac{T^3}{K^3}\right) d\frac{T}{K}\right] \text{J mol}^{-1}$$

$$= \left(19.175 - 1.215 - 32.452 + 14.835\right) \text{kJ mol}^{-1} = 2.77 \text{ kJ mol}^{-1}$$

$$\Delta H^{\circ}_{reaction}\left(298.15 \text{ K}\right) = 5\Delta H^{\circ}_f\left(N_2, g\right) + 6\Delta H^{\circ}_f\left(H_2O, g\right) - 4\Delta H^{\circ}_f\left(NH_3, g\right) - 6\Delta H^{\circ}_f\left(NO, g\right)$$

$$\Delta H^{\circ}_{reaction}\left(298.15 \text{ K}\right) = 6\times241.8 \text{ kJ mol}^{-1} + 4\times45.9 \text{ kJ mol}^{-1} - 6\times91.3 \text{ kJ mol}^{-1} = -1812 \text{ kJ mol}^{-1}$$

$$\Delta H^{\circ}_{reaction}\left(650 \text{ K}\right) = -1812 \text{ kJ mol}^{-1} + 2.77 \text{ kJ mol}^{-1} = -1815 \text{ kJ mol}^{-1}$$

The difference is small, not because the heat capacities of reactants and products are small, but because the difference in heat capacities of reactants and products is small.

P4.24) From the following data at 298.15 K as well as data in Table 4.1 (Appendix B, Data Tables), calculate the standard enthalpy of formation of $H_2S(g)$ and of $FeS_2(s)$:

ΔH_R° (kJ mol^{-1})

$Fe(s) + 2H_2S(g) \rightarrow FeS_2(s) + 2H_2(g)$ -137.0

$H_2S(g) + 3/2O_2(g) \rightarrow H_2O(l) + SO_2(g)$ 562.0

calculate the standard enthalpy of formation of $H_2S(g)$ and of FeS_2 (s).

	$\Delta H_{reaction}^\circ$ (kJ mol^{-1})
$H_2O(l) + SO_2(g) \rightarrow H_2S(g) + 3/2O_2(g)$	562.0
$S(s) + O_2(g) \rightarrow SO_2(g)$	-296.8
$H_2(g) + 1/2O_2(g) \rightarrow H_2O(l)$	-285.8

$H_2(g) + S(s) \rightarrow H_2S(g)$	$\Delta H_f^\circ = -20.6$ kJ mol^{-1}

	ΔH_R° (kJ mol^{-1})
$Fe(s) + 2H_2S(g) \rightarrow FeS_2(s) + 2H_2(g)$	ΔH_R° -137.0
$2H_2(g) + 2S(s) \rightarrow 2H_2S(g)$	ΔH_R° -2×20.6

$Fe(s) + 2S(s) \rightarrow FeS_2(s)$	$\Delta H_f^\circ = -178.2$ kJ mol^{-1}

P4.27) Calculate ΔH for the process in which $Cl_2(g)$ initially at 298.15 K at 1 bar is heated to 750 K at 1 bar. Use the temperature dependent heat capacities in the data tables. How large is the relative error if the molar heat capacity is assumed to be constant at its value of 298.15 K over the temperature interval?

$$\Delta H = \Delta H_f^\circ \left(Cl_2, g, 298.15 \text{ K}\right) + \int_{298.15}^{750} C_{P,m}\left(\frac{T}{K}\right) d\frac{T}{K}$$

$$= \left[\int_{298.15}^{750.} \left(22.85 + 0.06543\frac{T}{K} - 1.2517 \times 10^{-4}\frac{T^2}{K^2} + 1.1484 \times 10^{-7}\frac{T^3}{K^3} \right) d\frac{T}{K} \right] \text{J K}^{-1} \text{ mol}^{-1}$$

$$= (10324 + 15494 - 16496 + 8857) \text{J mol}^{-1} = 18.2 \text{ kJ mol}^{-1}$$

If it is assumed that the heat capacity is constant at its value at 298 K,

$$\Delta H^\circ \approx \left[\int_{298.15}^{750.} (33.95) d\frac{T}{K} \right] \text{J K}^{-1}\text{mol}^{-1} = 15.3 \text{ kJ mol}^{-1}$$

$$\text{Error} = 100 \times \frac{18.2 \text{ kJ mol}^{-1} - 15.3 \text{ kJ mol}^{-1}}{18.2 \text{ kJ mol}^{-1}} = 15.6\%$$

P4.28) From the following data at 298.15 K, calculate the standard enthalpy of formation of FeO(s) and of $Fe_2O_3(s)$:

ΔH_R° (kJ mol^{-1})

$Fe_2O_3(s) + 3C(graphite) \rightarrow 2Fe(s) + 3CO(g)$ 492.6

$FeO(s) + C(graphite) \rightarrow Fe(s) + CO(g)$ 155.8

$C(graphite) + O_2(g) \rightarrow CO_2(g)$ –393.51

$CO(g) + 1/2O_2(g) \rightarrow CO_2(g)$ –282.98

Calculate the standard enthalpy of formation of FeO(s) and of $Fe_2O_3(s)$.

	$\Delta H_{reaction}^\circ$ (kJ mol^{-1})
$Fe(s) + CO(g) \rightarrow FeO(s) + C(graphite)$	–155.8
$CO_2(s) \rightarrow CO(g) + 1/2O_2(g)$	282.98
$C(graphite) + O_2(g) \rightarrow CO_2(g)$	–393.51

$Fe(g) + 1/2O_2(g) \rightarrow FeO(s)$	$\Delta H_f^\circ = -266.3$ kJ mol^{-1}

	$\Delta H_{reaction}^\circ$ (kJ mol^{-1})
$2Fe(g) + 3CO(g) \rightarrow Fe_2O_3(s) + 3C(graphite)$	–492.6
$3C(graphite) + 3O_2(g) \rightarrow 3CO_2(g)$	-3×393.51
$3CO_2(g) \rightarrow 3CO(g) + 3/2O_2(g)$	3×282.98

$2Fe(g) + 3/2O_2(g) \rightarrow Fe_2O_3(s)$	$\Delta H_f^\circ = -824.2$ kJ mol^{-1}

P4.33) A camper stranded in snowy weather loses heat by wind convection. The camper is packing emergency rations consisting of 60.% sucrose, 30.% fat, and 10.% protein by weight. Using the data provided in Problem P4.32 and assuming the fat content of the rations can be treated with palmitic acid data and the protein content similarly by the protein data in Problem P4.32, how much emergency rations must the camper consume in order to compensate for a reduction in body temperature of 4.0 K? Assume the heat capacity of the body equals that of water. Assume the camper weighs 70. kg. State any additional assumptions.

At constant pressure $q = \Delta H$. The composition of the emergency rations means that 1 kg of the rations contains the following number of moles of sucrose, fat, and protein:

$$n_{succrose} = \frac{m}{M} = \frac{(0.60 \text{ kg})}{(342.3 \text{ g mol}^{-1})} = 1.753 \text{ mol}$$

$$n_{fat} = \frac{m}{M} = \frac{(0.30 \text{ kg})}{(256.43 \text{ g mol}^{-1})} = 1.17 \text{ mol}$$

$$n_{protein} = \frac{m}{M} = \frac{(0.10 \text{ kg})}{(88.30 \text{ g mol}^{-1})} = 1.13 \text{ mol}$$

Therefore, the enthalpy of combustion for 1 kg of rations is:

$$\Delta H^{\circ}_{combustion,1 \text{ } kg} = (1.753 \text{ mol}) \times (-5647 \text{ kJ mol}^{-1}) + (1.17 \text{ mol}) \times (-10035 \text{ kJ mol}^{-1}) + (1.13 \text{ mol}) \times (-22 \text{ kJ mol}^{-1})$$
$$= -21665 \text{ kJ}$$

The heat the stranded camper loses is given by:

$$q_{lost} = n_{H_2O} \, C_{p,m} \, \Delta T = \frac{m_{H_2O}}{M_{H_2O}} \times C_{p,m} \times \Delta T = \frac{(70. \text{ kg})}{(18.01 \text{ g mol}^{-1})} \times (75.3 \text{ J K}^{-1} \text{ mol}^{-1}) \times (4.0 \text{ K})$$
$$= 1170.7 \text{ kJ}$$

Finally, the mass of rations that needs to be consumed to produce the lost amount of heat assuming the body consists of 90% water is then:

$$m_{rations} = 0.9 \times \frac{(1 \text{ kg}) \times (1170.7 \text{kJ})}{(21665 \text{kJ})} = 0.0486 \text{ kg} = 49 \text{ g}$$

Chapter 5: Entropy and the Second and Third Laws of Thermodynamics

P5.5) One mole of $H_2O(l)$ is compressed from a state described by $P = 1.00$ bar and $T = 325$ K to a state described by $P = 660.$ bar and $T = 625$ K. In addition, $\beta = 2.07 \times 10^{-4}$ K^{-1} and the density of liquid water can be assumed to be constant at the value 997 kg m^{-3}. Calculate ΔS for this transformation, assuming that $\kappa = 0$.

From equation 5.24,

$$\Delta S = \int_{T_i}^{T_f} \frac{C_P}{T} dT - \int_{P_i}^{P_f} V \beta \, dP \approx nC_{P,m} \ln\frac{T_f}{T_i} - nV_{m,i}\beta\left(P_f - P_i\right)$$

$$= 1\,\text{mol} \times 75.3 \text{ J mol}^{-1}\text{K}^{-1} \times \ln\frac{625 \text{ K}}{325 \text{ K}}$$

$$- 1\,\text{mol} \times \frac{18.02 \times 10^{-3} \text{ kg mol}^{-1}}{997 \text{ kg m}^{-3}} \times 2.07 \times 10^{-4} \text{ K}^{-1} \times 659 \text{ bar} \times 10^5 \text{ Pa bar}^{-1}$$

$$= 49.2 \text{ J K}^{-1} - 0.247 \text{ J K}^{-1} = 49.0 \text{ J K}^{-1}$$

P5.8) The average heat evolved by the oxidation of foodstuffs in an average adult per hour per kilogram of body weight is 7.20 kJ kg^{-1} hr^{-1}. Suppose the heat evolved by this oxidation is transferred into the surroundings over a period lasting one day. Calculate the entropy change of the surroundings associated with this heat transfer. Assume the weight of an average adult is 70.0 kg. Assume also the surroundings are at $T = 293.0$ K.

$$q(\text{per day, } 70.0 \text{ kg}) = 7.20 \text{ kJ kg}^{-1} \text{ hr}^{-1} \times 24 \text{ h day}^{-1} \times 70.0 \text{ kg} = 1.21 \times 10^4 \text{ kJ day}^{-1}$$

$$\Delta S = \frac{q}{T} = \frac{1.21 \times 10^4 \text{ kJ day}^{-1}}{293 \text{ K}} = 41.3 \text{ kJ K}^{-1} \text{ day}^{-1}$$

P5.9) Calculate ΔS, ΔS_{total}, and $\Delta S_{surroundings}$ when the volume of 123 g of CO initially at 298 K and 1.00 bar increases by a factor of four in (a) an adiabatic reversible expansion, (b) an expansion against $P_{external} = 0$, and (c) an isothermal reversible expansion. Take $C_{P,m}$ to be constant at the value 29.14 J mol^{-1} K^{-1} and assume ideal gas behavior. State whether each process is spontaneous. The temperature of the surroundings is 298 K.

a) an adiabatic reversible expansion

$\Delta S_{surroundings} = 0$ because $q = 0$. $\Delta S = 0$ because the process is reversible.

$\Delta S_{total} = \Delta S + \Delta S_{surroundings} = 0$. The process is not spontaneous.

b) an expansion against $P_{external} = 0$

ΔT and $w = 0$. Therefore $\Delta U = q = 0$.

$$\Delta S = nR \ln \frac{V_f}{V_i} = \frac{123 \text{ g}}{28.01 \text{ g mol}^{-1}} \times 8.314 \text{ J mol}^{-1} \text{ K}^{-1} \times \ln 4 = 50.6 \text{ J K}^{-1}$$

$\Delta S_{total} = \Delta S + \Delta S_{surroundings} = 50.6 \text{ J K}^{-1} + 0 = 50.6 \text{ J K}^{-1}$. The process is spontaneous.

c) an isothermal reversible expansion

$\Delta T = 0$. Therefore $\Delta U = 0$.

$$w = -q = -nRT \ln \frac{V_f}{V_i} = -\frac{123 \text{ g}}{28.01 \text{ g mol}^{-1}} \text{ mol} \times 8.314 \text{ J mol}^{-1} \text{ K}^{-1} \times 298 \text{ K} \times \ln 4 = -15.1 \times 10^3 \text{ J}$$

$$\Delta S = \frac{q_{reversible}}{T} = \frac{15.1 \times 10^3 \text{ J}}{298 \text{ K}} = 50.6 \text{ J K}^{-1}$$

$$\Delta S_{surroundings} = \frac{-q}{T} = \frac{-15.1 \times 10^3 \text{ J}}{298 \text{ K}} = -50.6 \text{ J K}^{-1}$$

$\Delta S_{total} = \Delta S + \Delta S_{surroundings} = 50.6 \text{ J K}^{-1} - 50.6 \text{ J K}^{-1} = 0$. The system and surroundings are at equilibrium.

P5.10) The maximum theoretical efficiency of an internal combustion engine is achieved in a reversible Carnot cycle. Assume that the engine is operating in the Otto cycle and that $C_{V,m} = 5/2R$ for the fuel–air mixture initially at 298 K (the temperature of the cold reservoir). The mixture is compressed by a factor of 7.5 in the adiabatic compression step. What is the maximum theoretical efficiency of this engine? How much would the efficiency increase if the compression ratio could be increased to 25? Do you see a problem in doing so?

$$T_f = T_i \left(\frac{V_f}{V_i} \right)^{1-\gamma} = T_i \left(\frac{1}{7.5} \right)^{1-\frac{7}{5}} = 298 \text{ K} \times \left(\frac{1}{7.5} \right)^{-0.4} = 667 \text{ K}$$

$$\varepsilon = 1 - \frac{T_{cold}}{T_{hot}} = 1 - \frac{298 \text{ K}}{667 \text{ K}} = 0.553$$

$$T_f = T_i \left(\frac{V_f}{V_i}\right)^{1-\gamma} = T_i \left(\frac{1}{25}\right)^{1-\frac{7}{5}} = 298 \text{ K} \times \left(\frac{1}{25}\right)^{-0.4} = 1.079 \times 10^3 \text{ K}$$

$$\varepsilon = 1 - \frac{T_{cold}}{T_{hot}} = 1 - \frac{298 \text{ K}}{1079 \text{ K}} = 0.724$$

It would be difficult to avoid ignition of the fuel–air mixture before the compression was complete.

P5.13) Calculate ΔS for the isothermal compression of one mole of Cu(s) from 1.00 bar to 1000. bar at 298 K. $\Delta = 0.492 \times 10^{-4}$ K^{-1}, $\Delta = 0.78 \times 10^{-6}$ bar^{-1}, and the density is 8.92 g cm^{-3}. Repeat the calculation assuming that $\Delta = 0$.

$$\Delta S = -\int_{P_i}^{P_f} V_i (1 - \kappa P) \beta dP$$

$$= -\int_{10^5}^{10^8} \frac{63.55 \times 10^{-3} \text{ kg mol}^{-1}}{8.92 \times 10^3 \text{ kg m}^3} \left(1 - 0.780 \times 10^{-11} \text{ Pa}^{-1} \times P\right) 0.492 \times 10^{-4} \text{ K}^{-1} dP$$

$$= -0.0350 \text{ J K}^{-1}$$

Repeating the calculation for $\kappa = 0$, we see no change because $1 \gg \kappa P$

$$\Delta S = -\int_{P_i}^{P_f} V_i (1 - \kappa P) \beta dP$$

$$= -\int_{10^5}^{10^8} \frac{63.55 \times 10^{-3} \text{ kg mol}^{-1}}{8.92 \times 10^3 \text{ kg m}^3} \times 0.492 \times 10^{-4} \text{ K}^{-1} dP$$

$$= -0.0350 \text{ J K}^{-1}$$

P5.14) Calculate ΔS° for the reaction $3H_2(g) + N_2(g) \rightarrow 2NH_3(g)$ at 650. K. Omit terms in the temperature-dependent heat capacities higher than T^2/K^2.

From Table 2.4,

$$C_P^\circ (H_2, g) = 22.66 + 4.38 \times 10^{-2} \frac{T}{K} - 1.0835 \times 10^{-4} \frac{T^2}{K^2} \text{ J K}^{-1} \text{ mol}^{-1}$$

$$C_P^\circ (N_2, g) = 30.81 - 1.187 \times 10^{-2} \frac{T}{K} + 2.3968 \times 10^{-5} \frac{T^2}{K^2} \text{ J K}^{-1} \text{ mol}^{-1}$$

$$C_P^\circ (NH_3, g) = 29.29 + 1.103 \times 10^{-2} \frac{T}{K} + 4.2446 \times 10^{-5} \frac{T^2}{K^2} \text{ J K}^{-1} \text{ mol}^{-1}$$

$$\Delta C_P^\circ = 2\left(29.29 + 1.103\times10^{-2}\frac{T}{K} + 4.2446\times10^{-5}\frac{T^2}{K^2}\,\text{J K}^{-1}\,\text{mol}^{-1}\right)$$

$$-\left(30.81 - 1.187\times10^{-2}\frac{T}{K} + 2.3968\times10^{-5}\frac{T^2}{K^2}\,\text{J K}^{-1}\,\text{mol}^{-1}\right)$$

$$-3\left(22.66 + 4.38\times10^{-7}\frac{T}{K} - 1.0835\times10^{-4}\frac{T^2}{K^2}\,\text{J K}^{-1}\,\text{mol}^{-1}\right)$$

$$\Delta C_P^\circ = -40.21 - 0.0975\frac{T}{K} - 3.860\times10^{-4}\frac{T^2}{K^2}\,\text{J K}^{-1}\,\text{mol}^{-1}$$

$$\Delta S^\circ = 2S_{298.15}^\circ(NH_3, g) - S_{298.15}^\circ(N_2, g) - 3S_{298.15}^\circ(H_2, g)$$

$$= 2\times192.8\,\text{J K}^{-1}\,\text{mol}^{-1} - 191.6\,\text{J K}^{-1}\,\text{mol}^{-1} - 3\times130.7\,\text{J K}^{-1}\,\text{mol}^{-1}$$

$$= -198.1\,\text{J K}^{-1}\,\text{mol}^{-1}$$

$$\Delta S_T^\circ = \Delta S_{298.15}^\circ + \int_{298.15}^{T}\frac{\Delta C_p^\circ}{T'}dT'$$

$$= -198.1\,\text{J K}^{-1}\,\text{mol}^{-1} + \int_{298.15}^{650}\frac{\left(-40.21 - 0.0975\frac{T}{K} - 3.860\times10^{-4}\frac{T^2}{K^2}\right)}{\frac{T}{K}}d\frac{T}{K}\,\text{J K}^{-1}\,\text{mol}^{-1}$$

$$= -198.1\,\text{J K}^{-1}\,\text{mol}^{-1} - 31.34\,\text{J K}^{-1}\,\text{mol}^{-1} - 34.30\,\text{J K}^{-1}\,\text{mol}^{-1} + 64.38\,\text{J K}^{-1}\,\text{mol}^{-1}$$

$$= -199.4\,\text{J K}^{-1}\,\text{mol}^{-1}$$

P5.17) The interior of a refrigerator is typically held at 35°F and the interior of a freezer is typically held at 0.00°F. If the room temperature is 70.°F, by what factor is it more expensive to extract the same amount of heat from the freezer than from the refrigerator? Assume that the theoretical limit for the performance of a reversible refrigerator is valid in this case.

From equation 5.44

$$\eta_r = \frac{T_{cold}}{T_{hot} - T_{cold}}$$

$$T_{room} = \frac{5}{9}(70. - 32) + 273.15 = 294.3\ \text{K}$$

$$T_{freezer} = \frac{5}{9}(0.00 - 32) + 273.15 = 255.4\ \text{K}$$

$$T_{refrigerator} = \frac{5}{9}(35 - 32) + 273.15 = 274.8\ \text{K}$$

for the freezer $\eta_r = \dfrac{255.4 \text{ K}}{294.2 \text{ K} - 255.4 \text{ K}} = 6.57$

for the refrigerator $\eta_r = \dfrac{274.8 \text{ K}}{294.2 \text{ K} - 274.8 \text{ K}} = 14.1$

The freezer is more expansive to operate than the refrigerator by the ratio $14.1/6.57 = 2.15$.

P5.19) At the transition temperature of 95.4°C, the enthalpy of transition from rhombic to monoclinic sulfur is 0.38 kJ mol^{-1}.

a. Calculate the entropy of transition under these conditions.

b. At its melting point, 119°C, the enthalpy of fusion of monoclinic sulfur is 1.23 kJ mol^{-1}. Calculate the entropy of fusion.

c. The values given in parts (a) and (b) are for 1 mol of sulfur; however, in crystalline and liquid sulfur, the molecule is present as S_8. Convert the values of the enthalpy and entropy of fusion in parts (a) and (b) to those appropriate for S_8.

a) $\Delta S_{transition} = \dfrac{\Delta H_{transition}}{T_{transition}} = \dfrac{0.38 \text{ kJ mol}^{-1}}{(273.15 + 95.4) \text{ K}} = 1.0 \text{ J K}^{-1} \text{ mol}^{-1}$

b) $\Delta S_{fusion} = \dfrac{\Delta H_{fusion}}{T_{fusion}} = \dfrac{1.23 \text{ kJ mol}^{-1}}{(273.15 + 119) \text{ K}} = 3.14 \text{ J K}^{-1} \text{ mol}^{-1}$

c) Each of the ΔS in parts (a) and (b) should be multiplied by 8.

$\Delta S_{transition} = 8.2 \text{ J K}^{-1} \text{ mol}^{-1}$

$\Delta S_{fusion} = 25.1 \text{ J K}^{-1} \text{ mol}^{-1}$

P5.22) Calculate ΔH and ΔS if the temperature of one mole of Hg(l) is increased from 0.00°C to 100.°C at 1 bar. Over this temperature range, $C_{P,m} = 30.093 - 4.944 \times 10^{-3} T$ J mol^{-1} K^{-1}.

$$\Delta H = 1 \text{ mol} \times \int_{273.15 \text{ K}}^{373.15 \text{ K}} \left(30.093 - 4.944 \times 10^{-3} T\right) \text{ J mol}^{-1} \text{ K}^{-1} dT$$

$$= 30.093\left(T_f - T_i\right) - 2.472 \times 10^{-2} \left(T_f - T_i\right)^2 \text{ J}$$

$$= 3009 \text{ J} - 169 \text{ J} = 2.85 \times 10^3 \text{ J}$$

$$\Delta S = 1 \text{ mol} \times \int_{273.15 \text{ K}}^{373.15 \text{ K}} \left(\frac{30.093 - 4.944 \times 10^{-3} T \text{ J mol}^{-1} \text{ K}^{-1}}{T}\right) dT$$

$$= -4.944 \times 10^{-3} \left(T_f - T_i\right) + 30.093 \ln\frac{T_f}{T_i} \text{ J K}^{-1}$$

$$= -0.494 \text{ J K mol}^{-1} + 9.38 \text{ J K}^{-1} = 8.89 \text{ J K}^{-1}$$

P5.23) Calculate ΔS if the temperature of 1.75 mol of an ideal gas with $C_V = 3/2R$ is increased from 195 to 425 K under conditions of (a) constant pressure and (b) constant volume.

a) at constant pressure $\Delta S = nC_{P,m} \ln\frac{T_f}{T_i} = 1.75 \text{ mol} \times \left(\frac{3}{2} + 1\right) \times 8.314 \text{ J mol}^{-1} \text{ K}^{-1} \times \ln\frac{425 \text{ K}}{195 \text{ K}} = 28.3 \text{ J K}^{-1}$

b) at constant volume $\Delta S = nC_{V,m} \ln\frac{T_f}{T_i} = 1.75 \text{ mol} \times \frac{3}{2} \times 8.314 \text{ J mol}^{-1} \text{K}^{-1} \times \ln\frac{425 \text{ K}}{195 \text{ K}} = 17.0 \text{ J K}^{-1}$

P5.25) Calculate ΔS_R° for the reaction $H_2(g) + Cl_2(g) \rightarrow 2HCl(g)$ at 725 K. Omit terms in the temperature-dependent heat capacities higher than T^2/K^2.

From Table 2.4,

$$C_P^\circ(H_2, g) = 22.66 + 4.38 \times 10^{-2} \frac{T}{K} - 1.0835 \times 10^{-4} \frac{T^2}{K^2} \text{ J K}^{-1} \text{ mol}^{-1}$$

$$C_P^\circ(Cl_2, g) = 22.85 + 6.543 \times 10^{-2} \frac{T}{K} - 1.2517 \times 10^{-4} \frac{T^2}{K^2} \text{ J K}^{-1} \text{ mol}^{-1}$$

$$C_P^\circ(HCl, g) = 29.81 - 4.12 \times 10^{-3} \frac{T}{K} + 6.2231 \times 10^{-6} \frac{T^2}{K^2} \text{ J K}^{-1} \text{ mol}^{-1}$$

$$\Delta C_P^\circ = 2\left(29.81 - 4.12\times10^{-3}\frac{T}{K} + 6.2231\times10^{-6}\frac{T^2}{K^2}\ \text{J K}^{-1}\ \text{mol}^{-1}\right)$$

$$-\left(22.66 + 4.38\times10^{-2}\frac{T}{K} - 1.0835\times10^{-4}\frac{T^2}{K^2}\ \text{J K}^{-1}\ \text{mol}^{-1}\right)$$

$$-\left(22.85 + 6.543\times10^{-7}\frac{T}{K} - 1.2517\times10^{-4}\frac{T^2}{K^2}\ \text{J K}^{-1}\ \text{mol}^{-1}\right)$$

$$\Delta C_P^\circ = 14.11 - 0.117\frac{T}{K} + 2.460\times10^{-4}\frac{T^2}{K^2}\ \text{J K}^{-1}\ \text{mol}^{-1}$$

$$\Delta S^\circ = 2S_{298.15}^\circ(\text{HCl},g) - S_{298.15}^\circ(\text{Cl}_2,g) - S_{298.15}^\circ(\text{H}_2,g)$$

$$= 2\times186.9\ \text{J K}^{-1}\ \text{mol}^{-1} - 223.1\ \text{J K}^{-1}\ \text{mol}^{-1} - 130.7\ \text{J K}^{-1}\ \text{mol}^{-1}$$

$$= 20.0\ \text{J K}^{-1}\ \text{mol}^{-1}$$

$$\Delta S_T^\circ = \Delta S_{298.15}^\circ + \int_{298.15}^{T}\frac{\Delta C_P^\circ}{T'}dT'$$

$$= 20.0\ \text{J K}^{-1}\ \text{mol}^{-1} + \int_{298.15}^{725}\frac{\left(14.11 - 0.117\frac{T}{K} + 2.460\times10^{-4}\frac{T^2}{K^2}\ \text{J K}^{-1}\ \text{mol}^{-1}\right)}{\frac{T}{K}}d\frac{T}{K}\ \text{J K}^{-1}\ \text{mol}^{-1}$$

$$= 20.0\ \text{J K}^{-1}\ \text{mol}^{-1} + 12.53\ \text{J K}^{-1}\ \text{mol}^{-1} - 0.1040\ \text{J K}^{-1}\ \text{mol}^{-1} + 0.0002186\ \text{J K}^{-1}\ \text{mol}^{-1}$$

$$= 36.1\ \text{J K}^{-1}\ \text{mol}^{-1}$$

P5.28) The amino acid glycine dimerizes to form the dipeptide glycylglycine according to the reaction

$$2\text{Glycine}(s) \rightarrow \text{Glycylglycine}(s) + \text{H}_2\text{O}(l)$$

Calculate, ΔS, ΔS_{surr}, and ΔS_{total} at $T = 298$ K. Useful thermodynamic data are:

	Glycine	Glycylglycine	Water
ΔH_f^0 (kJ mol^{-1})	-537.2	-746.0	-285.8
S^0 (J K^{-1} mol^{-1})	103.5	190.0	70.0

$$\Delta S_R^\circ = 2\times103.5\ \text{J K}^{-1}\ \text{mol}^{-1} + 190.\ \text{J K}^{-1}\ \text{mol}^{-1} + 70.0\ \text{J K}^{-1}\ \text{mol}^{-1} = 53.0\ \text{J K}^{-1}\ \text{mol}^{-1}$$

$$\Delta H_R^\circ = 2\times537.2\ \text{kJ mol}^{-1} - 746\ \text{kJ mol}^{-1} - 285.8\ \text{kJ mol}^{-1} = 42.6\ \text{kJ mol}^{-1}$$

$$\Delta S_{surr} = \frac{-\Delta H_R^\circ}{T} = \frac{-42.6\ \text{kJ mol}^{-1}}{298.0\ \text{K}} = -143\ \text{J K}^{-1}\ \text{mol}^{-1}$$

$$\Delta S_{total} = -142.95\ \text{J K}^{-1}\ \text{mol}^{-1} + 53.0\ \text{J K}^{-1}\ \text{mol}^{-1} = -90.0\ \text{J K}^{-1}\ \text{mol}^{-1}$$

P5.34) Consider the reversible Carnot cycle shown in Figure 5.2 with 1.75 mol of an ideal gas with $C_{V,m} = 3/2R$ as the working substance. The initial isothermal expansion occurs at the hot reservoir temperature of $T_{hot} = 920.$ K from an initial volume of 4.00 L (V_a) to a volume of 11.50 L (V_b). The system then undergoes an adiabatic expansion until the temperature falls to $T_{cold} = 375$ K. The system then undergoes an isothermal compression and a subsequent adiabatic compression until the initial state described by $T_a = 920.$ K and $V_a = 4.00$ L is reached.

a. Calculate V_c and V_d.

b. Calculate w for each step in the cycle and for the total cycle.

c. Calculate ε and the amount of heat that is extracted from the hot reservoir to do 1.00 kJ of work in the surroundings.

a) We first calculate V_c and V_d.

$$\frac{T_c}{T_b} = \left(\frac{V_c}{V_b}\right)^{1-\gamma} ; \frac{V_c}{V_b} = \left(\frac{T_c}{T_b}\right)^{\frac{1}{1-\gamma}}$$

$$\frac{V_c}{V_b} = \left(\frac{375 \text{ K}}{920 \text{ K}}\right)^{\frac{1}{1-\frac{5}{3}}} = \left(\frac{375 \text{ K}}{920 \text{ K}}\right)^{-\frac{3}{2}} = 3.84; \quad V_c = 44.2 \text{ L}$$

$$\frac{T_d}{T_a} = \left(\frac{V_d}{V_a}\right)^{1-\gamma} ; \frac{V_d}{V_a} = \left(\frac{T_d}{T_a}\right)^{\frac{1}{1-\gamma}}$$

$$\frac{V_d}{V_a} = \left(\frac{375 \text{ K}}{920 \text{ K}}\right)^{\frac{1}{1-\frac{5}{3}}} = \left(\frac{375 \text{ K}}{920 \text{ K}}\right)^{-\frac{3}{2}} = 3.84; \quad V_d = 15.4 \text{ L}$$

b) We next calculate w for each step in the cycle, and for the total cycle.

$$w_{ab} = -nRT_a \ln\frac{V_b}{V_a} = -1.75 \text{ mol} \times 8.314 \text{ J mol}^{-1} \text{ K}^{-1} \times 920 \text{ K} \times \ln\frac{11.5 \text{ L}}{4.00 \text{ L}} = -1.41 \times 10^4 \text{ J}$$

$$w_{bc} = nC_{V,m}(T_c - T_b) = 1.75 \text{ mol} \times \frac{3}{2} \times 8.314 \text{ J mol}^{-1} \text{ K}^{-1} \times (375 \text{ K} - 920 \text{ K}) = -1.19 \times 10^4 \text{ J}$$

$$w_{cd} = -nRT_c \ln\frac{V_d}{V_c} = -1.75 \text{ mol} \times 8.314 \text{ J mol}^{-1} \text{ K}^{-1} \times 375 \text{ K} \times \ln\frac{15.4 \text{ L}}{44.2 \text{ L}} = +5.76 \times 10^3 \text{ J}$$

$$w_{da} = nC_{V,m}(T_a - T_d) = 1.75 \text{ mol} \times \frac{3}{2} \times 8.314 \text{ J mol}^{-1} \text{ K}^{-1} \times (920 \text{ K} - 375 \text{ K}) = +1.19 \times 10^4 \text{ J}$$

$$w_{total} = -1.41 \times 10^4 \text{ J} - 1.19 \times 10^4 \text{ J} + 5.76 \times 10^3 \text{ J} + 1.19 \times 10^4 \text{ J} = -8.37 \times 10^3 \text{ J}$$

$$\varepsilon = 1 - \frac{T_{cold}}{T_{hot}} = 1 - \frac{375 \text{ K}}{920 \text{ K}} = 0.592 \qquad q = \frac{w}{\varepsilon} = \frac{1.00 \times 10^3 \text{ J}}{0.592} = 1.69 \times 10^3 \text{ J}$$

Therefore, 1.69 kJ of heat must be extracted from the surroundings to do 1.00 kJ of work in the surroundings.

P5.35) Between 0°C and 100°C, the heat capacity of Hg(l) is given by

$$\frac{C_{P,m}(\text{Hg},l)}{\text{J K}^{-1}\text{mol}^{-1}} = 30.093 - 4.944\times10^{-3}\frac{T}{\text{K}}$$

Calculate ΔH and ΔS if 1.75 moles of Hg(l) are raised in temperature from 0.00° to 75.0°C at constant P.

$$\Delta H = n \int_{273.15}^{348.15} C_{P,m} d[T/K]$$

$$= 1.75 \text{ mol} \times \left[30.093\times(348.15-273.15) - 2.472\times10^{-3}\left(348.15^2 - 273.15^2\right)\right] \text{J mol}^{-1}$$

$$= 3.75\times10^3 \text{ J}$$

$$\Delta S =$$

$$n\int_{273.15}^{348.15}\frac{C_{P,m}}{[T/K]}d[T/K] = 1.75 \text{ mol}\times\left[30.093 \ln\frac{T_f}{T_i} - 4.944\times10^{-3}(348.15-273.15)\right]\text{J K}^{-1}\text{mol}^{-1}$$

$$= 12.1 \text{ J K}^{-1}$$

P5.36) Using the data in Problem P4.10 and the equation $\Delta H_{den} = \int_{T_1}^{T_2}\delta C_P^{trs}dT$, calculate ΔS_{den}. Assume

the denaturation occurs reversibly. Hint: Determine ΔH_{den} graphically, then determine T_m. You can perform the integration numerically using either a spread sheet program or a curve-fitting routine and a graphing calculator (see Example Problem 5.9).

We divide the area under the curve into rectangles of width 1.0 K, add the areas and multiply by the molecular mass of 14000. g mol^{-1}. The vertical axis is $C_{P,m}/T$ and the units for the areas in the figure below are J K^{-1} g^{-1}. The result is 190 kJ mol^{-1}.

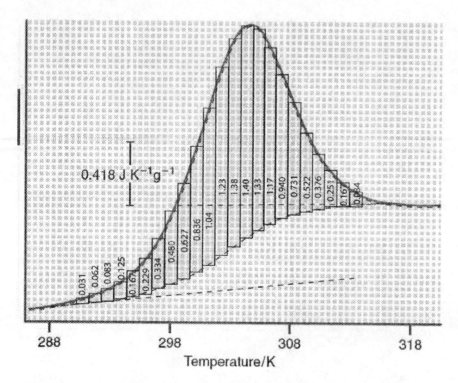

P5.39) The heat capacity of α-quartz is given by

$$\frac{C_{P,m}(\alpha\text{-quartz},s)}{\text{J K}^{-1}\text{ mol}^{-1}} = 46.94 + 34.31 \times 10^{-3}\frac{T}{\text{K}} - 11.30 \times 10^5 \frac{T^2}{\text{K}^2}$$

The coefficient of thermal expansion is given by $\beta = 0.3530 \times 10^{-4}\text{ K}^{-1}$ and $V_m = 22.6\text{ cm}^3\text{ mol}^{-1}$.

Calculate ΔS_m for the transformation α–quartz (38.0°C, 1 atm) $\rightarrow \alpha$–quartz (315°C, 850. atm). From equations 5.23 and 5.24

$$\Delta S_m = \int_{T_i}^{T_f} C_{P,m}\frac{dT}{T} - V\beta\left(P_f - P_i\right)$$

$$= \int_{311.15}^{588.15} \frac{\left(46.94 + 34.31 \times 10^{-3}\frac{T}{\text{K}} - 11.3 \times 10^{-5}\left(\frac{T}{\text{K}}\right)^2\right)}{T/\text{K}} d\frac{T}{\text{K}}\text{J K}^{-1}\text{ mol}^{-1} - V\beta\left(P_f - P_i\right)$$

$$= \left[46.94 \times \ln\frac{588.15\text{ K}}{311.15\text{ K}} + 34.31 \times 10^{-3} \times (498 - 298) - 5.65 \times 10^{-5} \times \left(588.15^2 - 311.15^2\right)\right]\text{J K}^{-1}\text{ mol}^{-1}$$

$$-22.6\text{ cm}^3\text{ mol}^{-1} \times \frac{1\text{ m}^3}{10^6\text{ cm}^3} \times 0.3530 \times 10^{-4}\text{ K}^{-1} \times 849\text{ atm} \times \frac{1.0125 \times 10^5\text{ Pa}}{\text{atm}}$$

$$= 29.9\text{ J K}^{-1}\text{ mol}^{-1} + 0.0218\text{ J K}^{-1}\text{ mol}^{-1} - 14.1\text{ J K}^{-1}\text{ mol}^{-1} - 0.0686\text{ J K}^{-1}\text{ mol}^{-1} = 25.2\text{ J K}^{-1}\text{ mol}^{-1}$$

P5.40)

a. Calculate ΔS if 1 mol of liquid water is heated from $0.00°$ to $10.0°C$ under constant pressure if $C_{P,m} = 75.3$ J K^{-1} mol^{-1}.

b. The melting point of water at the pressure of interest is $0.00°C$ and the enthalpy of fusion is 6.010 kJ mol^{-1}. The boiling point is $100.°C$ and the enthalpy of vaporization is 40.65 kJ mol^{-1}. Calculate ΔS for the transformation $H_2O(s, 0.00°C) \rightarrow H_2O(g, 100.°C)$.

a) The heat input is the same for a reversible and an irreversible process.

$$dq = dq_{reversible} = nC_{P,m}dT$$

$$\Delta S = n\int \frac{C_{P,m}}{T}dT = nC_{P,m}\ln\frac{T_f}{T_i}$$

$$= 1\ mol \times 75.3\ J\ mol^{-1}\ K^{-1}\ln\frac{283.15\ K}{273.15\ K}$$

$$= 2.71\ J\ K^{-1}$$

b)

$$\Delta S_{fusion} = \frac{\Delta H_{fusion}}{T_{fusion}} = \frac{1\ mol \times 6010\ J\ mol^{-1}}{273.15\ K} = 22.00\ J\ K^{-1}$$

$$\Delta S_{vaporization} = \frac{\Delta H_{vaporization}}{T_{vaporization}} = \frac{1\ mol \times 40650.\ J\ mol^{-1}}{373.15\ K} = 108.95\ J\ K^{-1}$$

$$\Delta S_{total} = \Delta S_{fusion} + \Delta S_{vaporization} + \Delta S_{heating} = (22.00 + 108.95 + 23.49)\ J\ K^{-1}$$

$$= 154\ J\ K^{-1}$$

P5.42) The following heat capacity data have been reported for L-alanine:

$T(K)$	10.	20.	40.	60.	80.	100.	140.	180.	220.	260.	300.
$C°_{P,m}$ (J K^{-1} mol^{-1})	0.49	3.85	17.45	30.99	42.59	52.50	68.93	83.14	96.14	109.6	122.7

By a graphical treatment, obtain the molar entropy of L-alanine at $T = 300.$ K. You can perform the integration numerically using either a spread sheet program or a curve-fitting routine and a graphing calculator (see Example Problem 5.9).

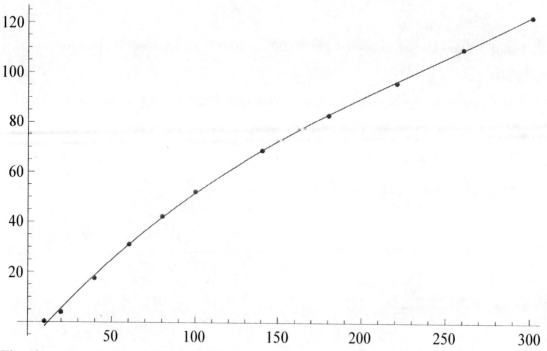

The above graph shows $C_{P,m}$ as a function of T. The line is the best fit to the data of a polynomial of the form $a + bx + cx^2 + dx^3$ given by

$-9.1602 + 0.77610\, x - 0.0019511\, x^2 + 2.76403 \times 10^{-6}\, x^3$.

$$S^\circ(300\text{ K}) = \int_{10\,\text{K}}^{300\,\text{K}} \frac{-9.1602 + 0.77610T - 0.0019511T^2 + 2.76403 \times 10^{-6} T^3}{T}\, dT = 135.2 \text{ J K}^{-1}\text{ mol}^{-1}$$

We have integrated from 10 K rather than 0 K because the integrand is not well defined at 0 K and because the contribution to S below 10 K is very small.

P5.43) The mean solar flux at the Earth's surface is ~2.00 J cm^{-2} min^{-1}. In a nonfocusing solar collector, the temperature can reach a value of 85.0°C. A heat engine is operated using the collector as the hot reservoir and a cold reservoir at 298 K. Calculate the area of the collector needed to produce one horsepower (1 hp = 746 watts). Assume that the engine operates at the maximum Carnot efficiency.

$$\varepsilon = 1 - \frac{T_{hot}}{T_{cold}} = 1 - \frac{298.2\text{ K}}{358.2\text{ K}} = 0.168$$

The area required for the solar panel is

$$\frac{\text{power}}{\text{efficiency} \times \text{flux}} = \frac{746 \text{ J s}^{-1}}{0.168 \times 2.00 \text{ J cm}^{-2}\text{ min}^{-1} \times 1 \text{ min}/60 \text{ s} \times 10^4 \text{cm}^2/\text{m}^2} = 13.3 \text{ m}^2$$

Chapter 6: Chemical Equilibrium

P6.1) Calculate $\Delta A_{reaction}^{\circ}$ and $\Delta G_{reaction}^{\circ}$ for the reaction $C_6H_6(l) + 15/2O_2(g) \rightarrow 6CO_2(g) + 3H_2O(l)$ at 298 K from the combustion enthalpy of benzene and the entropies of the reactants and products. All gaseous reactants and products are treated as ideal gases

$$\Delta G_{combustion}^{\circ} = \Delta H_{combustion}^{\circ} - T\Delta S_{combustion}^{\circ}$$

$$\Delta S_{combustion}^{\circ} = 6S^{\circ}(CO_2,g) + 3S^{\circ}(H_2O,l) - S^{\circ}(C_6H_6,l) - 15/2S^{\circ}(O_2,g)$$

$$= 6 \times 213.8 \text{ J mol}^{-1} \text{ K}^{-1} + 3 \times 70.0 \text{ J mol}^{-1} \text{ K}^{-1} - 173.4 \text{ J mol}^{-1} \text{ K}^{-1} - 15/2 \times 205.2 \text{ J mol}^{-1} \text{ K}^{-1}$$

$$= -219.6 \text{ J mol}^{-1} \text{ K}^{-1}$$

$$\Delta G_{combustion}^{\circ} = -3268 \times 10^3 \text{ kJ mol}^{-1} - 298.15 \text{ K} \times \left(-219.6 \text{ J mol}^{-1} \text{ K}^{-1}\right) = -3203 \times 10^3 \text{ kJ mol}^{-1}$$

$$\Delta A_{combustion}^{\circ} = \Delta U_{combustion}^{\circ} - T\Delta S_{combustion}^{\circ}$$

$$= \Delta H_{combustion}^{\circ} - \Delta(PV)_{combustion} - T\Delta S_{combustion}^{\circ}$$

$$= \Delta G_{combustion}^{\circ} + T\Delta S_{combustion}^{\circ} - \Delta(PV) - T\Delta S_{combustion}^{\circ}$$

$$= \Delta G_{combustion}^{\circ} - \Delta nRT$$

where Δn is the change in the number of moles of gas phase species in the reaction

$$\Delta A_{combustion}^{\circ} = -3203 \times 10^3 \text{ kJ mol}^{-1} + 1.5 \times 8.314 \text{ J mol}^{-1}\text{K}^{-1} \times 298.15 \text{ K}$$

$$= -3199 \times 10^3 \text{ kJ mol}^{-1}$$

The change in PV for the liquid can be neglected because liquids are essentially incompressible over the pressure range in this problem.

P6.3) A sample containing 3.50 moles of N_2 and 7.50 mol of H_2 is placed in a reaction vessel and brought to equilibrium at 35.0 bar and 750. K in the reaction $1/2N_2(g) + 3/2H_2(g) \rightarrow NH_3(g)$.

a. Calculate K_P at this temperature.

b. Set up an equation relating K_P and the extent of reaction as in Example Problem 6.10.

c. Using a numerical equation solver, calculate the number of moles of each species present at equilibrium.

a)

$$\Delta H^{\circ}_{reaction} = \Delta H^{\circ}_f\left(NH_3,g\right) = -45.9\times10^3 \text{ J mol}^{-1}$$

$$\Delta G^{\circ}_{reaction} = \Delta G^{\circ}_f\left(NH_3,g\right) = -16.5\times10^3 \text{ J mol}^{-1}$$

$$\ln K_P\left(T_f\right) = -\frac{\Delta G^{\circ}_{reaction}\left(298.15 \text{ K}\right)}{R\times298.15 \text{ K}} - \frac{\Delta H^{\circ}_{reaction}}{R}\left(\frac{1}{T_f} - \frac{1}{298.15 \text{ K}}\right)$$

$$\ln K_P\left(750.\text{ K}\right) = \frac{16.5\times10^3 \text{ J mol}^{-1}}{8.314 \text{ J mol}^{-1}\text{ K}^{-1}\times298.15 \text{ K}} + \frac{45.9\times10^3 \text{ J mol}^{-1}}{8.314 \text{ J mol}^{-1}\text{ K}^{-1}}\times\left(\frac{1}{750.\text{ K}} - \frac{1}{298.15 \text{ K}}\right)$$

$$= -4.4994$$

$$K_P\left(750.\text{ K}\right) = 1.11\times10^{-2}$$

b)

	$1/2 N_2(g) +$	$3/2 H_2(g) \leftrightarrow$	$NH_3(g)$
Initial number of moles	3.50	7.50	0
Moles present at equilibrium	$3.50 - \xi$	$7.50 - 3\xi$	2ξ
Mole fraction present at equilibrium	$\dfrac{3.50 - \xi}{11.00 - 2\xi}$	$\dfrac{7.50 - 3\xi}{11.00 - 2\xi}$	$\dfrac{2\xi}{11.00 - 2\xi}$
Partial pressure at Equilibrium, $P_i = x_i P$	$\dfrac{3.50 - \xi}{11.00 - 2\xi}P$	$\dfrac{7.50 - 3\xi}{11.00 - 2\xi}P$	$\left(\dfrac{2\xi}{11.00 - 2\xi}\right)P$

We next express K_P in terms of n_0, ξ, and P.

$$K_P\left(T\right) = \frac{\left(\dfrac{P^{eq}_{NH_3}}{P^{\circ}}\right)}{\left(\dfrac{P^{eq}_{N_2}}{P^{\circ}}\right)^{\frac{1}{2}}\left(\dfrac{P^{eq}_{H_2}}{P^{\circ}}\right)^{\frac{3}{2}}} = \frac{\left(\dfrac{2\xi}{11.00 - 2\xi}\right)\dfrac{P}{P^{\circ}}}{\left(\left(\dfrac{3.50 - \xi}{11.00 - 2\xi}\right)\dfrac{P}{P^{\circ}}\right)^{\frac{1}{2}}\left(\left(\dfrac{7.50 - 3\xi}{11.00 - 2\xi}\right)\dfrac{P}{P^{\circ}}\right)^{\frac{3}{2}}}$$

The following equation can be solved numerically using a program such as Mathematica

$$K_P\left(T\right) = \frac{\left(\dfrac{2\xi}{11.00 - 2\xi}\right)35.0}{\left(\left(\dfrac{3.50 - \xi}{11.00 - 2\xi}\right)35.0\right)^{\frac{1}{2}}\left(\left(\dfrac{7.50 - 3\xi}{11.00 - 2\xi}\right)35.0\right)^{\frac{3}{2}}} = 1.11\times10^{-2}$$

The physically meaningful root of the cubic equation is $\xi = 0.4964$. Therefore, there are 3.00 moles of $N_2(g)$, 6.01 moles of $H_2(g)$, and 0.99 moles of $NH_3(g)$ at equilibrium.

P6.4) Consider the equilibrium $NO_2(g) \rightleftharpoons NO(g) + 1/2O_2(g)$. One mole of $NO_2(g)$ is placed in a vessel and allowed to come to equilibrium at a total pressure of 1 bar. An analysis of the contents of the vessel gives the following results:

T	600. K	900. K
P_{NO}/P_{NO_2}	0.224	5.12

a. Calculate K_P at 600. and 900. K.

b. Calculate ΔG_R° and ΔH_R° for this reaction at 298.15 K, using only the data in the problem. Assume that ΔH_R° is independent of temperature.

c. Calculate ΔH_R° using the data tables and compare your answer with that obtained in part b.

a) $NO_2(g) \rightarrow NO(g) + \dfrac{1}{2}O_2(g)$

$$K_P = \frac{\left(P_{NO}/P^\circ\right)\left(P_{O_2}/P_O\right)^{1/2}}{P_{NO_2}/P^\circ}$$

At 600. K, $\dfrac{P_{NO}}{P_{NO_2}} = 0.224$ and $P_{O_2} = \dfrac{1}{2}P_{NO}$

$P_{total} = P_{NO} + P_{NO_2} + P_{O_2} = 1$ bar

1 bar $= 0.224\,P_{NO_2} + P_{NO_2} + 0.112\,P_{NO_2}$

1 bar $= 1.336\,P_{NO_2}$

$P_{NO_2} = 0.7485$ bar

$$K_P = \frac{(0.224 \times 0.7485) \times \sqrt{0.112 \times 0.7485}}{0.7485} = 0.0649$$

At 900. K, $\dfrac{P_{NO}}{P_{NO_2}} = 5.12$ and $P_{O_2} = \dfrac{1}{2}P_{NO}$

$P_{Total} = P_{NO} + P_{NO_2} + P_{O_2}$

1 bar $= 5.12\,P_{NO_2} + P_{NO_2} + 2.56\,P_{NO_2} = 8.78\,P_{NO_2}$

$P_{NO_2} = 0.1152$ bar

$$K_P = \frac{(5.12 \times 0.1152) \times \sqrt{2.56 \times 0.1152}}{0.1152} = 2.78$$

b) Assuming that $\Delta H^\circ_{reaction}$ is independent of temperature,

$$\ln \frac{K_P(900\text{ K})}{K_P(600\text{ K})} = \frac{-\Delta H^\circ_{reaction}}{R}\left(\frac{1}{900.\text{ K}} - \frac{1}{600.\text{ K}}\right)$$

$$\Delta H^\circ_{reaction} = -\frac{R \times \ln\left(\dfrac{K_P(900.\text{ K})}{K_P(600.\text{ K})}\right)}{\left(\dfrac{1}{900.\text{ K}} - \dfrac{1}{600.\text{ K}}\right)} - 56.2 \times 10^3 \text{ J mol}^{-1}$$

$$\ln K_P(298.15\text{ K}) = \ln K_P(600.\text{ K}) - \frac{\Delta H^\circ_{reaction}}{R}\times\left(\frac{1}{298.15\text{ K}} - \frac{1}{600.\text{ K}}\right) = -14.15$$

$$\Delta G^\circ_{reaction}(298.15\text{ K}) = -RT\ln K_P(298.15\text{ K})$$

$$= -8.314 \text{ J mol}^{-1}\text{ K}^{-1}\times298.15\text{ K}\times(-14.15)$$

$$= 35.1\times10^3 \text{ J mol}^{-1}$$

c. $\Delta H^\circ_{reaction} = \Delta H^\circ_f(NO, g) - \Delta H^\circ_f(NO_2, g)$

$$= 91.3\times10^3 \text{ J mol}^{-1} - 33.2\times10^3 \text{ J mol}^{-1} = 58.1\times10^3 \text{ J mol}^{-1}$$

P6.7) The pressure dependence of G is quite different for gases and condensed phases. Calculate ΔG_m for the processes (C, *solid, graphite*, 1 bar, 298.15 K)$\rightarrow$ (C, *solid, graphite*, 250. bar, 298.15 K) and (He, g, 1 bar, 298.15 K) $\rightarrow$ (He, g, 250. bar, 298.15 K). By what factor is ΔG_m greater for He than for graphite?

For a solid or liquid, we can assume that the volume is independent of pressure over a limited range in P.

$$\Delta G = \int_{P_i}^{P_f} VdP = V\left(P_f - P_i\right)$$

$$\Delta G_m(C, s, 250\text{ bar}) = \Delta G_m(C, s, 1\text{ bar}) + V_m\left(P_f - P_i\right) = G_m(C, s, 1\text{ bar}) + \frac{M}{\rho}\left(P_f - P_i\right)$$

$$= 0 + \frac{12.011\times10^{-3}\text{ kg}}{2250 \text{ kg m}^{-3}}\times249.0\times10^5 \text{ Pa} = 133 \text{ J mol}^{-1}$$

Treating He as an ideal gas,

$$G_m(He, g, 250.\text{ bar}) = G_m(He, g, 1\text{ bar}) + \int_{P_i}^{P_f} VdP$$

$$= 0 + RT\ln\frac{P_f}{P_i} = 1 \text{ mole}\times8.314 \text{ J mol}^{-1}\text{ K}^{-1}\times298.15\text{ K}\times\ln\frac{250.\text{ bar}}{1\text{ bar}} = 13.7\times10^3 \text{ J mol}^{-1}$$

This result is a factor of 103 greater than that for graphite.

P6.10) Calculate K_P at 725 K for the reaction $N_2O_4(l) \rightarrow 2NO_2(g)$ assuming that ΔH_R° is constant over the interval 298–725 K.

$$\Delta H_{reaction}^\circ = 2\Delta H_f^\circ(NO_2, g) - \Delta H_f^\circ(N_2O_4, l)$$
$$= 2 \times 33.2 \times 10^3 \text{ J mol}^{-1} + 19.5 \times 10^3 \text{ J mol}^{-1} = 85.9 \times 10^3 \text{ J mol}^{-1}$$

$$\Delta G_{reaction}^\circ = 2\Delta G_f^\circ(NO_2, g) - \Delta G_f^\circ(N_2O_4, l)$$
$$= 2 \times 51.3 \times 10^3 \text{ J mol}^{-1} - 97.5 \times 10^3 \text{ J mol}^{-1} = 5.1 \times 10^3 \text{ J mol}^{-1}$$

We use the Gibbs–Helmholtz equation to relate the Gibbs reaction energy at two temperatures.

$$\ln K_P(T_f) = -\frac{\Delta G_{reaction}^\circ(298.15 \text{ K})}{R \times 298.15 \text{ K}} - \frac{\Delta H_{reaction}^\circ}{R}\left(\frac{1}{T_f} - \frac{1}{298.15 \text{ K}}\right)$$

$$\ln K_P(725 \text{ K}) = -\frac{5.10 \times 10^3 \text{ J mol}^{-1}}{8.314 \text{ J K}^{-1}\text{mol}^{-1} \times 298.15 \text{ K}} - \frac{85.9 \times 10^3 \text{ J mol}^{-1}}{8.314 \text{ J K}^{-1}\text{ mol}^{-1}} \times \left(\frac{1}{725 \text{ K}} - \frac{1}{298.15 \text{ K}}\right)$$

$$\ln K_P(725 \text{ K}) = 18.35$$

$$K_P(725 \text{ K}) = 9.27 \times 10^7$$

P6.12) For the reaction $C(graphite) + H_2O(g) \rightleftharpoons CO(g) + H_2(g)$, $\Delta H_R^\circ = 131.28$ kJ mol^{-1} at 298.15 K. Use the values of $C_{P,m}^\circ$ at 298.15 K in the data tables to calculate ΔH_R° at 125.0°C.

$$\Delta H(T_2) = \Delta H(T_1) + \Delta C_{P,m}(T_2 - T_1)$$

$$\Delta H(398.15 \text{ K}) = 131.28 \text{ kJ mol}^{-1}$$

$$+100. \text{ K} \times \left(\begin{array}{c} 29.14 \text{ J K}^{-1}\text{ mol}^{-1} + 28.84 \text{ J K}^{-1}\text{ mol}^{-1} - 8.52 \text{ J K}^{-1}\text{ mol}^{-1} \\ -33.59 \text{ J K}^{-1}\text{ mol}^{-1} \end{array}\right)$$

$$= 132.9 \text{ kJ mol}^{-1}$$

P6.13) Ca(HCO$_3$)$_2$(s) decomposes at elevated temperatures according to the stoichiometric equation Ca(HCO$_3$)$_2$(s) $\rightarrow$ CaCO$_3$(s) + H$_2$O(g) + CO$_2$(g).

a. If pure Ca(HCO$_3$)$_2$(s) is put into a sealed vessel, the air is pumped out, and the vessel and its contents are heated, the total pressure is 0.235 bar. Determine K_P under these conditions.

b. If the vessel initially also contains 0.105 bar H$_2$O(g), what is the partial pressure of CO$_2$(g) at equilibrium?

a) $$Ca(HCO_3)_2(s) \leftrightarrow CaCO_3(s) + H_2O(g) \quad + \quad CO_2(g)$$

Partial pressure at

Equilibrium, $P_i = x_i P$ $\qquad\qquad\qquad \xi P \qquad\qquad\qquad \xi P$

The total pressure is made up of equal partial pressures of $H_2O(g)$ and $CO_2(g)$.

$$K_P = \frac{P_{H_2O}}{P^\circ}\frac{P_{CO_2}}{P^\circ} = \left(\frac{P_{H_2O}}{P^\circ}\right)^2 = \left(\frac{0.235}{2}\right)^2 = 0.0138$$

b) If one of the products is originally present

$$Ca(HCO_3)_2(s) \leftrightarrow CaCO_3(s) + H_2O(g) \quad + \quad CO_2(g)$$

Partial pressure at

Equilibrium, $P_i = x_i P$ $\qquad\qquad\qquad \xi P + P_i \qquad\qquad \xi P$

$$K_P = \frac{P_{H_2O}}{P^\circ}\frac{P_{CO_2}}{P^\circ} = \left(\frac{P + P_i}{P^\circ}\right)\left(\frac{P}{P^\circ}\right) = \left(\frac{P}{P^\circ} + 0.105\right)\left(\frac{P}{P^\circ}\right) = 0.0138$$

$$\frac{P}{P^\circ} = \frac{P_{CO_2}}{P^\circ} = 0.0762; \quad P_{CO_2} = 0.0762 \text{ bar}$$

P6.16) Collagen is the most abundant protein in the mammalian body. It is a fibrous protein that serves to strengthen and support tissues. Suppose a collagen fiber can be stretched reversibly with a force constant of $k = 10.0 \text{ N m}^{-1}$ and that the force, **F** (see Table 2.1), is given by $\mathbf{F} = -k\mathbf{l}$. When a collagen fiber is contracted reversibly, it absorbs heat $q_{rev} = 0.050 \text{ J}$. Calculate the change in the Helmholtz energy, ΔA, as the fiber contracts isothermally from $l = 0.20$ to 0.10 m. Calculate also the reversible work performed on w_{rev}, ΔS, and ΔU. Assume that the temperature is constant at $T = 310.$ K.

$$w_{rev} = -\int_{0.20m}^{0.10m} kl\,dl = \left[-\frac{1}{2}kl^2\right]_{0.20m}^{0.10m} = -10.0 \text{ N m}^{-1} \times \frac{(0.20 \text{ m})^2 - (0.10 \text{ m})^2}{2} = -0.15 \text{ J}$$

$$\Delta A = w_{rev} = -0.15 \text{ J}$$

$$\Delta S = \frac{q_{rev}}{T} = \frac{0.050 \text{ J}}{310.\text{ K}} = 1.61 \times 10^{-4} \text{ J K}^{-1}$$

$$\Delta U = \Delta A + T\Delta S = -0.15 \text{ J} + 0.050 \text{ J} = -0.10 \text{ J}$$

P6.17) Calculate $\mu_{N_2}^{mixture}(298.15 \text{ K}, 1 \text{bar})$ for nitrogen in air, assuming that the mole fraction of N_2 in air is 0.800. Use the conventional molar Gibbs energy defined in Section 6.17.

We calculate the conventional molar Gibbs energy as described in Example Problem 6.17.

$$\mu^{\circ}_{N_2}(T) = -TS^{\circ}(N_2,g,298.15\ K) = -298.15\ K \times 191.6\ J\ K^{-1}\ mol^{-1} = -57.1\ kJ\ mol^{-1}$$

$$\mu^{mixture}_{N_2}(T,P) = \mu^{\circ}_{N_2}(T) + RT\ln\frac{P}{P^{\circ}} + RT\ln x_{N_2}$$

$$= -57.1\ kJ\ mol^{-1} + RT\ln\frac{1\ bar}{1\ bar} + 8.314\ J\ mol^{-1}K^{-1} \times 298.15\ K \times \ln 0.800$$

$$= -57.7\times10^3\ J\ mol^{-1}$$

P6.18) Calculate the maximum nonexpansion work that can be gained from the combustion of benzene(*l*) and of $H_2(g)$ on a per gram and a per mole basis under standard conditions. Is it apparent from this calculation why fuel cells based on H_2 oxidation are under development for mobile applications?

$C_6H_6(l) + 15/2O_2(g) \leftrightarrow 6CO_2(g) + 3H_2O(l)$

$$w^{max}_{nonexpansion} = \Delta G^{\circ}_R = 3\Delta G^{\circ}_f(H_2O,l) + 6\Delta G^{\circ}_f(CO_2,g) - \frac{15}{2}\Delta G^{\circ}_f(O_2,g) - \Delta G^{\circ}_f(C_6H_6,l)$$

$$w^{max}_{nonexpansion} = 3\times(-237.1\ kJ\ mol^{-1}) + 6\times(-394.4\ kJ\ mol^{-1}) - \frac{15}{2}\times(0) - 124.5 - 124.5\ kJ\ mol^{-1}$$

$$= -3202\ kJ\ mol^{-1}$$

$$= -3202\ kJ\ mol^{-1} \times \frac{1\ mol}{78.18\ g} = -40.96\ kJ\ g^{-1}$$

$H_2(g) + 1/2O_2(g) \rightarrow H_2O(l)$

$$w^{max}_{nonexpansion} = \Delta G^{\circ}_R = \Delta G^{\circ}_f(H_2O,l) - \frac{1}{2}\Delta G^{\circ}_f(O_2,g) - \Delta G^{\circ}_f(H_2,g)$$

$$w^{max}_{nonexpansion} = -237.1\ kJ\ mol^{-1} - 0 - 0$$

$$= -237.1\ kJ\ mol^{-1}$$

$$= -237.1\ kJ\ mol^{-1} \times \frac{1\ mol}{2.016\ g} = -117.6\ kJ\ g^{-1}$$

On a per gram basis, nearly three times as much work can be extracted from the oxidation of hydrogen than benzene.

P6.20) Calculate ΔG for the isothermal expansion of 1.75 mol of an ideal gas at 298 K from an initial pressure of 7.5 bar to a final pressure of 1.5 bar.

$$dG = -SdT + VdP$$

At constant T, we consider the reversible process. Because G is a state function, any path between the same initial and final states will give the same result.

$$\Delta G = \int_{P_i}^{P_f} VdP = nRT \ln \frac{P_f}{P_i} = 1.75 \ mol \times 8.314 \ J \ mol^{-1} \ K^{-1} \times 298 \ K \times \ln \frac{1.50 \ bar}{7.50 \ bar}$$

$$= -6.98 \times 10^3 \ J$$

P6.24) Consider the reaction $FeO(s) + CO(g) \rightleftharpoons Fe(s) + CO_2(g)$ for which K_P is found to have the following values:

T	600.°C	1000.°C
K_P	0.900	0.396

a. Calculate $\Delta G^{\circ}_{reaction}$, $\Delta S^{\circ}_{reaction}$, and $\Delta H^{\circ}_{reaction}$ for this reaction at 600. °C. Assume that $\Delta H^{\circ}_{reaction}$ is independent of temperature.

b. Calculate the mole fraction of $CO_2(g)$ present in the gas phase at 600.°C.

a) $FeO(s) + CO(g) \rightarrow Fe(s) + CO_2(g)$

$$K_P = \frac{P_{CO_2} / P^{\circ}}{P_{CO} / P^{\circ}}$$

$$\ln \frac{K_P \left(1000.^{\circ}C\right)}{K_P \left(600.^{\circ}C\right)} = \frac{\Delta H^{\circ}_{reaction}}{R} \left(\frac{1}{1273.15 \ K} - \frac{1}{873.15 \ K}\right)$$

Assume that $\Delta H^{\circ}_{reaction}$ is independent of temperature.

$$\Delta H^{\circ}_{reaction} = \frac{-R \ln \frac{K_P \left(1000.^{\circ}C\right)}{K_P \left(600.^{\circ}C\right)}}{\left(\frac{1}{1273.15 \ K} - \frac{1}{873.15 \ K}\right)}$$

$$= \frac{-8.314 \ J \ mol^{-1} \ K^{-1} \times \frac{0.0396}{0.900}}{\left(\frac{1}{1273.15 \ K} - \frac{1}{873.15 \ K}\right)} = -19.0 \ kJ \ mol^{-1}$$

$$\Delta G^{\circ}_{reaction}\left(600.^{\circ}C\right) = -RT \ln K_P \left(600.^{\circ}C\right)$$

$$= -8.314 \ J \ mol^{-1} \ K^{-1} \times 873.15 \ K \times \ln(0.900) = 765 \ J \ mol^{-1}$$

$$\Delta S^{\circ}_{reaction}\left(600.^{\circ}C\right) = \frac{\Delta H^{\circ}_{reaction} - \Delta G^{\circ}_{reaction}\left(600.^{\circ}C\right)}{T} = \frac{-18960 \ J \ mol^{-1} - 765 \ J \ mol^{-1}}{873.15 \ K}$$

$$= -22.6 \ J \ mol^{-1} \ K^{-1}$$

b) because $K_P = P_{CO2}/P_{CO} = 0.900$

$K_P = K_x$ because $\Delta n = 0$

$\dfrac{x_{CO_2}}{x_{CO}} = 0.900$ and $x_{CO_2} + x_{CO} = 1$

$x_{CO_2} = 0.47 \quad x_{CO} = 0.53$

P6.27) A gas mixture with 3.00 mol of Ar, x moles of Ne, and y moles of Xe is prepared at a pressure of 1 bar and a temperature of 298 K. The total number of moles in the mixture is four times that of Ar. Write an expression for ΔG_{mixing} in terms of x. At what value of x does the magnitude of ΔG_{mixing} have its minimum value? Calculate ΔG_{mixing} for this value of x.

If the number of moles of Ne is x, the number of moles of Xe is $y = 8 - x$.

$$\Delta G_{mixing} = nRT \sum_i x_i \ln x_i$$

$$= nRT\left(-\frac{1}{4}\ln 4 + \frac{x}{12}\ln\frac{x}{12} + \frac{9-x}{12}\ln\frac{9-x}{12}\right)$$

$$\frac{d\Delta G_{mixing}}{dx} = nRT\left(\frac{1}{12}\ln\frac{x}{12} + \frac{x}{12}\frac{12}{x} - \frac{1}{12}\ln\frac{9-x}{12} + \frac{9-x}{12}\frac{12}{9-x}(-1)\right) = 0$$

$$= nRT\left(\frac{1}{12}\ln\frac{x}{12} + 1 - \frac{1}{12}\ln\frac{9-x}{12} - 1\right) = \frac{nRT}{12}\ln\frac{x}{9-x} = 0$$

$$\frac{x}{9-x} = 1; \quad x = \frac{9}{2}$$

$$\Delta G_{mixing} = nRT\left(-\frac{1}{4}\ln 4 + \frac{3}{8}\ln\frac{3}{8} + \frac{3}{8}\ln\frac{3}{8}\right)$$

$$= 12 \text{ mol} \times 8.314 \text{ J mol}^{-1}\text{ K}^{-1} \times 298.15 \text{ K} \times 6.054 = -32.2\times10^3 \text{ J}$$

P6.34) You have containers of pure O_2 and N_2 at 298 K and 1 atm pressure. Calculate ΔG_{mixing} relative to the unmixed gases of

a. a mixture of 12 mol of O_2 and 18 mol of N_2

b. a mixture of 15 mol of O_2 and 15 mol of N_2

c. Calculate ΔG_{mixing} if 12 mol of pure N_2 are added to the mixture of 15 mol of O_2 and 15 mole of N_2.

$$\Delta G_{mixing} = nRT(x_1 \ln x_1 + x_2 \ln x_2)$$

a) $\Delta G_{mixing} = 30 \text{ mol} \times 8.314 \text{ J K}^{-1} \text{ mol}^{-1} \times 298 \text{ K} \times \left(\dfrac{2}{5}\ln\dfrac{2}{5} + \dfrac{3}{5}\ln\dfrac{3}{5} \right) = -50.0 \text{ kJ}$

b) $\Delta G_{mixing} - 30 \text{ mol} \times 8.314 \text{ J K}^{-1} \text{ mol}^{-1} \times 298 \text{ K} \times \left(\dfrac{1}{2}\ln\dfrac{1}{2} + \dfrac{1}{2}\ln\dfrac{1}{2} \right) = -51.5 \text{ kJ}$

c) $\Delta G_{mixing} = \Delta G_{mixing} \ (\text{separate gases})$

$\quad\quad - \Delta G_{mixing} \ (15 \text{ mol A} + 15 \text{ mol B})$

$\quad\quad = 42 \text{ mol} \times 8.314 \text{ J K}^{-1} \text{ mol}^{-1} \times 298 \text{ K} \times \left(\dfrac{5}{14}\ln\dfrac{5}{14} + \dfrac{5}{14}\ln\dfrac{5}{14} \right) + 51.5$

$\quad\quad = -67.8 \text{ kJ} + 51.5 \text{ kJ} = -16.3 \text{ kJ}$

P6.36) Consider the equilibrium in the reaction $3O_2(g) \rightleftharpoons 2O_3(g)$, with $\Delta H_R^\circ = 285.4 \times 10^3 \text{ J mol}^{-1}$ at 298 K. Assume that ΔH_R° is independent of temperature.

a. Without doing a calculation, predict whether the equilibrium position will shift toward reactants or products as the pressure is increased.
b. Without doing a calculation, predict whether the equilibrium position will shift toward reactants or products as the temperature is increased.
c. Calculate K_P at 750. K.
d. Calculate K_x at 750. K and 1.50 bar.

a) The number of moles of products is fewer than the number of moles of reactants. Therefore, the equilibrium position will shift towards products as the pressure is increased.

b)

Because $\Delta H_{reaction}^\circ > 0$, the equilibrium position will shift towards products as the temperature is increased.

c)

$\Delta G_{reaction}^\circ = 2\Delta G_f^\circ (O_3, g) = 2 \times 163.2 \times 10^3 \text{ J mol}^{-1}$

$\Delta H_{reaction}^\circ = 2\Delta H_f^\circ (O_3, g) = 2 \times 142.7 \times 10^3 \text{ J mol}^{-1}$

$$\ln K_{P}\left(T_{f}\right)=-\frac{\Delta G_{reaction}^{\circ}\left(298.15\text{ K}\right)}{R\times298.15\text{ K}}-\frac{\Delta H_{reaction}^{\circ}}{R}\left(\frac{1}{T_{f}}-\frac{1}{298.15\text{ K}}\right)$$

$$\ln K_{P}\left(750.\text{ K}\right)=-\frac{2\times163.2\times10^{3}\text{J mol}^{-1}}{8.314\text{ J K}^{-1}\text{ mol}^{-1}\times298.15\text{ K}}-\frac{2\times142.7\times10^{3}\text{ J mol}^{-1}}{8.314\text{ J K}^{-1}\text{ mol}^{-1}}$$

$$\times\left(\frac{1}{750.\text{ K}}-\frac{1}{298.15\text{ K}}\right)$$

$$\ln K_{P}\left(750.\text{ K}\right)=-62.3103$$

$$K_{P}\left(750.\text{ K}\right)=8.69\times10^{-28}$$

d) Calculate K_X at 750 K and 1.50 bar.

$$K_{X}=K_{P}\left(\frac{P}{P^{\circ}}\right)^{-\Delta\nu}=8.69\times10^{-28}\times\left(\frac{1.50\text{ bar}}{1\text{ bar}}\right)^{+1}=1.30\times10^{-27}$$

P6.37) For a protein denaturation the entropy change is 2.31 J K^{-1} mol^{-1} at P = 1.00 atm and at the melting temperature T = 338 K. Calculate the melting temperature at a pressure of P = 1.00 × 10^{3} atm if the heat capacity change $\Delta C_{P,m}$ = 7.98 J K^{-1} mol^{-1} and if ΔV = 3.10 mL mol^{-1}. State any assumptions you make in the calculation.

At equilibrium for T = 338 K, $\Delta G_{R}=\Delta H_{R}-T\Delta S_{R}=0$

$\Delta H_{R}=T\Delta S_{R}=338\text{ K}\times2.31\text{ J K}^{-1}\text{ mol}^{-1}=780.8\text{ J mol}^{-1}$

The change in enthalpy as both the temperature and pressure change is given by

$\Delta H_{R}\left(T\right)=\Delta H_{R}\left(338\text{ K}\right)+\left(P_{T}-P_{338K}\right)\Delta V+\Delta C_{P}\left(T-338\text{ K}\right)$

$\Delta G_{R}\left(338\text{ K}\right)=\Delta H_{R}\left(338\text{ K}\right)-T\Delta S_{R}\left(338\text{ K}\right)=0$

$\Delta H_{R}\left(338\text{ K}\right)+\left(P_{T}-P_{338K}\right)\Delta V+\Delta C_{P}\left(T-338\text{ K}\right)=T\Delta S_{R}\left(338\text{ K}\right)$

$$T=\frac{\Delta H_{R}\left(338\text{ K}\right)+\left(P_{T}-P_{338K}\right)\Delta V-\Delta C_{P}\left(338\text{ K}\right)}{\Delta S_{R}\left(338\text{ K}\right)-\Delta C_{P}}$$

$$=\frac{\left(\begin{array}{l}780.8\text{ J mol}^{-1}+1.01325\times999\text{ bar}\times10^{5}\text{ Pa bar}^{-1}\times3.10\times10^{-6}\text{ m}^{3}\text{ mol}^{-1}\\-7.98\text{ J K}^{-1}\text{ mol}^{-1}\times338\text{ K}\end{array}\right)}{2.31\text{ J K}^{-1}\text{ mol}^{-1}-7.98\text{ J K}^{-1}\text{ mol}^{-1}}$$

$=283$ K

P6.39) Assume the internal energy of an elastic fiber under tension (see Problem 6.16) is given by $dU=T\,dS-P\,dV-F\,d\ell$. Obtain an expression for $\left(\partial G/\partial\ell\right)_{P,T}$ and calculate the maximum nonexpansion work obtainable when a collagen fiber contracts from ℓ = 20.0 to 10.0 cm at constant P

and T. Assume other properties as described in Problem 6.16.

$$dU = TdS - PdV - \gamma dl$$

$$dG = d(U + PV - TS) = TdS - PdV - \gamma dl + PdV + VdP - TdS - SdT$$

$$= -\gamma dl + VdP - SdT$$

$$\left(\frac{\partial G}{\partial \ell}\right)_{P,T} = -\gamma = kl$$

$$\Delta G = w_{rev} = \int_{0.20m}^{0.10m} kldl = \left[\frac{1}{2}kl^2\right]_{0.20m}^{0.10m} = -10 \text{ N m}^{-1} \times \frac{(0.20 \text{ m})^2 - (0.10 \text{ m})^2}{2} = -0.15 \text{ J}$$

Chapter 7: The Properties of Real Gases

P7.1) A van der Waals gas has a value of $z = 1.00035$ at 325 K and 1 bar and the Boyle temperature of the gas is 180. K. Because the density is low, you can calculate V_m from the ideal gas law. Use this information and the result of Problem 7.28, $z \approx 1 + (b - a/RT)(1/V_m)$, to estimate a and b.

$$z - 1 = \frac{1}{V_m}\left(b - \frac{a}{RT}\right); \quad T_B = \frac{a}{Rb}$$

$$z - 1 = \frac{b}{V_m}\left(1 - \frac{T_B}{T}\right)$$

$$b = \frac{z-1}{1 - \dfrac{T_B}{T}}\frac{RT}{P} = \frac{0.00035}{1 - \dfrac{180.\ K}{325\ K}} \times \frac{8.314 \times 10^{-2} \text{dm}^3 \ \text{bar}\ \text{mol}^{-1}\ \text{K}^{-1} \times 325\ \text{K}}{1\ \text{bar}}$$

$$= 0.0212\ \text{dm}^3\ \text{mol}^{-1} = 2.12 \times 10^{-5}\ \text{m}^3\ \text{mol}^{-1}$$

$$a = RbT_B = 8.314\,\text{J}\,\text{mol}^{-1}\ \text{K}^{-1} \times 3.59 \times 10^{-5}\,\text{m}^3\ \text{mol}^{-1} \times 180.\ \text{K} = 3.17 \times 10^{-2}\,\text{m}^6\ \text{Pa}\ \text{mol}^{-2}$$

P7.3) Assume that the equation of state for a gas can be written in the form $P(V_m - b(T)) = RT$. Derive an expression for $\beta = 1/V\,(\partial V/\partial T)_P$ and $\kappa = -1/V\,(\partial V/\partial P)_T$ for such a gas in terms of $b(T)$, $db(T)/dT$, P, and V_m.

$$P\left(\frac{V}{n} - b(T)\right) = RT; \quad \frac{V}{n} = \frac{RT}{P} + b(T)$$

$$V = nb(T) + \frac{nRT}{P}$$

$$\beta = \frac{1}{V}\left(\frac{\partial V}{\partial T}\right)_P = \frac{1}{V}\left(\frac{n\,db(T)}{dT} + \frac{nR}{P}\right) = \frac{1}{V_m}\left(\frac{db(T)}{dT} + \frac{R}{P}\right)$$

$$\kappa = -\frac{1}{V}\left(\frac{\partial V}{\partial P}\right)_T = -\frac{1}{V}\left(-\frac{nRT}{P^2}\right) = \frac{RT}{V_m P^2}$$

P7.6) For values of z near one, it is a good approximation to write $z(P) = 1 + (\partial z/\partial P)_T P$. If $z = 1.00131$ at 0°C and 1 bar, and the Boyle temperature of the gas is 180. K, calculate the values of a, b, and V_m for the van der Waals gas.

From Example Problem 7.2,

$$\left(\frac{\partial z}{\partial P}\right)_T = \frac{1}{RT}\left(b - \frac{a}{RT}\right)$$

We can write three equations in three unknowns:

$$z - 1 = \left(b - \frac{a}{RT}\right)\frac{P}{RT}$$

$$1.31 \times 10^{-3} = \left(b - \frac{a}{8.314 \times 10^{-2} \text{ L bar mol}^{-1} \text{ K}^{-1} \times 273.15 \text{ K}}\right)$$

$$\times \frac{1 \text{ bar}}{8.314 \times 10^{-2} \text{ L bar mol}^{-1} \text{ K}^{-1} \times 273.15 \text{ K}}$$

$$T_B = \frac{a}{Rb} = \frac{a}{8.314 \times 10^{-2} \text{ L bar mol}^{-1} \text{ K}^{-1} \times b} = 180. \text{ K}$$

$$P = \frac{RT}{V_m - b} - \frac{a}{V_m} = \frac{8.314 \times 10^{-2} \text{ L bar mol}^{-1} \text{ K}^{-1} \times 273.15 \text{ K}}{V_m - b} - \frac{a}{V_m} = 1 \text{ bar}$$

Using an equation solver, we obtain a and b by solving the first 2 equations simultaneously. We substitute these values into the third equation to obtain V_m. The results are:

$a = 1.31 \text{ L}^2 \text{ bar mol}^{-2}$, $b = 0.0872 \text{ L mol}^{-1}$, $V_m = 22.7 \text{ L mol}^{-1}$.

P7.8) The experimentally determined density of N_2 at 180. bar and 150. K is 544 g L^{-1}. Calculate z and V_m from this information. Compare this result with what you would have estimated from Figure 7.8. What is the relative error in using Figure 7.8 for this case?

$$V_m = \frac{M}{\rho} = \frac{28.02 \text{ g mol}^{-1}}{544 \text{ g L}^{-1}} = 5.15 \times 10^{-2} \text{ L mol}^{-1}$$

$$z = \frac{PV_m}{RT} = \frac{180. \text{ bar} \times 5.15 \times 10^{-2} \text{ L mol}^{-1}}{8.314 \times 10^{-2} \text{ L bar mol}^{-1} \text{ K}^{-1} \times 150. \text{ K}} = 0.743$$

Because $P_r = \dfrac{180. \text{ bar}}{33.98 \text{ bar}} = 5.30$ and $T_r = \dfrac{150. \text{ K}}{126.2 \text{ K}} = 1.19$, Figure 7.8 predicts $z = 0.85$. The relative error in z is 14%.

P7.14) Use the law of corresponding states and Figure 7.8 to estimate the molar volume of pentane at T = 725 K and P = 75.0 bar.

We use the values for the critical constants in Table 7.2.

$$T_r = \frac{725 \text{ K}}{469.7 \text{ K}} = 1.54 \quad P_r = \frac{75.0 \text{ bar}}{33.70 \text{ bar}} = 2.23. \text{ Therefore, } z \approx 0.84$$

$$\frac{PV_m}{RT} = 0.84; \quad V_m = 0.84\frac{RT}{P} = 0.8 \times \frac{8.314 \times 10^{-2} \ dm^3 \ bar\,K^{-1} \ mol^{-1} \times 725 \ K}{75.0 \ bar}$$

$$V_m = 0.674 \ L\,mol^{-1}$$

P7.15) Another equation of state is the Berthelot equation, $V_m = (RT/P) + b - (a/RT^2)$. Derive expressions for $\beta = 1/V \ (\partial V/\partial T)_P$ and $\kappa = -1/V \ (\partial V/\partial P)_T$ from the Berthelot equation in terms of V, T, and P.

$$V = \frac{nRT}{P} + nb - \frac{na}{RT^2}$$

$$\beta = \frac{1}{V}\left(\frac{\partial V}{\partial T}\right)_P = \frac{1}{V}\left(\frac{nR}{P} + \frac{2na}{RT^3}\right) = \frac{1}{V_m}\left(\frac{R}{P} + \frac{2a}{RT^3}\right)$$

$$\kappa = -\frac{1}{V}\left(\frac{\partial V}{\partial P}\right)_T = -\frac{1}{V}\left(-\frac{nRT}{P^2}\right) = \frac{nRT}{P^2 V} = \frac{RT}{P^2 V_m}$$

P7.16) Show that $P\kappa = 1 - P\left(\dfrac{\partial \ln z}{\partial P}\right)_T$ for a real gas where κ is the isothermal compressibility.

$$\kappa = -\frac{1}{V}\left(\frac{\partial V}{\partial P}\right)_T ; \quad P\kappa = -\frac{P}{V}\left(\frac{\partial V}{\partial P}\right)_T$$

$$z = \frac{V}{V_{ideal}} = \frac{PV}{nRT} ; \quad \ln z = \ln(PV) - \ln(nRT)$$

$$\left(\frac{\partial \ln z}{\partial P}\right)_T = \frac{1}{PV} \times \left[V + P\left(\frac{\partial V}{\partial P}\right)_T\right] = \frac{1}{P} + \frac{1}{V}\left(\frac{\partial V}{\partial P}\right)_T$$

$$\frac{1}{V}\left(\frac{\partial V}{\partial P}\right)_T = \left(\frac{\partial \ln z}{\partial P}\right)_T - \frac{1}{P}$$

Therefore $P\kappa = 1 - P\left(\dfrac{\partial \ln z}{\partial P}\right)_T$

P7.17) Calculate the van der Waals parameters of ethanol from the values of the critical constants. We use the values for the critical constants in Table 7.2.

$$b = \frac{RT_c}{8P_c} = \frac{8.314 \times 10^{-2} \text{ dm}^3 \text{ bar K}^{-1} \text{ mol}^{-1} \times 513.92 \text{ K}}{8 \times 61.37 \text{ bar}} = 0.0870 \text{ dm}^3 \text{ mol}^{-1}$$

$$a = \frac{27R^2T_c^2}{64P_c} = \frac{27 \times \left(8.314 \times 10^{-2} \text{ dm}^3 \text{ bar K}^{-1} \text{ mol}^{-1}\right)^2 \times \left(513.92 \text{ K}\right)^2}{64 \times 61.37 \text{ bar}}$$

$$= 12.55 \text{ dm}^6 \text{ bar mol}^{-2}$$

P7.21) At what temperature does the slope of the z versus P curve as $P \to 0$ have its maximum value for a van der Waals gas? What is the value of the maximum slope?

$$\left(\frac{\partial z}{\partial P}\right)_{T,P\to 0} = \frac{1}{RT}\left(b - \frac{a}{RT}\right) \text{ for a van der Waals gas}$$

$$\left(\frac{\partial}{\partial T}\left(\frac{\partial z}{\partial P}\right)_T\right)_{P\to 0} = -\frac{1}{RT^2}\left(b - \frac{a}{RT}\right) + \frac{1}{RT^3} = -\frac{1}{RT^2}\left(b - \frac{2a}{RT}\right)$$

Setting this derivative equal to zero gives

$$b - \frac{2a}{RT_{max}} = 0 \quad T_{max} = \frac{2a}{Rb}$$

The maximum slope is $\dfrac{1}{RT_{max}}\left(b - \dfrac{a}{RT_{max}}\right) = \dfrac{b}{2a}\left[b - a\left(\dfrac{b}{2a}\right)\right] = \dfrac{b^2}{4a}$

P7.22) Calculate the density of $O_2(g)$ at 415 K and 310. bar using the ideal gas and the van der Waals equations of state. Use a numerical equation solver to solve the van der Waals equation for V_m or use an iterative approach starting with V_m equal to the ideal gas result. Based on your result, does the attractive or repulsive contribution to the interaction potential dominate under these conditions?

$$V_m = \frac{RT}{P} = \frac{8.314 \times 10^{-2} \text{ L bar K}^{-1} \text{ mol}^{-1} \times 415 \text{ K}}{310. \text{ bar}} = 0.1113 \text{ L}$$

$$P_{vdW} = \frac{RT}{V_m - b} - \frac{a}{V_m^2} = \frac{8.314 \times 10^{-2} \text{ L bar K}^{-1} \text{ mol}^{-1} \times 415 \text{ K}}{V_m - 0.0319 \text{ L mol}^{-1}} - \frac{1.382 \text{ L}^2 \text{ bar mol}^{-2}}{\left(V_m\right)^2}$$

The three solutions to this equation are

$V_m = \left(0.0140 \pm 0.0323 \text{ } i\right) \text{L mol}^{-1}$ and $V_m = 0.1152 \text{ L mol}^{-1}$

Only the real solution is of significance.

$$\rho_{ideal \text{ } gas} = \frac{M}{V_m} = \frac{32.0 \text{ g mol}^{-1}}{0.1113 \text{ L mol}^{-1}} = 288 \text{ g L}^{-1}$$

$$\rho_{vdW} = \frac{M}{V_m} = \frac{32.0 \text{ g mol}^{-1}}{0.1152 \text{ L mol}^{-1}} = 278 \text{ g L}^{-1}$$

Because the van der Waals density is less than the ideal gas density, the repulsive part of the potential dominates.

P7.28) For a van der Waals gas, $z = V_m / (V_m - b) - a / RTV_m$. Expand the first term of this expression in a Taylor series in the limit $V_m \gg b$ to obtain $z \approx 1 + (b - a/RT)(1/V_m)$.

$$f(x) = f(0) + \left(\frac{df(x)}{dx} \right)_{x=0} x + \dots \text{ In this case, } f(x) = \frac{1}{1-x} \text{ and } x = \frac{b}{V_m}$$

$$z = \frac{V_m}{V_m - b} - \frac{a}{RTV_m} = \frac{1}{1 - \dfrac{b}{V_m}} - \frac{a}{RTV_m}$$

Because $\dfrac{1}{1-x} \approx 1 + x + \dfrac{x^2}{2} + \dots$

$$\frac{1}{1 - \dfrac{b}{V_m}} \approx 1 + \frac{b}{V_m}$$

$$z \approx 1 + \frac{b}{V_m} - \frac{a}{RTV_m} = 1 + \frac{1}{V_m}\left(b - \frac{a}{RT} \right)$$

Chapter 8: Phase Diagrams and the Relative Stability of Solids, Liquids, and Gases

P8.2) The vapor pressure of benzene (*l*) is given by

$$\ln\left(\frac{P}{\text{Pa}}\right) = 20.767 - \frac{2.7738 \times 10^3}{\dfrac{T}{\text{K}} - 53.08}$$

a. Calculate the standard boiling temperature.

b. Calculate $\Delta H_m^{vaporization}$ at 298 K and at the standard boiling temperature.

a.

$$\ln\left(\frac{P}{\text{Pa}}\right) = 20.767 - \frac{2.7738 \times 10^3}{\dfrac{T}{\text{K}} - 53.08} = \ln 10^5 \text{ at } T_b$$

$$T_b = \frac{2.7738 \times 10^3}{20.767 - \ln\left(10^5\right)} + 53.08 = 352.8 \text{ K}$$

b.

$$\Delta H_m^{vaporization}\left(298 \text{ K}\right) = RT^2 \frac{d\ln P}{dT} = \frac{8.314 \text{ J mol}^{-1} \text{ K}^{-1} \times \left(298 \text{ K}\right)^2 \times 2.7738 \times 10^3}{\left(298 - 53.08\right)^2}$$

$$= 34.1 \text{ kJ mol}^{-1} \text{ at } 298 \text{ K}$$

$$\Delta H_m^{vaporization}\left(352.8 \text{ K}\right) = RT^2 \frac{d\ln P}{dT} = \frac{8.314 \text{ J mol}^{-1} \text{ K}^{-1} \times \left(352.8 \text{ K}\right)^2 \times 2.7738 \times 10^3}{\left(352.8 - 53.08\right)^2}$$

$$= 32.0 \text{ kJ mol}^{-1} \text{ at } 335.9 \text{ K}$$

P8.5) Within what range can you restrict the values of *P* and *T* if the following information is known about CO_2? Use Figure 8.10 to answer this question.

a. As the temperature is increased, the solid is first converted to the liquid and subsequently to the gaseous state.

b. As the pressure on a cylinder containing pure CO_2 is increased from 65 to 80 atm, no interface delineating liquid and gaseous phases is observed.

c. Solid, liquid, and gas phases coexist at equilibrium.

d. An increase in pressure from 10 to 50 atm converts the liquid to the solid.

e.   An increase in temperature from –80° to 20°C converts a solid to a gas with no intermediate liquid phase.

a. As the temperature is increased, the solid is first converted to the liquid and subsequently to the gaseous state.

The temperature and pressure are greater than the values for the triple point, –56.6°C and 5.11 atm.

b. As the pressure on a cylinder containing pure CO_2 is increased from 65 to 80 atm, no interface delineating liquid and gaseous phases is observed.

The temperature is greater than the critical temperature, 31.0°C.

c. Solid, liquid, and gas phases coexist at equilibrium.

The system is at the triple point, –56.6°C and 5.11 atm.

d. An increase in pressure from 10 to 50 atm converts the liquid to the solid.

The temperature is slightly greater than the triple point value of –56.6°C.

e. An increase in temperature from –80° to 20°C converts a solid to a gas with no intermediate liquid phase.

The pressure is below the triple point pressure value of 5.11 atm.

P8.8) It has been suggested that the surface melting of ice plays a role in enabling speed skaters to achieve peak performance. Carry out the following calculation to test this hypothesis. At 1 atm pressure, ice melts at 273.15 K, $\Delta H_m^{fusion} = 6010$ J mol^{-1}, the density of ice is 920. kg m^{-3}, and the density of liquid water is 997 kg m^{-3}.

a.   What pressure is required to lower the melting temperature by 5.0°C?

b.   Assume that the width of the skate in contact with the ice has been reduced by sharpening to 25×10^{-3} cm, and that the length of the contact area is 15 cm. If a skater of mass 85 kg is balanced on one skate, what pressure is exerted at the interface of the skate and the ice?

c.   What is the melting point of ice under this pressure?

d.   If the temperature of the ice is – 5.0°C, do you expect melting of the ice at the ice–skate interface to occur?

a) What pressure is required to lower the melting temperature by 5.0°C?

$$\left(\frac{dP}{dT}\right)_{fusion} = \frac{\Delta S_m^{fusion}}{\Delta V_m^{fusion}} \approx \frac{\Delta S_m^{fusion}}{\dfrac{M}{\rho_{H_2O,l}} - \dfrac{M}{\rho_{H_2O,s}}} = \frac{22.0 \ \text{J mol}^{-1} \ \text{K}^{-1}}{\dfrac{18.02\times10^{-3} \ \text{kg}}{998 \ \text{kg m}^{-3}} - \dfrac{18.02\times10^{-3} \ \text{kg}}{920 \ \text{kg m}^{-3}}}$$

$$= -1.45\times10^{7} \ \text{Pa K}^{-1} = -145 \ \text{bar K}^{-1}$$

The pressure must be increased by 727 bar to lower the melting point by 5.0°C.

b) Assume that the width of the skate in contact with the ice is 25 x 10^{-3} cm, and that the length of the contact area is 15 cm. If a skater of mass 85 kg is balanced on one skate, what pressure is exerted at the interface of the skate and the ice?

$$P = \frac{F}{A} = \frac{85 \ \text{kg} \times 9.81 \ \text{ms}^{-2}}{15\times10^{-2} \ \text{m} \times 25\times10^{-5} \ \text{m}} = 2.2\times10^{7} \ \text{Pa} = 2.2\times10^{2} \ \text{bar}$$

c) What is the melting point of ice under this pressure?

$$\Delta T = \left(\frac{dT}{dP}\right)_{fusion} \Delta P = \frac{1°C}{144 \ \text{bar}} \times 2.20\times10^{2} \ \text{bar} = 1.5°C ; \ T_m = -1.5°C$$

d) If the temperature of the ice is –5.0°C, do you expect melting of the ice at the ice–skate interface to occur?

No, because the lowering of the melting temperature is less than the temperature of the ice.

P8.9) Answer the following questions using the *P-T* phase diagram for carbon sketched below.

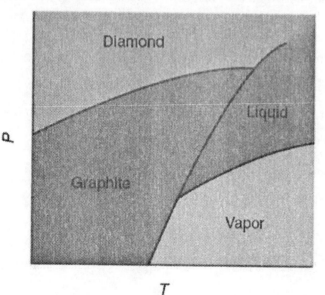

a) Which substance is denser, graphite or diamond? Explain your answer.

According to LeChatelier's principle, for a given temperature, the phase with the higher desnity will be found at higher pressure. Therefore diamond is more dense than graphite.

b) Which phase is denser, graphite or liquid carbon? Explain your answer.

Graphite is more dense because the slope of the graphite-liquid coexistence line is positive.

c) Why does the phase diagram have two triple points? Explain your answer.

Multiple triple points are possible only if there are more than a single solid phase or in rare cases more than one liquid phase.

P8.10) You have a compound dissolved in chloroform and need to remove the solvent by distillation. Because the compound is heat sensitive, you hesitate to raise the temperature above 0°C and decide on vacuum distillation. What pressure is required to boil chloroform at 0°C?

$$\ln P = a_1 - \frac{a_2}{T/K + a_3} = 20.907 - \frac{2696.1}{273.15 - 46.926} = \ln\left(10^5\right) = 8.9892$$

$$P = 8.02 \times 10^3 \text{ Pa}$$

P8.13) Autoclaves that are used to sterilize surgical tools require a temperature of 134°C to kill some bacteria. If water is used for this purpose, at what pressure must the autoclave operate? The normal boiling point of water is 373.15 K, and $\Delta H_m^{vaporization} = 40.650 \times 10^3 \text{ J mol}^{-1}$ at the normal boiling point.

$$\ln\frac{P_f}{P_i} = -\frac{\Delta H_m^{vaporization}}{R}\left(\frac{1}{T_f} - \frac{1}{T_i}\right)$$

$$\ln\frac{P_f}{P_i} = -\frac{40.656 \times 10^3 \text{ J mol}^{-1}}{8.314 \text{ J mol}^{-1}\text{K}^{-1}} \times \left(\frac{1}{407.15 \text{ K}} - \frac{1}{373.15 \text{ K}}\right) = 1.09349$$

$$\frac{P_f}{P_i} = 2.99; \quad P_f = 2.99 \text{ atm}$$

P8.20) The vapor pressure of liquid ethanol is 7615 Pa at 298.15 K, and

$\Delta H_m^{vaporization} = 38.56 \text{ kJ mol}^{-1}$. Calculate the normal and standard boiling points. Does your result for the normal boiling point agree with that in Table 8.3? If not, suggest a possible cause.

$$\ln\frac{P_f}{P_i} = -\frac{\Delta H_m^{vaporization}}{R}\left(\frac{1}{T_f} - \frac{1}{T_i}\right)$$

$$T_f = \frac{\Delta H_m^{vaporization}}{R\left(\dfrac{\Delta H_m^{vaporization}}{RT_i} - \ln\dfrac{P_f}{P_i}\right)}$$

At the normal boiling point, $P = 101325$ Pa.

$$T_{b,normal} = \frac{38.56\times10^3 \text{ J mol}^{-1}}{8.314 \text{ J mol}^{-1}\text{K}^{-1}\times\left(\dfrac{38.56\times10^3 \text{ J mol}^{-1}}{8.314 \text{ J mol}^{-1}\text{ K}^{-1}\times298.15\text{K}} - \ln\dfrac{101325}{7615}\right)} = 357.7 \text{ K}$$

At the standard boiling point, $P = 10^5$ Pa.

$$T_{b,standard} = \frac{38.56\times10^3 \text{ J mol}^{-1}}{8.314 \text{ J mol}^{-1}\text{K}^{-1}\times\left(\dfrac{38.56\times10^3 \text{ J mol}^{-1}}{8.314 \text{ J mol}^{-1}\text{ K}^{-1}\times298.15\text{K}} - \ln\dfrac{10^5}{7615}\right)} = 357.3 \text{ K}$$

The result for the normal boiling point is ~6 K higher than the value tabulated in Table 8.3. The most probable reason for this difference is that the calculation above has assumed that $\Delta H_m^{vaporization}$ is independent of T.

P8.21) Carbon disulfide, $CS_2(l)$, at 25°C has a vapor pressure of 0.4741 bar and an enthalpy of vaporization of 27.66 kJ mol^{-1}. The $C_{P,m}$ of the vapor and liquid phases at that temperature are 45.4 and 75.7 J K^{-1} mol^{-1}, respectively. Calculate the vapor pressure of $CS_2(l)$ at 115.0°C assuming

a. that the enthalpy of sublimation does not change with temperature.

b. that the enthalpy of sublimation at temperature T can be calculated from the equation $\Delta H_m^{sublimation}(T) = \Delta H_m^{sublimation}(T_0) + \Delta C_P(T-T_0)$ assuming that ΔC_P does not change with temperature.

a) If the enthalpy of vaporization is constant

$$\ln\frac{P_2}{P_1} = -\frac{\Delta H_m^{vaporization}}{R}\left(\frac{1}{T_2} - \frac{1}{T_1}\right)$$

$$\ln P_2 = \ln(0.4741) - \frac{27.66\times10^3 \text{ J mol}^{-1}}{8.314 \text{ J mol}^{-1}\text{ K}^{-1}}\times\left(\frac{1}{373.15 \text{ K}} - \frac{1}{298.15 \text{ K}}\right) = 1.4964$$

$$P_2 = 4.66 \text{ bar}$$

b) If the enthalpy of vaporization is given by $\Delta H_m^{vaporization}(T) = \Delta H_m^{vaporization}(T_0) + \Delta C_P(T-T_0)$

$$\int_{P_1}^{P_2} \frac{dP}{P} = \int_{T_1}^{T_2} \frac{\Delta H_m^{vaporization}}{RT^2} dT = \int_{T_1}^{T_2} \frac{\Delta H_m^{vaporization}(T_1) + \Delta C_P (T - T_1)}{RT^2} dT$$

$$\ln \frac{P_2}{P_1} = -\frac{\Delta H_m^{vaporization}(T_1)}{R}\left(\frac{1}{T_2} - \frac{1}{T_1}\right) + \frac{\Delta C_P T_1}{R}\left(\frac{1}{T_2} - \frac{1}{T_1}\right) + \frac{\Delta C_P}{R}\ln\frac{T_2}{T_1}$$

$$\ln P_2 = \ln(0.4741) - \frac{27.66 \times 10^3 \text{ J mol}^{-1}}{8.314 \text{ J mol}^{-1}\text{ K}^{-1}} \times \left(\frac{1}{373.15 \text{ K}} - \frac{1}{298.15 \text{ K}}\right)$$

$$+ \frac{(45.4 - 75.7) \text{ J mol}^{-1}\text{ K}^{-1} \times 298.15 \text{ K}}{8.314 \text{ J mol}^{-1}\text{ K}^{-1}}\left(\frac{1}{373.15 \text{ K}} - \frac{1}{298.15 \text{ K}}\right)$$

$$+ \frac{(45.4 - 75.7) \text{ J mol}^{-1}\text{ K}^{-1}}{8.314 \text{ J mol}^{-1}\text{ K}^{-1}}\ln\frac{373.15 \text{ K}}{298.15 \text{ K}}$$

$$\ln P_2 = 1.411$$

$$P_2 = 4.10 \text{ bar}$$

P8.22) Use the values for ΔG_f° (benzene, g) from Appendix B to calculate the vapor pressure of benzene at 298.15 K.

For the transformation C_6H_6 (*l*) $\rightarrow$ C_6H_6 (*g*)

$$\ln K_P = \ln\frac{P_{C_6H_6(g)}}{P^\circ} = -\frac{\Delta G_f^\circ(C_6H_6,g) - \Delta G_f^\circ(C_6H_6,l)}{RT}$$

$$= -\frac{129.7 \times 10^3 \text{ J mol}^{-1} + 124.5 \times 10^3 \text{ J mol}^{-1}}{8.314 \text{ J mol}^{-1}\text{ K}^{-1} \times 298.15 \text{ K}} = -2.0978$$

$$K_P = \frac{P_{C_6H_6(g)}}{1 \text{ bar}} = 0.1227 \text{ bar} = 1.22 \times 10^4 \text{ Pa}$$

P8.25) For water, $\Delta H_m^{vaporization}$ is 40.65 kJ mol^{-1}, and the normal boiling point is 373.15 K. Calculate the boiling point for water on the top of Mt. Rainier (elevation 4392 m), where the normal barometric pressure is 461 Torr.

$$\ln\frac{P_f}{P_i} = -\frac{\Delta H_m^{vaporization}}{R}\left(\frac{1}{T_f} - \frac{1}{T_i}\right)$$

$$T_f = \frac{\Delta H_m^{vaporization}}{R\left(\dfrac{\Delta H_m^{vaporization}}{RT_i} - \ln\dfrac{P_f}{P_i}\right)}$$

At the normal boiling point, $P = 760.$ Torr.

$$T_{b,normal} = \frac{40.656\times10^3\ \text{J mol}^{-1}}{8.314\ \text{J mol}^{-1}\ \text{K}^{-1}\times\left(\dfrac{40.656\times10^3\ \text{J mol}^{-1}}{8.314\ \text{J mol}^{-1}\ \text{K}^{-1}\times373.12\text{K}} - \ln\dfrac{461\ \text{Torr}}{760.\ \text{Torr}}\right)} = 359.4\ \text{K}$$

P8.28) Use the vapor pressures of $Cl_2(l)$ given in the following table to calculate the enthalpy of vaporization using a graphical method or a least squares fitting routine.

T (K)	P (atm)	T (K)	P (atm)
227.6	0.585	283.15	4.934
238.7	0.982	294.3	6.807
249.8	1.566	305.4	9.173
260.9	2.388	316.5	12.105
272.0	3.483	327.6	15.676

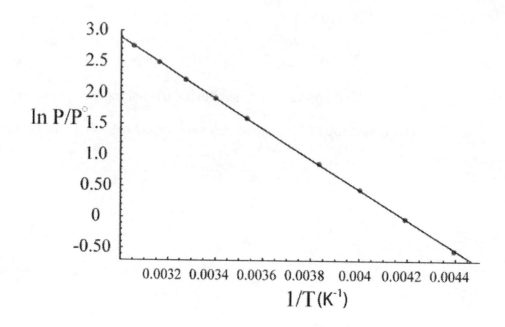

A least squares fit of $\ln P$ versus $1/T$ gives the result $\Delta H_m^{vaporization} = 20.32$ kJ mol^{-1}.

P8.36) The densities of a given solid and liquid of molar mass 147.2 g mol^{-1} at its normal melting temperature of 372.54 K are 992.6 and 933.4 kg m^{-3}, respectively. If the pressure is increased to 120. bar, the melting temperature increases to 375.88 K. Calculate ΔH_m^{fusion} and ΔS_m^{fusion} for this substance.

$$\frac{\Delta P}{\Delta T} \approx \frac{\Delta S}{\Delta V}; \quad \Delta S \approx \frac{\Delta P}{\Delta T}\Delta V$$

$$\Delta S_m^{fusion} = \frac{\Delta P}{\Delta T}M\left(\frac{1}{\rho_{liquid}} - \frac{1}{\rho_{solid}}\right)$$

$$\Delta S_m^{fusion} = \frac{119\times10^5\,\text{Pa}}{375.88\,\text{K} - 372.54\,\text{K}}\times147.2\times10^{-3}\text{kg mol}^{-1}\times\left(\frac{1}{933\ \text{kg m}^{-3}} - \frac{1}{992.6\ \text{kg m}^{-3}}\right)$$

$$= 33.5\ \text{J K}^{-1}\ \text{mol}^{-1}$$

$$\Delta H_m^{fusion} = T_{fusion}\Delta S_m^{fusion} = 372.54\ \text{K}\times33.5\ \text{J K}^{-1}\ \text{mol}^{-1} = 12.5\times10^3\ \text{J mol}^{-1}\ \text{at 1 bar}$$

P8.37) The variation of the vapor pressure of the liquid and solid forms of anthracene near the triple point are given in the Appendix A data tables. Calculate the temperature and pressure at the triple point. At the triple point, $P_{solid} = P_{liquid}$.

$$-11378\frac{\text{K}}{T} + 31.62 = -5873.3\frac{\text{K}}{T} + 21.965$$

$$31.62 - 21.965 = (11378 - 5873.3)\frac{\text{K}}{T}$$

$$T = \frac{(11378 - 5873.3)}{31.62 - 21.965} = 501\ \text{K}$$

$$\ln\frac{P_{tp}}{\text{Pa}} = \frac{-11378}{467.7} + 31.62 = 8.8673$$

$$\frac{P_{tp}}{\text{Pa}} = 7.24\times10^3$$

P8.41) Calculate the vapor pressure of a droplet of tetrachloromethane of radius 2.50×10^{-8} m at 45.0°C in equilibrium with its vapor. Use the tabulated value of the density and the surface tension at 298 K from Appendix B for this problem. (*Hint:* You need to calculate the vapor pressure of tetrachloromethane at this temperature.)

$$\ln \frac{P(T)}{\text{Pa}} = A(1) - \frac{A(2)}{\frac{T}{K} + A(3)} = 20.738 - \frac{2.7923 \times 10^3}{318.15 - 46.6667} = 10.4527$$

$$P = 3.46 \times 10^4 \text{ Pa}$$

Using Equation (8.29),

$$\Delta P = \frac{2\gamma}{r} = \frac{2 \times 26.43 \times 10^{-3} \text{ N m}^{-1}}{2.5 \times 10^{-8} \text{ m}} = 2.11 \times 10^6 \text{ Pa}$$

$$P_{inside} = P_{vapor} + \Delta P = 3.46 \times 10^4 \text{ Pa} + 2.11 \times 10^6 \text{ Pa} = 2.15 \times 10^6 \text{ Pa}$$

For a very large droplet, $\Delta P \to 0$, the vapor pressure is $3.46 \times 10^4 \text{Pa}$. For the small droplet, the vapor pressure is increased by the factor

$$\ln \left(\frac{P}{P_0} \right) = \frac{\frac{M}{\rho}(\mathbf{P} - P_0)}{RT} = \frac{\frac{153.82 \times 10^{-3} \text{ kg mol}^{-1}}{1594 \text{ kg m}^{-3}} \times \left(2.15 \times 10^6 - 3.46 \times 10^4\right) \text{Pa}}{8.314 \text{ J mol}^{-1} \text{ K}^{-1} \times 318.15 \text{ K}}$$

$$= 7.7138 \times 10^{-2}$$

$$P = 1.080 \, P_0 = 3.74 \times 10^4$$

P8.42) Solid iodine, $I_2(s)$, at 25°C has an enthalpy of sublimation of 56.30 kJ mol^{-1}. The $C_{P,m}$ of the vapor and solid phases at that temperature are 36.9 and 54.4 J K^{-1} mol^{-1}, respectively. The sublimation pressure at 25.00°C is 0.30844 Torr. Calculate the sublimation pressure of the solid at the melting point (113.6°C) assuming

a. that the enthalpy of sublimation and the heat capacities do not change with temperature.

b. that the enthalpy of sublimation at temperature T can be calculated from the equation

$$\Delta H^{\circ}_{sublimation}(T) = \Delta H^{\circ}_{sublimation}(T_0) + \Delta C_P (T - T_0)$$

a) If the enthalpy of sublimation is constant

$$\ln \frac{P_2}{P_1} = -\frac{\Delta H_m^{sublimation}}{R}\left(\frac{1}{T_2}-\frac{1}{T_1}\right)$$

$$\ln P_2 = \ln 0.30844 - \frac{56.30\times10^3 \text{ J mol}^{-1}}{8.314 \text{ J mol}^{-1}\text{ K}^{-1}}\times\left(\frac{1}{386.8 \text{ K}}-\frac{1}{298.15 \text{ K}}\right)$$

$$P_2 = 56.1 \text{ Torr}$$

b) If the enthalpy of sublimation is given by

$$\Delta H_{sublimation}^{\circ}(T) = \Delta H_{sublimation}^{\circ}(T_0)+\Delta C_P(T-T_0)$$

$$\int_{P_1}^{P_2}\frac{dP}{P} = \int_{T_1}^{T_2}\frac{\Delta H_m^{vaporization}}{RT^2}dT = \int_{T_1}^{T_2}\frac{\Delta H_m^{vaporization}(T_1)+\Delta C_P(T-T_1)}{RT^2}dT$$

$$\ln \frac{P_2}{P_1} = -\frac{\Delta H_m^{vaporization}(T_1)}{R}\left(\frac{1}{T_2}-\frac{1}{T_1}\right)+\frac{\Delta C_P T_1}{R}\left(\frac{1}{T_2}-\frac{1}{T_1}\right)+\frac{\Delta C_P}{R}\ln\frac{T_2}{T_1}$$

$$\ln P_2 = \ln 0.30844 - \frac{56.30\times10^3 \text{ J mol}^{-1}}{8.314 \text{ J K}^{-1}\text{ mol}^{-1}}\times\left(\frac{1}{386.8 \text{ K}}-\frac{1}{298.15 \text{ K}}\right)$$

$$+\frac{(36.9-54.4)\text{J K}^{-1}\text{ mol}^{-1}\times 298.15 \text{ K}}{8.314 \text{ J K}^{-1}\text{ mol}^{-1}}\times\left(\frac{1}{386.8 \text{ K}}-\frac{1}{298.15 \text{ K}}\right)$$

$$+\frac{(36.9-54.4)\text{J K}^{-1}\text{ mol}^{-1}}{8.314 \text{ J K}^{-1}\text{ mol}^{-1}}\times\ln\frac{386.8 \text{ K}}{298.15 \text{ K}}$$

$$\ln P_2 = 3.964$$

$$P_2 = 52.5 \text{ Torr}$$

P8.43) Consider the transition between two forms of solid tin, $\text{Sn}(s,\text{gray})\rightleftharpoons\text{Sn}(s,\text{white})$. The two phases are in equilibrium at 1 bar and 18°C. The densities for gray and white tin are 5750 and 7280 kg m^{-3}, respectively, and the molar entropies for gray and white tin are 44.14 and 51.18 J K^{-1} mol^{-1}, respectively. Calculate the temperature at which the two phases are in equilibrium at 200 bar.

In going from 1atm, 18°C to 200 atm, T

$$\Delta G^{gray} = V_m^{gray} \Delta P - S^{gray} \Delta T$$

$$\Delta G^{white} = V_m^{white} \Delta P - S^{white} \Delta T$$

At equilibrium

$$\Delta G^{gray} - \Delta G^{white} = 0 = \left(V_m^{gray} - V_m^{white} \right) \Delta P - \left(S^{gray} - S^{white} \right) \Delta T$$

$$\Delta T = \frac{\left(V_m^{gray} - V_m^{white} \right) \Delta P}{\left(S^{gray} - S^{white} \right)} = \frac{M_{Sn} \left(\dfrac{1}{\rho_{gray}} - \dfrac{1}{\rho_{white}} \right) \Delta P}{\Delta S_{transition}}$$

$$= \frac{118.71 \times 10^{-3} \ \text{kg mol}^{-1} \times \left(\dfrac{1}{5750 \ \text{kg m}^{-3}} - \dfrac{1}{7280 \ \text{kg m}^{-3}} \right) \times 199 \times 10^{5} \ \text{Pa}}{-7.04 \ \text{J K}^{-1} \ \text{mol}^{-1}} = -12.3°\text{C}$$

$$T_f = 5.7°\text{C}$$

Chapter 9: Ideal and Real Solutions

P9.2) At a given temperature, a nonideal solution of the volatile components A and B has a vapor pressure of 832 Torr. For this solution, $y_A = 0.404$. In addition, $x_A = 0.285$, $P_A^* = 591$ Torr, and $P_B^* = 503$ Torr. Calculate the activity and activity coefficient of A and B.

$$P_A = y_A P_{total} = 0.404 \times 832 \text{ Torr} = 336 \text{ Torr}$$
$$P_B = 832 \text{ Torr} - 336 \text{ Torr} = 496 \text{ Torr}$$
$$a_A = \frac{P_A}{P_A^*} = \frac{336 \text{ Torr}}{591 \text{ Torr}} = 0.569$$
$$\gamma_A = \frac{a_A}{x_A} = \frac{0.569}{0.285} = 2.00$$
$$a_B = \frac{P_B}{P_B^*} = \frac{496 \text{ Torr}}{503 \text{ Torr}} = 0.986$$
$$\gamma_B = \frac{a_B}{x_B} = \frac{0.986}{0.715} = 1.38$$

P9.4) At 365 K, pure toluene and hexane have vapor pressures of 5.82×10^4 Pa and 1.99×10^5 Pa, respectively.

a. Calculate the mole fraction of hexane in the liquid mixture that boils at 365 K at a pressure of 760. Torr.

b. Calculate the mole fraction of hexane in the vapor that is in equilibrium with the liquid of part (a).

a) $P_{total} = x_{hex} P_{hex}^* + (1 - x_{hex}) P_{tol}^*$

$1.01325 \times 10^5 \text{ Pa} = 1.99 \times 10^5 \text{ Pa } x_{hex} + 5.82 \times 10^4 \text{ Pa} (1 - x_{hex})$

$x_{hex} = 0.306$

b) $y_B = \dfrac{x_{hex} P_{hex}^*}{P_{tol}^* + (P_{hex}^* - P_{tol}^*) x_{hex}}$

$= \dfrac{0.306 \times 1.99 \times 10^5 \text{ Pa}}{5.82 \times 10^4 \text{ Pa} + 0.306 \times (1.99 \times 10^5 \text{ Pa} - 5.82 \times 10^4 \text{ Pa})}$

$= 0.602$

P9.7) The osmotic pressure of an unknown substance is measured at 298 K. Determine the molar mass if the concentration of this substance is 31.2 kg m^{-3} and the osmotic pressure is 6.10×10^4 Pa. The density of the solution is 997 kg m^{-3}.

$$\pi = \frac{n_{solute} RT}{V} = \frac{c_{solute} \rho_{solution} RT}{M_{solute}}; \quad M_{solute} = \frac{\rho_{solution} c_{solute} RT}{\pi}$$

$$M_{solute} = \frac{997 \text{ kg m}^{-3} \times 31.2 \text{ kg m}^{-3} \times 8.314 \text{ J mol}^{-1} \text{ K}^{-1} \times 298 \text{ K}}{6.10 \times 10^4 \text{ Pa}} = 1.26 \times 10^3 \text{ kg mol}^{-1}$$

P9.9) An ideal solution is made up of the volatile liquids A and B, for which $P_A^* = 172$ Torr and $P_B^* = 57.6$ Torr. As the pressure is reduced, the first vapor is observed at a total pressure of 70.0 Torr. Calculate x_A.

The first vapor is observed at a pressure of

$$P_{total} = x_A P_A^* + (1 - x_A) P_B^*$$

$$x_A = \frac{P_{total} - P_B^*}{P_A^* - P_B^*} = \frac{70.0 \text{ Torr} - 57.6 \text{ Torr}}{172 \text{ Torr} - 57.6 \text{ Torr}} = 0.108$$

P9.15) At 39.9°C, a solution of ethanol ($x_1 = 0.9006$, $P_1^* = 130.4$ Torr) and isooctane ($P_2^* = 43.9$ Torr) forms a vapor phase with $y_1 = 0.6667$ at a total pressure of 185.9 Torr.

a. Calculate the activity and activity coefficient of each component.

b. Calculate the total pressure that the solution would have if it were ideal.

a) The activity and activity coefficient for ethanol are given by

$$a_1 = \frac{y_1 P_{total}}{P_1^*} = \frac{0.6667 \times 185.9 \text{ Torr}}{130.4 \text{ Torr}} = 0.9504$$

$$\gamma_1 = \frac{a_1}{x_1} = \frac{0.9504}{0.9006} = 1.055$$

Similarly, the activity and activity coefficient for isooctane are given by

$$a_2 = \frac{(1 - y_1) P_{total}}{P_2^*} = \frac{0.3333 \times 185.9 \text{ Torr}}{43.9 \text{ Torr}} = 1.411$$

$$\gamma_2 = \frac{a_2}{x_2} = \frac{1.411}{1 - 0.9006} = 14.20$$

b) If the solution were ideal, Raoult's law would apply.

$$P_{Total} = x_1 P_1^* + x_2 P_2^*$$
$$= 0.9006 \times 130.4 \text{ Torr} + (1 - 0.9006) \times 43.9 \text{ Torr}$$
$$= 121.8 \text{ Torr}$$

P9.16) Calculate the solubility of CO in 1 L of water if its pressure above the solution is 55.0 bar. The density of water at this temperature is 997 kg m^{-3}.

$$x_{CO} = \frac{n_{CO}}{n_{CO} + n_{H_2O}} \approx \frac{n_{CO}}{n_{H_2O}} = \frac{P_{CO}}{k_{CO}^H} = \frac{55.0 \text{ bar}}{5.84 \times 10^3 \text{ bar}} = 9.42 \times 10^{-3}$$

$$n_{H_2O} = \frac{\rho_{H_2O} V}{M_{H_2O}} = \frac{10^{-3} \text{m}^3 \times 997 \text{ kg m}^{-3}}{18.02 \times 10^{-3} \text{kg mol}^{-1}} = 55.3$$

$$n_{CO} = x_{CO} n_{H_2O} = 9.42 \times 10^{-3} \times 55.3 = 0.521 \text{ mol}$$

P9.19) A and B form an ideal solution. At a total pressure of 0.605 bar, $y_A = 0.360$ and $x_A = 0.700$. Using this information, calculate the vapor pressure of pure A and of pure B.

$$P_{total} = x_A P_A^* + y_B P_{total}$$

$$P_A^* = \frac{P_{total} - y_B P_{total}}{x_A} = \frac{0.605 \text{ bar} \times (1 - 0.5560)}{0.700} = 0.311 \text{ bar}$$

$$P_B^* = \frac{P_A^* (x_A y_A - x_A)}{(x_A - 1) y_A} = \frac{0.311 \text{ bar} \times (0.700 \times 0.360 - 0.700)}{(0.700 - 1) \times 0.360} = 1.29 \text{ bar}$$

P9.24) An ideal solution is formed by mixing liquids A and B at 298 K. The vapor pressure of pure A is 136 Torr and that of pure B is 93.7 Torr. If the mole fraction of A in the vapor is 0.320, what is the mole fraction of A in the solution?

$$x_A = \frac{y_A P_B^*}{P_A^* + (P_B^* - P_A^*) y_A} = \frac{0.320 \times 93.7 \text{ Torr}}{136 \text{ Torr} + (93.7 \text{ Torr} - 136 \text{ Torr}) \times 0.320} = 0.245$$

P9.25) A solution is prepared by dissolving 89.6 g of a nonvolatile solute in 175 g of water. The vapor pressure above the solution is 20.62 Torr and the vapor pressure of pure water is 23.76 Torr at this temperature. What is the mass of the solute?

$$x_{H_2O} = \frac{P_{H_2O}}{P_{H_2O}^*} = \frac{20.62 \text{ Torr}}{23.76 \text{ Torr}} = 0.868$$

$$x_{solute} = 0.132 = \frac{n_{solute}}{n_{solute} + n_{H_2O}};$$

$$n_{solute} = \frac{x_{solute} \dfrac{m_{H_2O}}{M_{H_2O}}}{x_{H_2O}} = \frac{0.132 \times \dfrac{175 \text{ g}}{18.02 \text{ g mol}^{-1}}}{0.868} = 1.48 \text{ mol}$$

$$M = \frac{89.6 \text{ g}}{1.48 \text{ mol}} = 60.6 \text{ g mol}^{-1}$$

P9.28) The vapor pressures of 1-bromobutane and 1-chlorobutane can be expressed in the form

$$\ln \frac{P_{bromo}}{\text{Pa}} = 17.076 - \frac{1584.8}{\dfrac{T}{\text{K}} - 111.88}$$

and

$$\ln \frac{P_{chloro}}{\text{Pa}} = 20.612 - \frac{2688.1}{\dfrac{T}{\text{K}} - 55.725}$$

Assuming ideal solution behavior, calculate x_{bromo} and y_{bromo} at 290.0 K and a total pressure of 7325 Pa. At 300.0K, $P_{bromo}^\circ = 5719$ Pa and $P_{chloro}^\circ = 14877$ Pa.

$$P_{total} = x_{bromo} P_{bromo}^\circ + (1 - x_{bromo}) P_{chloro}^\circ$$

$$x_{bromo} = \frac{P_{total} - P_{chloro}^\circ}{P_{bromo}^\circ - P_{chloro}^\circ} = \frac{7325 \text{ Pa} - 9301 \text{ Pa}}{3564 \text{ Pa} - 9301 \text{ Pa}} = 0.344$$

$$y_{bromo} = \frac{x_{bromo} P_{bromo}^\circ}{P_{total}} = \frac{0.344 \times 3564 \text{ Pa}}{7325 \text{ Pa}} = 0.168$$

P9.29) In an ideal solution of A and B, 2.75 mol are in the liquid phase and 4.255 mol are in the gaseous phase. The overall composition of the system is $Z_A = 0.420$ and $x_A = 0.310$. Calculate y_A

$$n_{liq}^{tot}\left(Z_B - x_B\right) = n_{vapor}^{tot}\left(y_B - Z_B\right)$$

$$y_B = \frac{n_{liq}^{tot}\left(Z_B - x_B\right) + n_{vapor}^{tot}Z_B}{n_{vapor}^{tot}} = \frac{2.75\ \text{mol}\times\left(0.580 - 0.690\right) + 4.255\ \text{mol}\times 0.580}{4.255\ \text{mol}}$$

$$= 0.509$$

$$y_A = 1 - 0.509 = 0.491$$

P9.33) The dissolution of 5.25 g of a substance in 565 g of benzene at 298 K raises the boiling point by 0.625°C. Note that K_f = 5.12 K kg mol^{-1}, K_b = 2.53 K kg mol^{-1}, and the density of benzene is 876.6 kg m^{-3}. Calculate the freezing point depression, the ratio of the vapor pressure above the solution to that of the pure solvent, the osmotic pressure, and the molar mass of the solute. $P_{benzene}^{*}$ =103 Torr at 298 K.

$$\Delta T_b = K_b m_{solute};\quad m_{solute} = \frac{\Delta T_b}{K_b} = \frac{0.625\ \text{K}}{2.53\ \text{K kg mol}^{-1}} = 0.247\ \text{mol kg}^{-1}$$

$$M = \frac{5.25\ \text{g}}{0.247\ \text{mol kg}^{-1}\times 0.565\ \text{kg}} = 37.6\ \text{g mol}^{-1}$$

$$\Delta T_f = -K_f m_{solute} = -5.12\ \text{K kg mol}^{-1}\times 0.247\ \text{mol kg}^{-1} = -1.26\ \text{K}$$

$$\frac{P_{benzene}}{P_{benzene}^{*}} = x_{benzene} = \frac{n_{benzene}}{n_{benzene} + n_{solute}}$$

$$= \frac{\dfrac{565\ \text{g}}{78.11\ \text{g mol}^{-1}}}{\dfrac{565\ \text{g}}{78.11\ \text{g mol}^{-1}} + 0.247\ \text{mol kg}^{-1}\times 0.565\ \text{kg}} = 0.981$$

$$\pi = \frac{n_{solute}RT}{V} = \frac{\dfrac{5.25\times 10^{-3}\ \text{kg}}{37.6\times 10^{-3}\text{kg mol}^{-1}}\times 8.314\ \text{J mol}^{-1}\ \text{K}^{-1}\times 298\ \text{K}}{\dfrac{565\times 10^{-3}\ \text{kg}}{876.6\ \text{kg m}^{-3}}} = 5.37\times 10^{5}\ \text{Pa}$$

Chapter 10: Electrolyte Solutions

P10.2) Calculate $\Delta S^\circ_{reaction}$ for the reaction $AgNO_3(aq) + KCl(aq) \rightarrow AgCl(s) + KNO_3(aq)$

$$\Delta S^\circ_{reaction} = S^\circ(AgCl, s) - S^\circ(Ag^+, aq) - S^\circ(Cl^-, aq)$$

$$\Delta S^\circ_{reaction} = 96.3 \; J \; K^{-1} \; mol^{-1} - 72.7 \; J \; K^{-1} \; mol^{-1} - 56.5 \; J \; K^{-1} \; mol^{-1} = -32.9 \; J \; K^{-1} \; mol^{-1}$$

P10.7) At 25°C, the equilibrium constant for the dissociation of acetic acid, K_a, is 1.75×10^{-5}. Using the Debye–Hückel limiting law, calculate the degree of dissociation in $0.100m$ and $1.00m$ solutions. Compare these values with what you would obtain if the ionic interactions had been ignored.

$$CH_3COOH\,(aq) \rightarrow CH_3COO^-(aq) + H^+(aq)$$

For 0.100 m

$$\frac{m^2 \gamma_\pm^2}{0.100 \; mol\,kg^{-1} - m} = 1.75 \times 10^{-5}$$

when $\gamma_\pm = 1$

$$m = 1.314 \times 10^{-3} \; mol \; kg^{-1}$$

$$I = \frac{m}{2}(2) = m = 1.314 \times 10^{-3} \; mol \; kg^{-1}$$

$$\ln \gamma_\pm - 1.173 \times 1 \times \sqrt{1.314 \times 10^{-3}} = -0.04252$$

$$\gamma_\pm = 0.9584$$

when $\gamma_\pm = 0.9584$

$$\frac{\left(\dfrac{m}{m^\circ}\right)^2 \gamma_\pm^2}{0.100 - \dfrac{m}{m^\circ}} = 1.75 \times 10^{-5}$$

$$m = 1.371 \times 10^{-3} \; mol\,kg^{-1}$$

We iterate several times

when $m = 1.371 \times 10^{-3} \; mol\,kg^{-1}$

$$I = \frac{m}{2}(2) = m = 1.371 \times 10^{-3} \; mol\,kg^{-1}$$

$$\ln \gamma_\pm = -1.173 \times 1 \times \sqrt{1.371 \times 10^{-3}} = -0.04343$$

$$\gamma_\pm = 0.9575$$

when $\gamma_\pm = 0.9575$

$$\frac{\left(\dfrac{m}{m^\circ}\right)^2 \gamma_\pm^2}{0.100 - \dfrac{m}{m^\circ}} = 1.75 \times 10^{-5}$$ This result has converged sufficiently to calculate the degree of

$m = 1.372 \times 10^{-3} \text{ mol kg}^{-1}$

dissociation.

$m = 1.372 \times 10^{-3} \text{ mol kg}^{-1}$

$$\frac{1.372 \times 10^{-3} \text{ mol kg}^{-1}}{0.100 \text{ mol kg}^{-1}} \times 100\% = 1.37\%$$

for $1.00 \, m$

$$\frac{\left(\dfrac{m}{m^\circ}\right)^2 \gamma_\pm^2}{1.00 - \dfrac{m}{m^\circ}} = 1.75 \times 10^{-5}$$

when $\gamma_\pm = 1$

$m = 4.174 \times 10^{-3} \text{ mol kg}^{-1}$

$I = m = 4.174 \times 10^{-3} \text{ mol kg}^{-1}$

$\ln \gamma_\pm = -0.07579$

$\gamma_\pm = 0.9270$

when $\gamma_\pm = 0.9270$

$$\frac{\left(\dfrac{m}{m^\circ}\right)^2 \gamma_\pm^2}{1.00 - \dfrac{m}{m^\circ}} = 1.75 \times 10^{-5}$$

$m = 4.503 \times 10^{-3} \text{ mol kg}^{-1}$

We iterate several times

when $m = 4.503 \times 10^{-3} \text{ mol kg}^{-1}$

$$I = \frac{m}{2}(2) = m = 4.503 \times 10^{-3} \text{ mol kg}^{-1}$$

$$\ln \gamma_\pm = -1.173 \times 1 \times \sqrt{4.503 \times 10^{-3}} = -0.07871$$

$$\gamma_\pm = 0.9243$$

when $\gamma_{\pm} = 0.9243$

$$\frac{\left(\frac{m}{m^\circ}\right)^2 \gamma_{\pm}^2}{1.00 - \frac{m}{m^\circ}} = 1.75 \times 10^{-5}$$

$m = 4.516 \times 10^{-3} \text{ mol kg}^{-1} \rightarrow 4.52 \times 10^{-3} \text{ mol kg}^{-1}$

This result has converged sufficiently to calculate the degree of dissociation.

$m = 4.52 \times 10^{-3} \text{ mol kg}^{-1}$

$$\frac{4.52 \times 10^{-3} \text{ mol kg}^{-1}}{1.00 \text{ mol kg}^{-1}} \times 100\% = 0.452\%$$

If ionic interactions are ignored

For 0.100 m

$$\frac{\left(\frac{m}{m^\circ}\right)^2}{0.100 - \frac{m}{m^\circ}} = 1.75 \times 10^{-5}$$

$m = 1.314 \times 10^{-3} \text{ mol kg}^{-1}$

$$\frac{1.314 \times 10^{-3}}{0.100} \times 100\% = 1.31\%$$

for 1.00 m

$$\frac{\left(\frac{m}{m^\circ}\right)^2}{1.00 - \frac{m}{m^\circ}} = 1.75 \times 10^{-5}$$

$m = 4.174 \times 10^{-3} \text{ mol kg}^{-1}$

$$\frac{4.174 \times 10^{-3}}{1.00} \times 100\% = 0.417\%$$

P10.13) Calculate the ionic strength in a solution that is 0.0050m in K$_2$SO$_4$, 0.0010m in Na$_3$PO$_4$, and 0.0025m in MgCl$_2$

$$I_{K_2SO_4} = \frac{m}{2}\left(v_+ z_+^2 + v_- z_-^2\right)$$

$$= \frac{0.0050}{2}(2 + 4) = 0.0150 \text{ mol kg}^{-1}$$

$$I_{Na_3PO_4} = \frac{m}{2}\left(v_+ z_+^2 + v_- z_-^2\right)$$

$$= \frac{0.0010}{2}(3+9) = 0.0060 \text{ mol kg}^{-1}$$

$$I_{MgCl_2} = \frac{m}{2}\left(v_+ z_+^2 + v_- z_-^2\right)$$

$$= \frac{0.0025}{2}(4+2) = 0.0075 \text{ mol kg}^{-1}$$

total ionic strength

$$I = (0.0150 + 0.0060 + 0.0075) \text{ mol kg}^{-1}$$

$$= 0.0285 \text{ mol kg}^{-1}$$

P10.14) Calculate I, $\gamma_\pm$, and $a_\pm$ for a 0.0075m solution of Na_2SO_4 at 298 K. Assume complete

dissociation.

$$Na_2SO_4 \Rightarrow v_+ = 1,\ v_- = 2,\ z_+ = 1,\ z_- = 2$$

$$I = \frac{m}{2}\left(v_+ z_+^2 + v_- z_-^2\right)$$

$$I = \frac{0.0075}{2}(1+8) = 0.023 \text{ mol kg}^{-1}$$

$$\ln \gamma_\pm = -1.173\left|z_+ z_-\right|\sqrt{\frac{I}{\text{mol kg}^{-1}}} = -1.173 \times 2 \times \sqrt{0.0225} = -0.3519$$

$$\gamma_\pm = 0.70$$

$$m_\pm^{(v_+ + v_-)} = m_+^{v_+} m_-^{v_-}$$

$$m_\pm^3 = (0.015)^2 (0.0075)^1 = 1.6875 \times 10^{-6}$$

$$m_\pm = 0.012 \text{ mol kg}^{-1}$$

$$a_\pm = \left(\frac{m_\pm}{m^\circ}\right)\gamma_\pm$$

$$a_\pm = 0.012 \times 0.70 = 0.0084$$

P10.15) Express $\mu_\pm$ in terms of μ_+ and μ_- for (a) NaCl, (b) MgBr₂, (c) Li₃PO₄, and (d) Ca(NO₃)₂.

Assume complete dissociation.

a) NaCl $\quad \mu_\pm = \dfrac{\mu_{solute}}{v} = \dfrac{v_+ \mu_+ + v_- \mu_-}{v} = \dfrac{\mu_+ + \mu_-}{2}$

b) MgBr₂ $\quad \mu_\pm = \dfrac{\mu_{solute}}{v} = \dfrac{v_+ \mu_+ + v_- \mu_-}{v} = \dfrac{\mu_+ + 2\mu_-}{3}$

c) Li_3PO_4 $\mu_\pm = \dfrac{\mu_{solute}}{\nu} = \dfrac{\nu_+\mu_+ + \nu_-\mu_-}{\nu} = \dfrac{3\mu_+ + \mu_-}{4}$

d) $Ca(NO_3)_2$ $\mu_\pm = \dfrac{\mu_{solute}}{\nu} = \dfrac{\nu_+\mu_+ + \nu_-\mu_-}{\nu} = \dfrac{\mu_+ + 2\mu_-}{3}$

P10.17) Calculate the solubility of $BaSO_4$ ($K_{sp} = 1.08 \times 10^{-10}$) (a) in pure H_2O and (b) in an aqueous solution with $I = 0.0010$ mol kg^{-1}. For part (a), do an iterative calculation of $\gamma_\pm$ and the solubility until the answer is constant in the second decimal place. Do you need to repeat this procedure in part (b)?

a) $BaSO_4(s) \rightarrow Ba^{2+}(aq) + SO_4^{2-}(aq)$

$\quad \nu_+ = 1, \qquad \nu_- = 1$

$\quad z_+ = 2, \qquad z_- = 2$

$\quad K_{sp} = \left(\dfrac{c_{Ba^{2+}}}{c^\circ}\right)\left(\dfrac{c_{SO_4^{2-}}}{c^\circ}\right)\gamma_\pm^2 = 1.08 \times 10^{-10}$

$\quad c_{Ba^{2+}} = c_{SO_4^{2-}}$

$\quad K_{sp} = \left(\dfrac{c_{Ba^{2+}}}{c^\circ}\right)^2 \gamma_\pm^2 = 1.08 \times 10^{-10}$

when $\gamma_\pm = 1$ $\qquad c_{Ba^{2+}} = 1.039 \times 10^{-5}$ mol L^{-1}

$I = \dfrac{m}{2}\sum\left(\nu_+ z_+^2 + \nu_- z_-^2\right)$

$\quad = \dfrac{1.039 \times 10^{-5}}{2} \times (4 + 4) = 4.157 \times 10^{-5}$ mol kg^{-1}

$\ln \gamma_\pm = -1.173 \times 4 \times \sqrt{4.157 \times 10^{-5}} = -0.03025$

$\quad \gamma_\pm = 0.97020$

when $\gamma_\pm = 0.9702$ $\qquad c_{Ba^{2+}} = 1.0711 \times 10^{-5}$ mol L^{-1}

$I = \dfrac{1.0711 \times 10^{-5}}{2} \times (8) = 4.2846 \times 10^{-5}$ mol kg^{-1}

$\ln \gamma_\pm = -1.173 \times 4 \times \sqrt{4.2846 \times 10^{-5}} = -0.03071$

$\quad \gamma_\pm = 0.9698$

when $\gamma_\pm = 0.9698$ $\qquad c_{Ba^{2+}} = 1.0716 \times 10^{-5}$ mole L^{-1}

$$I = \frac{1.0716 \times 10^{-5}}{2} \times (8) = 4.2866 \times 10^{-5} \text{ mol kg}^{-1}$$

$$\ln \gamma_{\pm} = -1.173 \times 4 \times \sqrt{4.2866 \times 10^{-5}} = -0.03072$$

$$\gamma_{\pm} = 0.9697$$

$$\text{solubility} = 1.07 \times 10^{-5} \text{ mol L}^{-1}$$

b) $I = 0.0010 \text{ mol kg}^{-1}$

$$\ln \gamma_{\pm} = 1.173 \times 4 \times \sqrt{0.0010} = -0.148374$$

$$\gamma_{\pm} = 0.8621$$

$$K_{sp} = \left(\frac{c_{Ba^{2+}}}{c^{\circ}} \right)^2 \times (0.8621)^2 = 1.08 \times 10^{-10}$$

$$c_{Ba^{2+}} = 1.21 \times 10^{-5} \text{ mol L}^{-1}$$

There is no need to repeat the calculation because the ionic strength in the solution is not influenced by the dissociation of the BaSO₄.

P10.21) The equilibrium constant for the hydrolysis of dimethylamine,

$$(CH_3)_2NH(aq) + H_2O(aq) \rightarrow CH_3NH_3^+(aq) + OH^-(aq)$$

is 5.12×10^{-4}. Calculate the extent of hydrolysis for (a) a $0.125m$ solution of $(CH_3)_2NH$ in water and (b) a solution that is also $0.045m$ in $NaNO_3$.

a) $(CH_3)_2 NH (aq) + H_2O(l) \rightarrow CH_3NH_3^+(aq) + OH^-(aq)$

$$K = \frac{\left(\dfrac{m}{m^{\circ}} \right)^2 \gamma_{\pm}^2}{0.125 - \dfrac{m}{m^{\circ}}} = 5.12 \times 10^{-4}$$

if $\gamma_{\pm} = 1$,

$m = 7.75 \times 10^{-3} \text{ mol kg}^{-1}$

when $m = 7.75 \times 10^{-3} \text{ mol kg}^{-1}$

$$I = \frac{m}{2}(2) = m = 7.75 \times 10^{-3} \text{ mol kg}^{-1}$$

$$\ln \gamma_{\pm} = -1.173 \times 1 \times \sqrt{7.75 \times 10^{-3}} = -0.1033$$

$$\gamma_{\pm} = 0.9019$$

when $\gamma_\pm = 0.9019$

$$\frac{\left(\dfrac{m}{m^\circ}\right)^2 \gamma_\pm^2}{0.125 - \left(\dfrac{m}{m^\circ}\right)} = 5.12 \times 10^{-4}$$

$m = 8.561 \times 10^{-3} \ \text{mol kg}^{-1}$

We iterate several times

when $m = 8.561 \times 10^{-3} \ \text{mol kg}^{-1}$

$$I = \frac{m}{2}(2) = m = 8.561 \times 10^{-3} \ \text{mol kg}^{-1}$$

$$\ln \gamma_\pm = -1.173 \times 1 \times \sqrt{8.561 \times 10^{-3}} = -0.1085$$

$$\gamma_\pm = 0.8972$$

when $\gamma_\pm = 0.8972$

$$\frac{\left(\dfrac{m}{m^\circ}\right)^2 \gamma_\pm^2}{0.125 - \dfrac{m}{m^\circ}} = 5.12 \times 10^{-4}$$

$m = 8.605 \times 10^{-3} \ \text{mol kg}^{-1}$

when $m = 8.605 \times 10^{-3} \ \text{mol kg}^{-1}$

$$I = \frac{m}{2}(2) = m = 8.605 \times 10^{-3} \ \text{mol kg}^{-1}$$

$$\ln \gamma_\pm = -1.173 \times 1 \times \sqrt{8.605 \times 10^{-3}} = -0.1088$$

$$\gamma_\pm = 0.8969$$

when $\gamma_\pm = 0.8969$

$$\frac{\left(\dfrac{m}{m^\circ}\right)^2 \gamma_\pm^2}{0.125 - \dfrac{m}{m^\circ}} = 5.12 \times 10^{-4}$$

$m = 8.607 \times 10^{-3} \ \text{mol kg}^{-1} \rightarrow 8.61 \times 10^{-3} \ \text{mol kg}^{-1}$

The degree of hydrolysis is

$$\frac{0.00861}{0.125} \times 100\% = 6.88\%$$

b) $NaNO_3$

$$Na^+ \quad NO_3^- \quad v_+ = 1 \quad v_- = 1$$
$$0.045 \quad 0.045 \quad z_+ = 1 \quad z_- = 1$$
$$I = \frac{0.045}{2}(1+1) = 0.045 \text{ mol kg}^{-1}$$

Add to this the ionic strength from the last iteration of part a).

$$I_{total} = 0.045 + 8.605 \times 10^{-3} = 0.0536 \text{ mol kg}^{-1}$$

$$\ln \gamma_{\pm} = -1.173 \times 1 \times \sqrt{0.0536} = -0.2716$$
$$\gamma_{\pm} = 0.7622$$
$$K = \frac{\left(\frac{m}{m^\circ}\right)^2 (\gamma_{\pm})^2}{0.125} = \frac{\left(\frac{m}{m^\circ}\right)^2 (0.7622)^2}{0.125} = 5.12 \times 10^{-4}$$
$$m = 0.01006 \text{ mol kg}^{-1}$$

Carrying out another iteration

$$I_{total} = 0.045 + 0.01006 = 0.0551 \text{ mol kg}^{-1}$$

$$\ln \gamma_{\pm} = -1.173 \times 1 \times \sqrt{0.0551} = -0.2753$$
$$\gamma_{\pm} = 0.7593$$
$$K = \frac{\left(\frac{m}{m^\circ}\right)^2 (\gamma_{\pm})^2}{0.125} = \frac{\left(\frac{m}{m^\circ}\right)^2 (0.7593)^2}{0.125} = 5.12 \times 10^{-4}$$
$$m = 0.01010 \text{ mol kg}^{-1}$$

The degree of hydrolysis is

$$\frac{0.01010}{0.125} \times 100\% = 8.08\%$$

P10.23) Calculate the Debye–Hückel screening length $1/\kappa$ at 298 K in a $0.0025m$ solution of Na_2HPO_4.

$$\kappa = 3.29 x 10^9 \sqrt{I} \text{ m}^{-1}$$
$$I = \frac{m}{2}\left(v_+ z_+^2 + v_- z_-^2\right) = \frac{0.0025}{2}\left(2 \times 1^1 + 2^2\right) = 0.0075 \text{ mol kg}^{-1}$$
$$\kappa = 3.29 \times 10^9 \sqrt{0.0075} \text{ m}^{-1} = 2.85 \times 10^8 \text{ m}^{-1}$$
$$\frac{1}{\kappa} = 3.5 \times 10^{-9} \text{ m} = 3.5 \text{ nm}$$

P10.26) Calculate the mean ionic molality, $m_\pm$, in $0.0500m$ solutions of (a) $Ca(NO_3)_2$, (b) NaOH, (c) $MgSO_4$, and (d) $AlCl_3$.

$$m_\pm^v = m_+^{v_+} m_-^{v_-}$$

a) $Ca(NO_3)_2 \qquad v_+ = 1, \quad v_- = 2, \quad v = 3$

$\quad m_\pm^3 = (0.0500)(0.0500 \times 2)^2 = 5.00 \times 10^{-4}$

$\quad m_\pm = 0.0794 \ mol \, kg^{-1}$

b) NaOH $\qquad v_+ = 1, \quad v_- = 1, \quad v = 2$

$\quad m_\pm^2 = (0.0500)(0.0500)$

$\quad m_\pm = 0.0500 \ mol \, kg^{-1}$

c) $MgSO_4 \qquad v_+ = 1, \quad v_- = 1, \quad v = 2$

$\quad m_\pm^2 = (0.0500)(0.0500)$

$\quad m_\pm = 0.0500 \ mol \, kg^{-1}$

d) $AlCl_3 \qquad v_+ = 1, \quad v_- = 3, \quad v = 4$

$\quad m_\pm^4 = (0.0500)(0.0500 \times 3)^3 = 1.6875 \times 10^{-4}$

$\quad m_\pm = 0.1140 \ mol \, kg^{-1}$

P10.28) Calculate $\Delta H^\circ_{reaction}$ and $\Delta G^\circ_{reaction}$ for the reaction $AgNO_3(aq) + KCl(aq) \rightarrow AgCl(s) + KNO_3(aq)$.

$\Delta G^\circ_{reaction} = \Delta G^\circ_f (AgCl, s) + \Delta G^\circ_f (K^+, aq) + \Delta G^\circ_f (NO_3^-, aq) - \Delta G^\circ_f (Ag^+, aq)$

$- \Delta G^\circ_f (NO_3^-, aq) - \Delta G^\circ_f (K^+, aq) - \Delta G^\circ_f (Cl^-, aq)$

$\Delta G^\circ_{reaction} = \Delta G^\circ_f (AgCl, s) - \Delta G^\circ_f (Ag^+, aq) - \Delta G^\circ_f (Cl^-, aq) = -109.8 \ kJ \, mol^{-1} - 77.1 \ kJ$

$mol^{-1} + 131.2 \ kJ \, mol^{-1} = -55.7 \ kJ \, mol^{-1}$

$\Delta H^\circ_{reaction} = \Delta H^\circ_f (AgCl, s) - \Delta H^\circ_f (Ag^+, aq) - \Delta H^\circ_f (Cl^-, aq)$

$\Delta H^\circ_{reaction} = -127.0 \ kJ \, mol^{-1} - 105.6 \ kJ \, mol^{-1} + 167.2 \ kJ \, mol^{-1} = -65.4 \ kJ \, mol^{-1}$

Chapter 11: Electrochemical Cells, Batteries, and Fuel Cells

P11.3) For the half-cell reaction $Hg_2Cl_2(s) + 2e^- \rightarrow 2Hg(l) + 2Cl^-(aq)$, $E^\circ = +0.27$ V. Using this result and $\Delta G^\circ_f(Hg_2Cl_2,s) = -210.7$ kJ mol^{-1}, determine ΔG°_f (Cl^-, aq).

$$\Delta G^\circ_R = -nFE^\circ = 2\Delta G^\circ_f\left(Cl^-,aq\right) - \Delta G^\circ_f\left(Hg_2Cl_2,s\right)$$

$$\Delta G^\circ_f\left(Cl^-,aq\right) = \frac{-\Delta G^\circ_f\left(Hg_2Cl_2,s\right) - nFE^\circ}{2}$$

$$\Delta G^\circ_f\left(Cl^-,aq\right) = \frac{-210.7 \text{ kJ mol}^{-1} - 2 \text{ mol} \times 96485 \text{ C mol}^{-1} \times 0.26808 \text{ V}}{2} = -131.2 \text{ kJ mol}^{-1}$$

P11.9) Consider the half-cell reaction $AgCl(s) + e^- \rightleftharpoons Ag(s) + Cl^-(aq)$. If $\mu^\circ (AgCl, s) = -109.71$ kJ mol^{-1}, and if $E^\circ = +0.222$ V for this half-cell, calculate the standard Gibbs energy of formation of Cl^- (aq).

$$\Delta G^\circ_1 = -109.71 \text{ kJ mol}^{-1}$$

$$\Delta G^\circ_2 = -1 \times 96,485 \text{ C mol}^{-1} \times 0.222 \text{ V} = 21.4 \text{ kJ mol}^{-1}$$

$$\Delta G^\circ_3 = \Delta G^\circ_f\left(Cl^-\left(aq\right)\right) = \Delta G^\circ_1 + \Delta G^\circ_2 = -131.1 \text{ kJ mol}^{-1}$$

P11.11) Consider the cell $Hg(l)|Hg_2SO_4(s)|FeSO_4(aq, a = 0.0100)|Fe(s)$.

a. Write the cell reaction.

b. Calculate the cell potential, the equilibrium constant for the cell reaction, and ΔG° at 25°C.

a) Oxidation: $2Hg(l) + SO_4^{2-}(aq) \rightarrow Hg_2SO_4(s) + 2e^-$ $E^\circ = -0.6125$ V

    Reduction: $Fe^{2+}(aq) + 2e^- \rightarrow Fe(s)$ $E^\circ = -0.447$ V

Cell reaction:

$$2Hg(l) + Fe^{2+}(aq) + SO_4^{2-}(aq) \rightarrow Hg_2SO_4(s) + Fe(s)$$

$$E^\circ_{cell} = -0.6125 + (-0.447) = -1.0595 \text{ V}$$

$$E_{cell} = E_{cell}^{\circ} - \frac{RT}{nF} \ln\left(\frac{1}{a_{Fe^{2+}} a_{SO_4^{2-}}}\right)$$

$$= -1.0595 - \frac{8.3145 \text{ J mol}^{-1} \text{ K}^{-1} \times 298.15 \text{ K}}{2 \times 96,485 \text{ C mol}^{-1}} \ln\left(1/(0.0100)^2\right)$$

$$= -1.178 \text{ V}$$

$$\Delta G_{reaction}^{\circ} = -nFE^{\circ} = -2 \times 96,485 \text{ C mol}^{-1} \times (-1.0595 \text{ V})$$

$$= 204.5 \text{ kJ mol}^{-1}$$

$$K = e^{-\Delta G_{reaction}^{\circ}/RT} = 1.51 \times 10^{-36}$$

P11.13) a) Calculate $\Delta G_{reaction}^{\circ}$ and the equilibrium constant, K, at 298.15 K for the reaction

$Hg_2Cl_2(s) \rightarrow 2Hg(l) + Cl_2(g)$. b) Calculate K using Table 4.1. c) What value of ΔG_R would make the value of K the same as calculated from the half cell potentials?

a)

$Hg_2Cl_2(s) + 2e^- \rightarrow 2Hg(l) + 2Cl^-(aq)$	$E^{\circ} = +0.26808 \text{ V}$
$2Cl^-(aq) \rightarrow Cl_2(g) + 2e^-$	$E^{\circ} = -1.35827 \text{ V}$
$Hg_2Cl_2(s) \rightarrow 2Hg(l) + Cl_2(g)$	$E^{\circ} = -1.09019 \text{ V}$

$$\Delta G_{reaction}^{\circ} = -nFE^{\circ} = -2 \times 96485 \text{ C mol}^{-1} \times 1.09019 \text{ V} = 210.4 \text{ kJ mol}^{-1}$$

$$\ln K = \frac{nF}{RT} E^{\circ} = -\frac{2 \times 96485 \text{ C mol}^{-1} \times 1.09019 \text{ V}}{8.314 \text{ J K}^{-1}\text{mol}^{-1} \times 298.15 \text{ K}}$$

$$= -84.8686$$

$$K = 1.39 \times 10^{-37}$$

b)

$$\Delta G_{reaction}^{\circ} = -nFE^{\circ} = -2 \times 96485 \text{ C mol}^{-1} \times 1.09019 \text{ V} = 210.4 \text{ kJ mol}^{-1}$$

$$\ln K = -\frac{\Delta G_{reaction}^{\circ}}{RT} = -\frac{210700 \text{ J mol}^{-1}}{8.314 \text{ J K}^{-1} \text{ mol}^{-1} \times 298.15 \text{ K}}$$

$$= -85.0002$$

$$K = 1.22 \times 10^{-37}$$

c) The results would agree exactly if $\Delta G_{reaction}^{\circ} = 210374 \text{ J mol}^{-1}$

P11.20) Determine E° for the reaction $Cr^{2+}(aq) + 2e^- \rightarrow Cr(s)$ from the one-electron reduction potential for Cr^{3+} and the three-electron reduction potential for Cr^{3+} given in Table 11.1 (see Appendix B).

$$Cr^{3+}(aq) + 3e^- \rightarrow Cr(s) \qquad \Delta G^{\circ} = -nFE^{\circ} = -3 \times 96485 \text{ C mol}^{-1} \times (-0.744 \text{ V}) = 215.4 \text{ kJ mol}^{-1}$$

$$Cr^{2+}(aq) \rightarrow Cr^{3+}(aq) + e^{-} \qquad \Delta G^{\circ} = -nFE^{\circ} = -1 \times 96485 \text{ C mol}^{-1} \times 0.407 \text{ V} = -39.27 \text{ kJ mol}^{-1}$$

$$Cr^{2+}(aq) + 2e^{-} \rightarrow Cr(s) \qquad \Delta G = 215.4 \text{ kJ mol}^{-1} - 39.27 \text{ kJ mol}^{-1} = 176.1 \text{ kJ mol}^{-1}$$

$$E^{\circ}_{Cr^{2+}/Cr} = -\frac{\Delta G^{\circ}}{nF} = \frac{-176.1 \times 10^{3} \text{ J mol}^{-1}}{2 \times 96485 \text{ C mol}^{-1}} = -0.913 \text{ V}$$

P11.23) Consider the half-cell reaction $O_2(g) + 4H^{+}(aq) + 4e^{-} \rightarrow 2H_2O(l)$. By what factor are n, Q, E, and E° changed if all the stoichiometric coefficients are multiplied by the factor two? Justify your answers.

n is proportional to the number of electrons transferred, and increases by the factor two.

Q is squared if all stoichiometric factors are doubled. The factor by which it is increased depends on the activities of O_2 and H^{+}.

$E^{\circ} = \dfrac{\Delta G^{\circ}}{nF}$ is unchanged because both ΔG° and n are doubled.

$E = E^{\circ} - \dfrac{RT}{nF} \ln Q$ is unchanged because the squaring of Q is offset exactly by the doubling of n.

P11.25) The half-cell potential for the reaction $O_2(g) + 4H^{+}(aq) + 4e^{-} \rightarrow 2H_2O(l)$ is +1.03 V at 298.15 K when $a_{O_2} = 1$. Determine $a_{H^{+}}$.

$$E = E^{\circ} - \frac{RT}{nF} \ln \frac{1}{a_{O_2} a_{H^{+}}^{4}}$$

$$1.03 \text{ V} = 1.23 \text{ V} - \frac{0.05916 \text{ V}}{4} \log_{10} \frac{1}{a_{H^{+}}^{4}}$$

$$\log_{10} a_{H^{+}} = \frac{1.03 \text{ V} - 1.23 \text{ V}}{0.05916 \text{ V}} = -3.381$$

$$a_{H^{+}} = 4.16 \times 10^{-4}$$

P11.26) Using half-cell potentials, calculate the equilibrium constant at 298.15 K for the reaction 2 $H_2O(l) \rightarrow 2H_2(g) + O_2(g)$. Compare your answer with that calculated using ΔG_f° values from Table 11.1 (see Appendix B). What is the value of E° for the overall reaction that makes the two methods agree exactly?

$$2H_2O(l) + 2e^- \rightarrow H_2(g) + 2OH^-(aq) \qquad E^\circ = -0.8277 \text{ V}$$

$$4OH^-(aq) \rightarrow O_2(g) + 4e^- + 2H_2O(l) \qquad E^\circ = -0.401 \text{ V}$$

$$2H_2O(l) \rightarrow 2H_2(g) + O_2(g) \qquad E^\circ = -1.2287 \text{ V}$$

$$\ln K = \frac{nF}{RT} E^\circ = -\frac{4 \times 96485 \text{ C mol}^{-1} \times 1.2287 \text{ V}}{8.314 \text{ J K}^{-1} \text{ mol}^{-1} \times 298.15 \text{ K}}$$

$$= -191.303$$

$$K = 8.28 \times 10^{-84}$$

$$\ln K = -\frac{2\Delta G_f^\circ (H_2O, l)}{RT} = -\frac{2 \text{ mol} \times 237.1 \times 10^3 \text{ J mol}^{-1}}{8.3145 \text{ J mol}^{-1} \text{ K}^{-1} \times 298.15 \text{ K}} = -191.301$$

$$K = 8.30 \times 10^{-84}$$

For the two results to agree, E° must be given by

$$E^\circ = -\frac{191.325 \times 8.3145 \text{ J mol}^{-1} \text{ K}^{-1} \times 298.15 \text{ K}}{4 \times 96485 \text{ C mol}^{-1}} = -1.22869 \text{ V}$$

This value lies within the error limits of the determination of E°.

P11.29) Determine K_{sp} for AgBr at 298.15 K using the electrochemical cell described by

$$Ag(s)\big|AgBr(s)\big|Br^-(aq, a_{Br^-})\big\|Ag^+(aq, a_{Ag^+})\big|Ag(s)$$

The half cell and overall reactions are

$$AgBr(s) + e^- \rightarrow Ag(s) + Br^-(aq) \qquad E^\circ = +0.07133 \text{ V}$$

$$Ag(s) \rightarrow Ag^+(aq) + e^- \qquad E^\circ = -0.7996 \text{ V}$$

$$AgBr(s) \rightarrow Ag^+(aq) + Br^-(aq) \qquad E^\circ = -0.72827 \text{ V}$$

$$\log_{10} K_{sp} = -\frac{nE^\circ}{0.05916 \text{ V}} = -\frac{0.729 \text{ V}}{0.05916 \text{ V}} = -12.310$$

$$K_{sp} = 4.89 \times 10^{-13}$$

Chapter 12: From Classical to Quantum Mechanics

P12.5) Calculate the highest possible energy of a photon that can be observed in the emission spectrum of H.

The highest energy photon corresponds to a transition from $n = \infty$ to $n = 1$.

$$\tilde{\nu} = 109677\left(\frac{1}{1} - \frac{1}{\infty^2}\right) = 109677 \text{ cm}^{-1}$$

$$E = hc\tilde{\nu} = 2.17871 \times 10^{-18} \text{ J}$$

P12.8) What speed does a N_2 molecule have if it has the same momentum as a photon of wavelength 180. nm?

$$p = \frac{h}{\lambda} = m_{H_2} v_{H_2}$$

$$v_{H_2} = \frac{h}{m_{H_2}\lambda} = \frac{6.626 \times 10^{-34} \text{ J s}}{28.02 \text{ amu} \times 1.661 \times 10^{-27} \text{ kg} (\text{amu})^{-1} \times 180. \times 10^{-9} \text{ m}} = 0.0791 \text{ m s}^{-1}$$

P12.12) Electrons have been used to determine molecular structure by diffraction. Calculate the speed of an electron for which the wavelength is equal to a typical bond length, namely, 0.175 nm.

$$v = \frac{p}{m} = \frac{h}{m\lambda} = \frac{6.626 \times 10^{-34} \text{ J s}}{9.109 \times 10^{-31} \text{ kg} \times 0.175 \times 10^{-9} \text{ m}} = 4.16 \times 10^{6} \text{ m s}^{-1}$$

P12.13) For a monatomic gas, one measure of the "average speed" of the atoms is the root mean square speed, $v_{rms} = \langle v^2 \rangle^{1/2} = \sqrt{3kT/m}$, in which m is the molecular mass and k is the Boltzmann constant. Using this formula, calculate the de Broglie wavelength for H and Ne atoms at 250. and at 750. K.

$$\lambda = \frac{h}{mv_{rms}} = \frac{h}{\sqrt{3kTm}} = \frac{6.626 \times 10^{-34} \text{ J s}}{\sqrt{3 \times 1.381 \times 10^{-23} \text{ J K}^{-1} \times 250.\text{K} \times 1.008 \text{ amu} \times 1.661 \times 10^{-27} \text{ kg amu}^{-1}}}$$

$$= 1.59 \times 10^{-10} \text{ m}$$

or H at 750. K. $\lambda = 9.19 \times 10^{-11}$ m for He. For Ne, $\lambda = 3.56 \times 10^{-11}$ m and 2.05×10^{-11} m at 250. K and 750. K, respectively.

P12.16) If an electron passes through an electrical potential difference of 1 V, it has an energy of 1 electron-volt. What potential difference must it pass through in order to have a wavelength of 0.225 nm?

$$E = \frac{1}{2}m_e v^2 = \frac{1}{2}m_e \times \left(\frac{h}{m_e \lambda}\right)^2 = \frac{h^2}{2m_e \lambda^2}$$

$$= \frac{\left(6.626\times10^{-34}\text{ J s}\right)^2}{2\times9.109\times10^{-31}\text{ kg}\times\left(2.25\times10^{-10}\text{ m}\right)^2} \times \frac{1\text{eV}}{1.602\times10^{-19}\text{ J}} = 29.7\text{ eV}$$

The electron must pass through an electrical potential of 29.7 V.

P12.20) The power (energy per unit time) radiated by a black body per unit area of surface expressed in units of W m^{-2} is given by $P = \sigma T^4$ with σ= 5.67 × 10^{-8} W m^{-2} K^{-4}. The radius of the sun is 6.95 × 10^5 km and the surface temperature is 5750. K. Calculate the total energy radiated per second by the sun. Assume ideal blackbody behavior.

$$E = PA = \sigma T^4 \times 4\pi r^2$$

$$= 5.67\times10^{-8}\text{ W m}^{-2}\text{K}^{-4} \times \left(5750.\text{ K}\right)^4 \times 4\pi \times \left(6.95\times10^8\text{ m}\right)^2$$

$$= 3.76\times10^{26}\text{ W}$$

P12.21) The work function of tungsten is 4.50 eV. What is the minimum frequency of light required to observe the photoelectric effect on W? If light with a 225-nm wavelength is absorbed by the surface, what is the velocity of the emitted electrons?

a) For electrons to be emitted, the photon energy must be greater than the work function of the surface.

$$E = h\nu \geq 4.50\text{ eV} \times \frac{1.602\times10^{-19}\text{ J}}{\text{eV}} = 9.05\times10^{-19}\text{ J}$$

$$\nu \geq \frac{E}{h} \geq \frac{9.05\times10^{-19}\text{ J}}{6.626\times10^{-34}\text{ J s}} \geq 1.09\times10^{15}\text{ s}^{-1}$$

b) The outgoing electron must first surmount the barrier arising from the work function, so not all the photon energy is converted to kinetic energy.

$$E_e = h\nu - \phi = \frac{hc}{\lambda} - \phi$$

$$= \frac{6.626\times10^{-34}\text{ J s}\times2.998\times10^8\text{ m s}^{-1}}{225\times10^{-9}\text{ m}} - 7.21\times10^{-19}\text{ J} = 1.62\times10^{-19}\text{ J}$$

$$v = \sqrt{\frac{2E_e}{m_e}} = \sqrt{\frac{2\times1.62\times10^{-19}\text{ J}}{9.11\times10^{-31}\text{ kg}}} = 5.96\times10^5\text{ m s}^{-1}$$

P12.23) Calculate the longest and the shortest wavelength observed in the Lyman series.

For the longest wavelength, the transition is from $n = 2$ to $n = 1$.

$$\tilde{v} = 109677\left(\frac{1}{1} - \frac{1}{2^2}\right) = 82257.8 \text{ cm}^{-1}$$

$$\lambda = \frac{1}{\tilde{v}} = 1.21569 \times 10^{-5} \text{ cm}$$

For the shortest wavelength, the transition is from $n = \infty$ to $n = 1$.

$$\tilde{v} = 109677\left(\frac{1}{1} - \frac{1}{\infty^2}\right) = 109677 \text{ cm}^{-1}$$

$$\lambda = \frac{1}{\tilde{v}} = 9.11768 \times 10^{-6} \text{ cm}$$

P12.24) A 1000-W gas discharge lamp emits 5.25 W of ultraviolet radiation in a narrow range centered near 325 nm. How many photons of this wavelength are emitted per second?

$$n' = \frac{E_{total}}{E_{photon}} = \frac{3.00 \text{ W} \times 1 \text{ J s}^{-1} \text{ W}^{-1}}{\frac{hc}{\lambda}} = \frac{5.25 \text{ W} \times 1 \text{ J s}^{-1} \text{ W}^{-1}}{\frac{6.626 \times 10^{-34} \text{ J s} \times 2.998 \times 10^8 \text{ m s}^{-1}}{325 \times 10^{-9} \text{ m}}} = 8.59 \times 10^{18} \text{ s}^{-1}$$

P12.27) Pulsed lasers are powerful sources of nearly monochromatic radiation. Lasers that emit photons in a pulse of 10-ns duration with a total energy in the pulse of 0.10 J at 1000 nm are commercially available.

a. What is the average power (energy per unit time) in units of watts (1 W = 1 J/s) associated with such a pulse?

b. How many 1000-nm photons are emitted in such a pulse?

a) $P = \dfrac{\Delta E}{\Delta t} = \dfrac{0.10 \text{ J}}{1.0 \times 10^{-8} \text{ s}} = 1.0 \times 10^7 \text{ J s}^{-1}$

b) $N = \dfrac{E_{pulse}}{E_{photon}} = \dfrac{E_{pulse}}{h\dfrac{c}{\lambda}} = \dfrac{0.10 \text{ J}}{6.626 \times 10^{-34} \text{ J s}^{-1} \times \dfrac{2.998 \times 10^8 \text{ m s}^{-1}}{1.000 \times 10^6 \text{ m}}} = 5.0 \times 10^{17}$

Chapter 13: The Schrödinger Equation

P13.1) A wave traveling in the z direction is described by the wave function $\Psi(z,t) = A_1\ \mathbf{x}\ \sin(kz-\omega t + \varnothing_1) + A_2\ \mathbf{y}\ \sin(kz-\omega t + \varnothing_2)$, where $\mathbf{x}$ and $\mathbf{y}$ are vectors of unit length along the x and y axes, respectively. Because the amplitude is perpendicular to the propagation direction, $\Psi(z,t)$ represents a transverse wave.

a) What requirements must A_1 and A_2 satisfy for a plane polarized wave in the x-z plane? The amplitude of a plane polarized wave is non-zero only in one plane

b) What requirements must A_1 and A_2 satisfy for a plane polarized wave in the y-z plane?

c) What requirements must A_1 and A_2 and ϕ_1 and ϕ_2 satisfy for a plane polarized wave in a plane oriented at 45° to the xz plane?

d) What requirements must A_1 and A_2 and ϕ_1 and ϕ_2 satisfy for a circularly polarized wave? The phases of the two components of a circularly polarized wave differ by $\pi/2$.

a) The amplitude along the x axis must oscillate, and the amplitude along the y axis must vanish. Therefore $A_1 \neq 0$ and $A_2 = 0$.

b) The amplitude along the y axis must oscillate, and the amplitude along the x axis must vanish. Therefore $A_1 = 0$ and $A_2 \neq 0$.

c) The amplitude along both the x and y axes must oscillate. Therefore $A_1 \neq 0$ and $A_2 \neq 0$. Because they must oscillate in phase, $\phi_1 = \phi_2$.

d) The amplitude along both the x and y axes must oscillate with the same amplitude. Therefore $A_1 = A_2 \neq 0$. For a circularly polarized wave, the x and y components must be out of phase by $\pi/2$.

Therefore $\phi_1 = \phi_2 \pm \dfrac{\pi}{2}$. This can be seen by comparing the x and y amplitudes for the positive sign.

$$\Psi(z,t) = A_1\ \mathbf{x}\ \sin(kz - \omega t + \phi_1) + A_1\ \mathbf{y}\ \sin(kz - \omega t + \phi_1 + \frac{\pi}{2})$$

let $kz + \phi = kz'$

$$\Psi(z,t) = A_1\ \mathbf{x}\ \sin(kz' - \omega t) + A_1\ \mathbf{y}\ \sin(kz' - \omega t + \frac{\pi}{2})$$

$$= A_1\ \mathbf{x}\ \sin(kz' - \omega t) + A_1\ \mathbf{y}\left[\sin(kz' - \omega t)\cos\frac{\pi}{2} + \cos(kz' - \omega t)\sin\frac{\pi}{2}\right]$$

$$= A_1\ \mathbf{x}\ \sin(kz' - \omega t) + A_1\ \mathbf{y}\ \cos(kz' - \omega t)$$

The x and y amplitudes are $\pi/2$ out of phase and the sum of the squares of their amplitudes is a constant as required for a circle.

P13.2) Because $\int_0^d \cos(n\pi x/d)\cos(m\pi x/d)dx = 0$, $m \neq n$, the functions $\cos(n\pi x/d)$ for $n = 1, 2, 3, ...$ form an orthogonal set. What constant must these functions be multiplied by to form an orthonormal set?

$$1 = N^2 \int_0^d \cos\left(\frac{m\pi x}{d}\right)\cos\left(\frac{m\pi x}{d}\right)dx = N^2\left[\frac{x}{2} + \frac{d}{4m\pi}\sin\left(\frac{2m\pi x}{d}\right)\right]_0^d$$

where we have used the standard integral $\int\left(\cos^2 ax\right)dx = \frac{1}{2}x + \frac{1}{4a}\sin 2ax$

$$1 = N^2\left[\frac{d}{2} + \frac{d}{4m\pi}\sin\left(2m\pi\right) - \frac{0}{2} - \frac{d}{4m\pi}\sin\left(0\right)\right] = \frac{d}{2}N^2 = 1$$

$$N = \sqrt{\frac{2}{d}}$$

P13.6) Carry out the following coordinate transformations:

a. Express the point $x = 4$, $y = 2$, and $z = 3$ in spherical coordinates.

b. Express the point $r = 7$, $\theta = \frac{\pi}{8}$, and $\phi = \frac{5\pi}{8}$ in Cartesian coordinates.

a) $r = \sqrt{x^2 + y^2 + z^2} = \sqrt{4^2 + 2^2 + 3^2} = \sqrt{29}$

$$\theta = \cos^{-1}\frac{z}{\sqrt{x^2 + y^2 + z^2}} = \cos^{-1}\frac{2}{\sqrt{29}} = 0.980 \text{ radians}$$

$$\phi = \tan^{-1}\frac{y}{x} = \tan^{-1}\frac{2}{4} = 0.464 \text{ radians}$$

b) $x = r\sin\theta\cos\phi = 7\sin\frac{\pi}{8}\cos\frac{5\pi}{8} = -1.03$

$$y = r\sin\theta\sin\phi = 7\sin\frac{\pi}{8}\sin\frac{5\pi}{8} = 2.75$$

$$z = r\cos\theta = 7\cos\frac{\pi}{8} = 6.47$$

P13.8) Show that

$$\frac{a+ib}{c+id} = \frac{ac+bd+i(bc-ad)}{c^2+d^2}$$

$$\frac{a+ib}{c+id} = \left(\frac{a+ib}{c+id}\right)\left(\frac{c-id}{c-id}\right) = \frac{ac+bd+ibc-iad}{c^2+d^2} = \frac{ac+bd+i(bc-ad)}{c^2+d^2}$$

P13.9) Express the following complex numbers in the form $re^{i\theta}$.

a. $7-3i$

c. $\dfrac{7-i}{5+3i}$

b. $-5i$

d. $\dfrac{4+i}{1-2i}$

In the notation $re^{i\theta}$, $r = |z| = \sqrt{a^2+b^2}$ and $\theta = \sin^{-1}\left(\dfrac{\text{Im } z}{|z|}\right)$.

a) $7-3i = \sqrt{58}\exp\left(i\sin^{-1}\dfrac{-3}{\sqrt{58}}\right) = \sqrt{58}\exp(-0.129i\pi)$

b) $-5i = 5\exp\left(i\sin^{-1}\dfrac{-5}{\sqrt{25}}\right) = 5\exp\left(-\dfrac{\pi}{2}\right)$

c)

$$\frac{7-i}{3+5i} = \frac{7-i}{3+5i} \times \frac{3-5i}{3-5i} = \frac{21-38i-5}{34} = \frac{8}{17} - \frac{19i}{17}$$

We next calculate the magnitude of the complex number

$$\sqrt{\left(\frac{8}{17}\right)^2 + \left(-\frac{19}{17}\right)^2} = \sqrt{\frac{425}{289}} = \sqrt{\frac{25}{17}} = \frac{5}{\sqrt{17}}$$

$$= \frac{5}{\sqrt{17}}\exp\left(i\sin^{-1}\frac{19/17}{5/\sqrt{17}}\right) = \frac{5}{\sqrt{17}}\exp(-0.217i\pi)$$

d)

$$\frac{4+i}{1-2i} = \frac{4+i}{1-2i} \times \frac{1+2i}{1+2i} = \frac{4+9i-2}{5} = \frac{2}{5} + \frac{9i}{5}$$

We next calculate the magnitude of the complex number

$$\sqrt{\left(\frac{2}{5}\right)^2 + \left(\frac{9}{5}\right)^2} = \sqrt{\frac{85}{25}} = \sqrt{\frac{17}{5}}$$

$$= \sqrt{\frac{17}{5}} \exp\left(i\sin^{-1}\frac{9/5}{\sqrt{17/5}}\right) = \sqrt{\frac{17}{5}} \exp(0.430i\pi)$$

P13.14) Determine in each of the following cases if the function in the first column is an eigenfunction of the operator in the second column. If so, what is the eigenvalue?

a. $3\cos^2\theta - 1$ $\qquad\qquad \frac{1}{\sin\theta}\frac{d}{d\theta}\left(\sin\theta\frac{d}{d\theta}\right)$

b. $e^{-(x^2/2)}$ $\qquad\qquad \frac{d^2}{dx^2} - x^2$

c. $e^{-4i\phi}$ $\qquad\qquad \frac{d^2}{d\phi^2}$

a)

$$\frac{1}{\sin\theta}\frac{d}{d\theta}\left(\sin\theta\frac{d\left(3\cos^2\theta - 1\right)}{d\theta}\right) = \frac{1}{\sin\theta}\frac{d}{d\theta}\left(-6\cos\theta\sin^2\theta\right)$$

$$= \frac{1}{\sin\theta}\left(6\sin^3\theta - 12\cos^2\theta\sin\theta\right) = 6\sin^2\theta - 12\cos^2\theta$$

$$= 6 - 18\cos^2\theta = -6\left(3\cos^2\theta - 1\right)$$

Eigenfunction with eigenvalue –6.

b)

$$\frac{d^2 e^{-\frac{1}{2}x^2}}{dx^2} - x^2 e^{-\frac{1}{2}x^2} = -e^{-\frac{1}{2}x^2}$$

Eigenfunction with eigenvalue –1.

c)

$$\frac{d^2 e^{-4i\phi}}{d\phi^2} = -16 e^{-4i\phi}$$

Eigenfunction with eigenvalue –16.

P13.15) Show by carrying out the integration that $\sin(m\pi x/a)$ and $\cos(m\pi x/a)$, where m is an integer, are orthogonal over the interval $0 \le x \le a$. Would you get the same result if you used the interval $0 \le x \le 3a/4$? Explain your result..

$$\int_0^a \cos\left(\frac{m\pi x}{a}\right) \sin\left(\frac{m\pi x}{a}\right) dx = \left[\frac{a}{2m\pi} \sin^2\left(\frac{m\pi x}{a}\right)\right]_0^a = \frac{a}{2m\pi}\left[\sin^2(m\pi) - 0\right] = 0$$

$$\int_0^{\frac{3a}{4}} \cos\left(\frac{m\pi x}{a}\right) \sin\left(\frac{m\pi x}{a}\right) dx = \int_0^{\frac{3a}{4}} \cos\left(\frac{m\pi x}{a}\right) \sin\left(\frac{m\pi x}{a}\right) dx$$

$$= \left[\frac{a}{2m\pi} \sin^2\left(\frac{m\pi x}{a}\right)\right]_0^{\frac{3a}{4}} = \frac{a}{2m\pi}\left[\sin^2\left(\frac{3m\pi}{4}\right) - 0\right] \ne 0$$

except for the special case $\frac{3m}{4} = n$ where n is an integer. The length of the integration interval must be n periods (for n an integer) to make the integral zero.

P13.19) Is the function $3x^2 - 1$ an eigenfunction of the operator $-(1-x^2)(d^2/dx^2) + 2x(d/dx)$? If so, what is the eigenvalue?

$$-(1-x^2)\left(d^2\left[3x^2-1\right]/dx^2\right) + 2x\left(d\left[3x^2-1\right]/dx\right)$$
$$= -6(1-x^2) + 12x^2 = 18x^2 - 6 = 6\left[3x^2 - 1\right]$$

Eigenfunction with eigenvalue 6

P13.20) Find the result of operating with $d^2/dx^2 - 4x^2$ on the function e^{-ax^2}. What must the value of a be to make this function an eigenfunction of the operator?

$$\frac{d^2 e^{-ax^2}}{dx^2} - 4x^2 e^{-ax^2} = -2ae^{-ax^2} - 4x^2 e^{-ax^2} + 4a^2 x^2 e^{-ax^2} = -2ae^{-ax^2} + 4(a^2 - 1)x^2 e^{-ax^2}$$

For the function to be an eigenfunction of the operator, the terms containing $x^2 e^{-ax^2}$ must vanish. This is the case if $a = \pm 1$.

P13.21) Determine in each of the following cases if the function in the first column is an eigenfunction of the operator in the second column. If so, what is the eigenvalue?

a. $\sin\theta\cos\phi \quad \partial/\partial\phi$

b. $e^{(-x^2/2)} \quad (1/x)d/dx$

c. $\sin\theta \quad (\sin\theta/\cos\theta)d/d\theta$

a)

$\dfrac{\partial}{\partial\phi}\sin\theta\cos\phi = -\sin\theta\sin\phi.$ \qquad Not an eigenfunction

b)

$\dfrac{1}{x}\dfrac{d}{dx}e^{-\frac{1}{2}x^2} = -e^{-\frac{1}{2}x^2}$ \qquad Eigenfunction with eigenvalue -1

c)

$\dfrac{\sin\theta}{\cos\theta}\dfrac{d}{d\theta}\sin\theta = \sin\theta$ \qquad Eigenfunction with eigenvalue $+1$

P13.24) If two operators act on a wave function as indicated by $\hat{A}\hat{B}f(x)$, it is important to carry out the operations in succession with the first operation being that nearest to the function. Mathematically, $\hat{A}\hat{B}f(x) = \hat{A}(\hat{B}f(x))$ and $\hat{A}^2 f(x) = \hat{A}(\hat{A}f(x))$. Evaluate the following successive operations $\hat{A}\hat{B}f(x)$. The operators $\hat{A}$ and $\hat{B}$ are listed in the first and second columns and $f(x)$ is listed in the third column.

a. $\dfrac{d}{dx} \qquad \dfrac{d}{dx} \qquad x^2 + e^{ax^2}$

b. $\dfrac{\partial^2}{\partial y^2} \qquad \dfrac{\partial}{\partial x} \qquad \cos 3y \sin^2 x$

c. $\dfrac{\partial}{\partial\theta} \qquad \dfrac{\partial^2}{\partial\phi^2} \qquad \dfrac{\cos\phi}{\sin\theta}$

a) $\dfrac{d}{dx}\left[\dfrac{d\left(x^2+e^{ax^2}\right)}{dx}\right]=\dfrac{d}{dx}\left[2x+2a\,x\,e^{ax^2}\right]=2+4a^2x^2e^{ax^2}+2a\,e^{ax^2}$

b) $\dfrac{\partial^2}{\partial y^2}\left[\dfrac{\partial\left(\cos 3y\sin^2 x\right)}{\partial x}\right]=\dfrac{\partial^2}{\partial y^2}\left[2\cos 3y\sin x\cos x\right]=18\cos 3y\sin x\cos x$

c) $\dfrac{\partial}{\partial\theta}\left[\dfrac{\partial^2\left(\dfrac{\cos\phi}{\sin\theta}\right)}{\partial\phi^2}\right]=\dfrac{\partial}{\partial\theta}\left[-\dfrac{\cos\phi}{\sin\theta}\right]=\dfrac{\cos\phi\cos\theta}{\sin^2\theta}$

P13.26) Consider a two-level system with $\varepsilon_1 = 3.10\times10^{-21}$ J and $\varepsilon_2 = 6.10\times10^{-21}$ J. If $g_2 = g_1$, what value of T is required to obtain $n_2/n_1 = 0.225$? What value of T is required to obtain $n_2/n_1 = 0.875$?

$$\dfrac{n_2}{n_1}=\dfrac{g_2}{g_1}\exp\left[\dfrac{-\left(\varepsilon_2-\varepsilon_1\right)}{kT}\right]$$

$$\ln\left(\dfrac{n_2}{n_1}\right)=\ln\left(\dfrac{g_2}{g_1}\right)-\dfrac{\left(\varepsilon_2-\varepsilon_1\right)}{kT}$$

$$\dfrac{1}{T}=\dfrac{k}{\left(\varepsilon_2-\varepsilon_1\right)}\left[\ln\left(\dfrac{g_2}{g_1}\right)-\ln\left(\dfrac{n_2}{n_1}\right)\right]$$

$$T=\dfrac{\left(\varepsilon_2-\varepsilon_1\right)}{k\left[\ln\left(\dfrac{g_2}{g_1}\right)-\ln\left(\dfrac{n_2}{n_1}\right)\right]}$$

for $n_2/n_1 = 0.225$ $T=\dfrac{3.00\times10^{-21}\text{ J}}{1.381\times10^{-23}\text{ J K}^{-1}\times\left[\ln(1)-\ln(0.225)\right]}=145$ K

for $n_2/n_1 = 0.875$ $T=\dfrac{3.00\times10^{-21}\text{J}}{1.381\times10^{-23}\text{J K}^{-1}\times\left[\ln(1)-\ln(0.875)\right]}=1.63\times10^3$ K

P13.28) Normalize the set of functions $\phi_n(\theta)=e^{in\theta}$, $0\le\theta\le2\pi$. To do so, you need to multiply the functions by a so-called normalization constant N so that the integral

$$N\,N^*\int_0^{2\pi}\phi_m^*(\theta)\phi_n(\theta)d\theta=1\quad\text{for } m=n.$$

$$NN^* \int_0^{2\pi} e^{-in\theta} e^{in\theta} d\theta = NN^* \int_0^{2\pi} d\theta = 2\pi NN^* = 1 \quad \text{This is satisfied for} \quad N = \frac{1}{\sqrt{2\pi}} \quad \text{and the normalized}$$

functions are $\phi_n(\theta) = \frac{1}{\sqrt{2\pi}} e^{in\theta}, \quad 0 \le \theta \le 2\pi.$

P13.30) Operate with a) $\dfrac{\partial}{\partial x} + \dfrac{\partial}{\partial y} + \dfrac{\partial}{\partial z}$ and b) $\dfrac{\partial^2}{\partial x^2} + \dfrac{\partial^2}{\partial y^2} + \dfrac{\partial^2}{\partial z^2}$ on the function

$A\cos(k_1 x)\cos(k_2 y)\cos(k_3 z)$. Is the function an eigenfunction of either operator? If so, what is the eigenvalue?

$$\left[\frac{\partial}{\partial x} + \frac{\partial}{\partial y} + \frac{\partial}{\partial z}\right] A\cos k_1 x \cos k_2 y \cos k_3 z$$

$$= -Ak_1 \sin k_1 x \cos k_2 y \cos k_3 z - Ak_2 \cos k_1 x \sin k_2 y \cos k_3 z - Ak_3 \cos k_1 x \cos k_2 y \sin k_3 z$$

$$= -A\left[k_1 \sin k_1 x \cos k_2 y \cos k_3 z + k_2 \cos k_1 x \sin k_2 y \cos k_3 z + k_3 \cos k_1 x \cos k_2 y \sin k_3 z\right]$$

not an eigenfunction

$$\left[\frac{\partial^2}{\partial x^2} + \frac{\partial^2}{\partial y^2} + \frac{\partial^2}{\partial z^2}\right] A\cos k_1 x \cos k_2 y \cos k_3 z$$

$$= -A\left[k_1^2 + k_2^2 + k_3^2\right]\cos k_1 x \cos k_2 y \cos k_3 z$$

eigenfunction with eigenvalue $-\left[k_1^2 + k_2^2 + k_3^2\right]$

P13.36) Which of the following wave functions are eigenfunctions of the operator d/dx? If they are eigenfunctions, what is the eigenvalue?

a. $ae^{-3x} + be^{-3ix}$

d. $\cos ax$

b. $\sin^2 x$

e. e^{-ix^2}

c. e^{-ix}

a) $\dfrac{d\left(ae^{-3x} + be^{-3ix}\right)}{dx} = -3ae^{-3x} - 3ibe^{-3ix}$ Not an eigenfunction

b) $\dfrac{d\sin^2 x}{dx} = 2\sin x \cos x$ Not an eigenfunction

c) $\dfrac{de^{-ix}}{dx} = -ie^{-ix}$ Eigenfunction with eigenvalue $-i$

d) $\dfrac{d \cos a\, x}{d\, x} = -a \sin a\, x$ Not an eigenfunction

e) $\dfrac{d\, e^{-i x^2}}{d\, x} = -2i\, x\, e^{-i x^2}$ Not an eigenfunction

Chapter 15: Using Quantum Mechanics on Simple Systems

P15.3) Normalize the total energy eigenfunctions for the three-dimensional box in the interval

$0 \le x \le a, 0 \le y \le b, 0 \le z \le c$.

$$1 = \int_0^a \int_0^b \int_0^c \psi^*(x,y,z)\psi(x,y,z)\,dx\,dy\,dz$$

$$= N^2 \int_0^a \sin^2\left(\frac{n_x \pi x}{a}\right)dx \int_0^b \sin^2\left(\frac{n_y \pi y}{b}\right)dy \int_0^c \sin^2\left(\frac{n_z \pi z}{c}\right)dz$$

Using the standard integral $\int \sin^2 \alpha x\,dx = \dfrac{x}{2} - \dfrac{\sin(2\alpha x)}{4\alpha}$

$$1 = N^2 \int_0^a \sin^2\left(\frac{n_x \pi x}{a}\right)dx \int_0^b \sin^2\left(\frac{n_y \pi y}{b}\right)dy \int_0^c \sin^2\left(\frac{n_z \pi z}{c}\right)dz$$

$$1 = N^2 \left[\frac{a}{2} - \frac{a}{4n_x \pi}(\sin n_x \pi - \sin 0)\right] \times \left[\frac{b}{2} - \frac{b}{4n_y \pi}(\sin n_y \pi - \sin 0)\right] \times \left[\frac{c}{2} - \frac{c}{4n_z \pi}(\sin n_z \pi - \sin 0)\right] = N^2 \frac{abc}{8}$$

$$N = \sqrt{\frac{8}{abc}} \text{ and } \psi(x,y) = \sqrt{\frac{8}{abc}} \sin\left(\frac{n_x \pi x}{a}\right)\sin\left(\frac{n_y \pi y}{b}\right)\sin\left(\frac{n_z \pi z}{c}\right)$$

P15.4) Is the superposition wave function for the free particle an eigenfunction of the momentum

operator? Is it an eigenfunction of the total energy operator? Explain your result.

$$-i\hbar \frac{d}{dx}\left(A_+ e^{+i\sqrt{\frac{2mE}{\hbar^2}}x} + A_- e^{-i\sqrt{\frac{2mE}{\hbar^2}}x}\right) = -(i)^2 \hbar\sqrt{\frac{2mE}{\hbar^2}}A_+ e^{+i\sqrt{\frac{2mE}{\hbar^2}}x} + (i)^2 \hbar\sqrt{\frac{2mE}{\hbar^2}}A_- e^{-i\sqrt{\frac{2mE}{\hbar^2}}x}$$

$$= \hbar k\, A_+ e^{+i\sqrt{\frac{2mE}{\hbar^2}}x} - \hbar k\, A_- e^{-i\sqrt{\frac{2mE}{\hbar^2}}x}$$

This function is not an eigenfunction of the momentum operator, because the operation does not return

the original function multiplied by a constant.

$$-\frac{\hbar^2}{2m}\frac{d^2}{dx^2}\left(A_+ e^{+i\sqrt{\frac{2mE}{\hbar^2}}x} + A_- e^{-i\sqrt{\frac{2mE}{\hbar^2}}x}\right) = -(i)^2 \frac{\hbar^2}{2m}\frac{2mE}{\hbar^2}A_+ e^{+i\sqrt{\frac{2mE}{\hbar^2}}x} - (-i)^2 \frac{\hbar^2}{2m}\frac{2mE}{\hbar^2}A_- e^{-i\sqrt{\frac{2mE}{\hbar^2}}x}$$

$$= E\left(A_+ e^{+i\sqrt{\frac{2mE}{\hbar^2}}x} + A_- e^{-i\sqrt{\frac{2mE}{\hbar^2}}x}\right)$$

This function is an eigenfunction of the total energy operator. Because the energy is proportional to p^2, the difference in sign of the momentum of these two components does not affect the energy.

P15.8) Evaluate the normalization integral for the eigenfunctions of $\hat{H}$ for the particle in the box $\psi_n(x) = A\sin(n\pi x/a)$ using the trigonometric identity $\sin^2 y = (1-\cos 2y)/2$.

$$1 = \int_0^a A^2 \sin^2\left(\frac{n\pi x}{a}\right) dx$$

let $y = \dfrac{n\pi x}{a}$; $dx = \dfrac{a}{n\pi} dy$

$$1 = A^2 \frac{a}{n\pi} \int_0^{n\pi} \sin^2 y \, dy = A^2 \frac{a}{n\pi} \int_0^{n\pi} \frac{1-\cos 2y}{2} dy = A^2 \frac{a}{n\pi}\left[\frac{y}{2} - \frac{\sin 2y}{4}\right]_0^{n\pi}$$

$$= \frac{A^2}{2}\frac{a}{n\pi}n\pi - \frac{A^2}{4}\frac{a}{n\pi}(\sin 2n\pi - \sin 0) = \frac{A^2 a}{2}$$

$$A = \sqrt{\frac{2}{a}}$$

P15.10) What is the solution of the time-dependent Schrödinger equation $\Psi(x,t)$ for the total energy eigenfunction $\psi_4(x) = \sqrt{2/a}\sin(4\pi x/a)$ in the particle in the box model? Write $\omega = E/\hbar$ explicitly in terms of the parameters of the problem.

$$\psi(x,t) = \psi(x)e^{-i\omega t} = \psi(x)e^{-i\frac{Et}{\hbar}}$$

Because $E = \dfrac{n^2 h^2}{8ma^2} = \dfrac{16 h^2}{8ma^2}$,

$$\psi(x,t) = \sqrt{\frac{2}{a}}\sin\left(\frac{4\pi x}{a}\right)e^{-i\frac{4\pi ht}{ma^2}}$$

P15.13) Show that the energy eigenvalues for the free particle, $E = \hbar^2 k^2/2m$, are consistent with the classical result $E = (1/2)mv^2$.

$$E = \frac{1}{2}m\mathrm{v}^2 = \frac{p^2}{2m}$$

From the de Broglie relation, $p = \dfrac{h}{\lambda}$

$$E = \frac{1}{2m}\left(\frac{h}{\lambda}\right)^2 = \frac{\hbar^2 k^2}{2m}, \text{ showing consistency between the classical and quantum result.}$$

P15.15) Calculate the wavelength of the light emitted when an electron in a one-dimensional box of length 3.0 nm makes a transition from the $n = 5$ state to the $n = 4$ state.

$$E = h\nu = \frac{h^2}{8ma^2}\left(n_2^2 - n_1^2\right) \quad \nu = \frac{h}{8ma^2}\left(n_2^2 - n_1^2\right)$$

$$= \frac{6.26 \times 10^{-34} \text{ J s}}{8 \times 9.109 \times 10^{-31} \text{ kg} \times 9.0 \times 10^{-18} \text{ m}^2}\left(5^2 - 4^2\right) = 9.1 \times 10^{13}\,\mathrm{s}^{-1}$$

$$\lambda = \frac{c}{\nu} = \frac{2.998 \times 10^8 \text{ m s}^{-1}}{9.1 \times 10^{13} \text{ s}^{-1}} = 3.3 \times 10^{-6} \text{ m}$$

P15.20) Calculate a) the zero point energy of a CO molecule in a one-dimensional box of length 1.00 cm and b) the ratio of the zero point energy to kT at 300.K.

$$E_1 = \frac{h^2}{8ma^2} = \frac{\left(6.26 \times 10^{-34} \text{ J s}\right)^2}{8 \times 28.01 \text{ amu} \times 1.661 \times 10^{-27} \text{ kg amu}^{-1} \times 1.00 \times 10^{-4} \text{ m}^2}$$

$$= 1.18 \times 10^{-37} \text{ J}$$

$$\frac{E_1}{kT} = \frac{1.18 \times 10^{-37} \text{ J}}{1.381 \times 10^{-23} \text{ J K}^{-1} \times 300. \text{ K}} = 2.85 \times 10^{-18}$$

P15.21) Normalize the total energy eigenfunction for the rectangular two-dimensional box, in the interval $0 \le x \le a, 0 \le y \le b$.

$$\psi_{n_x, n_y}(x, y) = N \sin\left(\frac{n_x \pi x}{a}\right)\sin\left(\frac{n_y \pi y}{b}\right)$$

$$1 = \int_0^a \int_0^b \psi^*(x,y)\psi(x,y)\,dx\,dy = N^2 \int_0^a \sin^2\left(\frac{n_x \pi x}{a}\right)dx \int_0^b \sin^2\left(\frac{n_y \pi y}{b}\right)dy$$

Using the standard integral $\int \sin^2(\alpha x)\,dx = \dfrac{x}{2} - \dfrac{\sin(2\alpha x)}{4\alpha}$

$$N^2 \int_0^a \sin^2\left(\frac{n_x \pi x}{a}\right)dx \int_0^b \sin^2\left(\frac{n_y \pi y}{b}\right)dy$$

$$= N^2\left[\frac{a}{2} - \frac{a}{4n_x \pi}(\sin n_x \pi - \sin 0)\right] \times \left[\frac{b}{2} - \frac{b}{4n_y \pi}(\sin n_y \pi - \sin 0)\right] = N^2 \frac{ab}{4}$$

$$N = \sqrt{\frac{4}{ab}} \text{ and } \psi(x,y) = \sqrt{\frac{4}{ab}}\sin\left(\frac{n_x \pi x}{a}\right)\sin\left(\frac{n_y \pi y}{b}\right)$$

P15.24) What are the energies of the lowest 5 energy levels in a three-dimensional box with $a = b = c$? What is the degeneracy of each level?

$n_x^2 + n_y^2 + n_z^2 = 3 : (111)$ degeneracy 1

$n_x^2 + n_y^2 + n_z^2 = 6 : (112), (121), (211)$ degeneracy 3

$n_x^2 + n_y^2 + n_z^2 = 9 : (212), (122), (221)$ degeneracy 3

$n_x^2 + n_y^2 + n_z^2 = 11 : (311), (131), (113)$ degeneracy 3

$n_x^2 + n_y^2 + n_z^2 = 12 : (222)$ degeneracy 1

P15.25) In discussing the Boltzmann distribution in Chapter 13, we used the symbols g_i and g_j to indicate the degeneracies of the energy levels i and j. By degeneracy, we mean the number of distinct quantum states (different quantum numbers), all of which have the same energy.

a. Using your answer to Problem P15.19a, what is the degeneracy of the energy level $5h^2/8ma^2$ for the square two-dimensional box of edge length a?

b. Using your answer to Problem P15.14b, what is the degeneracy of the energy level $9h^2/8ma^2$ for a three-dimensional cubic box of edge length a?

a) The only pairs n_x, n_y that satisfy the equation $n_x^2 + n_y^2 = 5$ are 2, 1 and 1, 2. Therefore the degeneracy of this energy level is 2.

b) The only trios of nonzero numbers n, q, r that satisfy the equation $n_x^2 + n_y^2 + n_z^2 = 9$ are 2, 2, 1; 2, 1, 2; and 1, 2, 2. Therefore the degeneracy of this energy level is 3.

P15.27) Two wave functions are distinguishable if they lead to a different probability density. Which of the following wave functions are distinguishable from $\sin kx$?

a. $\left(e^{ikx} - e^{-ikx}\right)/2$

b. $e^{i\theta} \sin kx$, θ a constant

c. $\cos\left(kx - \pi/2\right)$

d. $i\cos\left(kx + \pi/2\right)\left(\sin\theta + i\cos\theta\right)\left(-\dfrac{\sqrt{2}}{2} + i\dfrac{\sqrt{2}}{2}\right)$

Two wave functions ψ_1 and ψ_2 are indistinguishable if $\psi_1^*\psi_1 = \psi_2^*\psi_2$. For the wave function in the problem, $\left(\sin kx\right)\left(\sin kx\right) = \sin^2 kx$.

a)

$$\frac{\left(e^{ikx} - e^{-ikx}\right)}{2i}\frac{\left(e^{-ikx} - e^{ikx}\right)}{2i} = \frac{-i^2\left(\cos kx + i\sin kx - \left[\cos kx - i\sin kx\right]\right)^2}{4}$$

$$= \left(2\sin kx\right)^2/4 = \sin^2 kx \quad \text{indistinguishable}$$

b) $\left(e^{i\theta}\sin kx\right)\left(e^{-i\theta}\sin kx\right) = e^{-i\theta}e^{i\theta}\sin^2 kx = \sin^2 kx$ indistinguishable

c)

$$\cos\left(kx - \pi/2\right)\cos\left(kx - \pi/2\right) = \left(\left[\cos kx \cos \pi/2 + \sin kx \sin \pi/2\right]\right)^2$$
$$= \left(\left[\cos kx \times 0 + \sin kx \times 1\right]\right)^2 = \sin^2 kx \quad \text{indistinguishable}$$

indistinguishable

d)

$$i\cos\left(kx + \pi/2\right)\left(\sin\theta + i\cos\theta\right)\left(-\frac{\sqrt{2}}{2} + i\frac{\sqrt{2}}{2}\right)$$

$$\times(-i)\cos\left(kx + \pi/2\right)\left(\sin\theta - i\cos\theta\right)\left(-\frac{\sqrt{2}}{2} - i\frac{\sqrt{2}}{2}\right)$$

$$= \left[\cos\left(kx + \pi/2\right)\right]^2 \left(\sin^2\theta + \cos^2\theta\right)\left(\frac{1}{2} + \frac{1}{2}\right)$$

$$= \left[\cos\left(kx + \pi/2\right)\right]^2 = -\left[\cos\left(kx\right)\cos\left(\pi/2\right) + \sin\left(kx\right)\sin\left(\pi/2\right)\right]^2$$
$$= \sin^2\left(kx\right)$$

indistinguishable

P15.28) Is the superposition wave function $\psi(x) = \sqrt{2/a}\left[\sin(n\pi x/a) + \sin(m\pi x/a)\right]$ an eigenfunction

of the total energy operator for the particle in the box?

$$\hat{H}\psi(x) = -\frac{\hbar^2}{2m}\frac{d^2}{dx^2}\left[\sqrt{\frac{2}{a}}\sin\left(\frac{n\pi x}{a}\right) + \sqrt{\frac{2}{a}}\sin\left(\frac{m\pi x}{a}\right)\right]$$

$$= \frac{h^2 n^2}{8ma^2}\sqrt{\frac{2}{a}}\sin\left(\frac{n\pi x}{a}\right) + \frac{h^2 m^2}{8ma^2}\sqrt{\frac{2}{a}}\sin\left(\frac{m\pi x}{a}\right)$$

Because the result is not the wave function multiplied by a constant, the superposition wave function is

not an eigenfunction of the total energy operator.

P15.34) Calculate the probability that a particle in a one-dimensional box of length a is found between $0.18a$ and $0.22a$ when it is described by the following wave functions:

a. $\sqrt{\dfrac{2}{a}}\sin\left(\dfrac{\pi x}{a}\right)$

b. $\sqrt{\dfrac{2}{a}}\sin\left(\dfrac{5\pi x}{a}\right)$

What would you expect for a classical particle? Compare your results in the two cases with the classical result.

a)

Using the standard integral $\int \sin^2(by)\,dy = \dfrac{y}{2} - \dfrac{1}{4b}\sin(2by)$

$$P = \frac{2}{a}\int_{0.18a}^{0.22a}\sin^2\left(\frac{\pi x}{a}\right)dx = \frac{2}{a}\left[\frac{x}{2} - \frac{a}{4\pi}\sin\left(\frac{2\pi x}{a}\right)\right]_{0.18a}^{0.22a}$$

$$= \frac{2}{a}\left[\frac{0.22a}{2} - \frac{a}{4\pi}\sin(0.44\pi) - \frac{0.18a}{2} + \frac{a}{4\pi}\sin(0.36\pi)\right]$$

$$= 0.04 + \frac{1}{2\pi}\left[-\sin(0.44\pi) + \sin(0.36\pi)\right] = 0.028$$

b)

Using the standard integral $\int \sin^2(by)\,dy = \dfrac{y}{2} - \dfrac{1}{4b}\sin(2by)$

$$P = \frac{2}{a}\int_{0.18a}^{0.22a}\sin^2\left(\frac{5\pi x}{a}\right)dx = \frac{2}{a}\left[\frac{0.22a}{2} - \frac{a}{20\pi}\sin(2.2\pi) - \frac{0.31a}{2} + \frac{a}{20\pi}\sin(1.8\pi)\right]$$

$$= 0.04 + \frac{1}{10\pi}\left[-\sin(2.2\pi) + \sin(1.8\pi)\right] = 0.0026$$

Because a classical particle is equally likely to be in any given interval, the probability will be 0.04 independent of the energy. In the ground state, the interval chosen is near the maximum of the wave function so that the quantum mechanical probability is greater than the classical probability. For the $n = 3$ state, the interval chosen is near a node of the wave function so that the quantum mechanical probability is much less than the classical probability.

Chapter 16: The Particle in the Box and the Real World

Computational Problems

Before solving the computational problems, it is recommended that students work through Tutorials 1–3 under the Help menus in Spartan Student Edition to gain familiarity with the program.

Computational Problem 16.1: Build (a) ethylene, (b) the trans conformation for 1,3 butadiene, and (c) all trans hexatriene and calculate the ground-state (singlet) energy of these molecules using the B3LYP method with the 6-311+G** basis set. Repeat your calculation for the triplet state, which corresponds to the excitation of an electron from the highest filled energy level to the lowest unoccupied energy level. Use a nonplanar input geometry for the triplet states. Compare the energy difference from these calculations to literature values of the maximum in the UV-visible absorption spectrum.

Your solution to this problem should answer the questions in bold type in Steps 5 and 10.

Step 1: Create a new file, and build structures for 1,3-butadiene, 1,3,5-hexatriene, 1,3,5,7-octatetraene. Create each one as a new molecule within the same file using the "New Molecule" command in the File menu.

Step 2: Go to Setup > Calculations. Set the calculation type to "Equilibrium Geometry" and the method to B3LYP/6-31G*. The meaning of the method (Density Functional Theory w/B3LYP functional) and basis set (6-31G*) will be explained later in the course, consider this to be a "black box" for now. "Total Charge" should be neutral, and "Global Calculations" should be on. Click "OK." This will write the parameters for the calculation to the file, but will NOT start running the calculation. You will submit the calculation in Step 4.

Step 3: You will now calculate molecular orbitals for butadiene. Use the left and right navigation buttons at the bottom of the window to select your Butadiene model, then go to Setup → Surfaces. Click "Add," then select "LUMO" from the "Surfaces" menu. (NOT |LUMO| from the "Properties" menu!). Click "OK." Repeat for the HOMO, and HOMO-1. Make sure "Apply Globally" is NOT checked.

Step 4: Go to Setup → Submit. Your calculations should start running. The calculation will most likely take a few minutes on most current computers.

Step 5: While on butadiene, go to Display → Surfaces. A dialog box will appear. **Examine the surfaces one at a time, and comment on the degree of electron delocalization. Is there any evidence that the system can be modeled as particles in a 1-dimensional box?**

Step 6: Turn off the surfaces, close the dialog box, and then go to Display → Properties. Record the HOMO and LUMO energies in your spreadsheet program of choice. Use the navigation buttons on the bottom of the screen to advance to the next molecule, and repeat until you have recorded data for all 4 molecules.

The method used in this problem is reasonably accurate in calculating HOMO-LUMO gaps. Quantitatively accurate excited state calculations require advanced methods that are beyond the scope of this course. However, for this exercise, you will be looking for trends in the energy, not exact values, so use of a simpler method is acceptable.

Step 7: Calculate the energy gap E_g (in eV) by subtracting the HOMO energy from the LUMO energy. This is the energy corresponding to the longest-wavelength absorption of light by the conjugated molecule. Calculate the wavelength of absorption in another column using equation (1). Be careful about units here!

$$E = \frac{hc}{\lambda} \quad (1)$$

Step 8: Go through each molecule, measure the carbon-carbon bond lengths using the "Distance" tool in the tool palette, and add them up to obtain a total conjugation length for each molecule. Record this in your spreadsheet.

Step 9: Perform a 1-dimensional particle in-a-box calculation for the HOMO-LUMO transition energy for all three of your molecules, using the lengths you measured. Keep track of your units! Example values for 1, 3, 5, 7, 9-decapentaene have been provided to help get you started. The table below does not include wavelengths, however, you should calculate wavelengths for all four molecules. **If you are stuck here, please reread Section 16.3 of your book.**

Molecule	n HOMO	L (m)	B3LYP E_{HOMO} (eV)	B3LYP E_{LUMO} (eV)	B3LYP E_g (eV)	1D Box E_g (eV)
Butadiene						
Hexatriene						
Octatetraene						
Decapentaene	5	1.253×10^{-9}	-5.116	-1.796	3.32	2.64

Step 10: **Generate a table** containing the B3LYP and 1D Box predicted energy values, and **graph them** with # of double bonds on the x-axis and energy in eV on the y-axis. Look at the shape of the curves—**how does the first absorption energy change as a function of double bonds?** While you may use regression techniques if you would like, a qualitative answer is acceptable.

Also, compare your calculated values with the experimental value for butadiene and with some values that are calculated using a higher level theory. **What trends do you see?**

A UV/Visible absorption spectrum for butadiene is below. Find the strongest peak and use its energy in your comparison. Results for E_g obtained using a higher level theory for all four molecules are below.

B3LYP/6-31G* TDDFT results for

Butadiene: 5.27 eV

Hexatriene: 4.34 eV

Octatetraene: 3.72 eV

Decapentaene: 3.27 eV

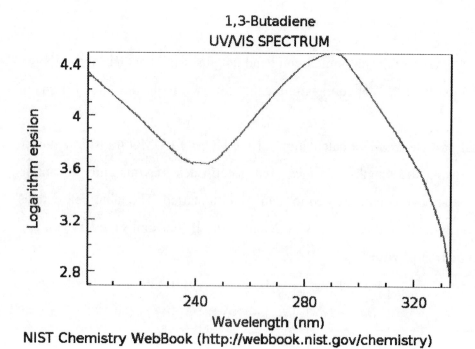

NIST Chemistry WebBook (http://webbook.nist.gov/chemistry)

Chapter 17: Commuting and Noncommuting Operators and the Surprising Consequences of Entanglement

P17.4) a) Show that $\psi(x) = e^{-x^2/2}$ is an eigenfunction of $\hat{A} = x^2 - \partial^2/\partial x^2$. Show that $\hat{B}\psi(x)$ where $\hat{B} = x - \partial/\partial x$ is another eigenfunction of $\hat{A}$.

$$\hat{A}\psi(x) = x^2 e^{-x^2/2} - \partial^2 e^{-x^2/2}/\partial x^2 = x^2 e^{-x^2/2} - \partial\left(-xe^{-x^2/2}\right)/\partial x$$

$$= e^{-x^2/2} + x^2 e^{-x^2/2} - x^2 e^{-x^2/2} = e^{-x^2/2}$$

$$\hat{B}\psi(x) = xe^{-x^2/2} - \partial e^{-x^2/2}/\partial x = xe^{-x^2/2} + xe^{-x^2/2} = 2xe^{-x^2/2}$$

$$\hat{A}\left(\hat{B}\psi(x)\right) = x^2\left(2xe^{-x^2/2}\right) - \partial^2\left(2xe^{-x^2/2}\right)/\partial x^2$$

$$= 2x^3 e^{-x^2/2} - \partial\left(2e^{-x^2/2} - 2x^2 e^{-x^2/2}\right)/\partial x$$

$$= 2x^3 e^{-x^2/2} + 2xe^{-x^2/2} + 4xe^{-x^2/2} - 2x^3 e^{-x^2/2}$$

$$= 6xe^{-x^2/2}$$

P17.5) Another important uncertainty principle is encountered in time-dependent systems. It relates the lifetime of a state Δt with the measured spread in the photon energy ΔE associated with the decay of this state to a stationary state of the system. "Derive" the relation $\Delta E\,\Delta t \geq \hbar/2$ in the following steps.

a. Starting from $E = p_x^2/2m$ and $\Delta E = (dE/dp_x)\Delta p_x$, show that $\Delta E = v_x \Delta p_x$.

b. Using $v_x = \Delta x/\Delta t$, show that $\Delta E\,\Delta t = \Delta p_x \Delta x \geq \hbar/2$.

c. Estimate the width of a spectral line originating from the decay of a state of lifetime 1.0×10^{-9} s and 1.0×10^{-11} s in inverse seconds and inverse centimeters.

a)

$$\frac{dE}{dp_x}\Delta p_x = \frac{p_x}{m}\Delta p_x = v_x \Delta p_x$$

b)

$$\Delta E\Delta t = \frac{\Delta x}{\Delta t}\Delta p_x \Delta t = \Delta x\Delta p_x \geq \frac{\hbar}{2}$$

c)

$$\Delta E \geq \frac{\hbar}{2\Delta t} = h\Delta v$$

$$\Delta v = \frac{1}{4\pi\Delta t} = \frac{1}{4\pi\left(1.0\times10^{-9}\text{s}\right)} = 8.0\times10^{7} \text{ s}^{-1}$$

$$\Delta v\left(\text{cm}^{-1}\right) = \frac{\Delta v\left(\text{s}^{-1}\right)}{c} = \frac{8.0\times10^{7} \text{ s}^{-1}}{2.998\times10^{10} \text{ cm s}^{-1}} = 0.00265 \text{ cm}^{-1}$$

The corresponding answers for 1.0×10^{-11} s are 8.0×10^{9} s^{-1} and 0.265 cm^{-1}, respectively.

P17.6) Evaluate the commutator $[x(\partial/\partial y), y]$ by applying the operators to an arbitrary function $f(x,y)$.

$$\left[x\frac{\partial}{\partial y}, y\right]f(x,y) = x\frac{\partial yf(x,y)}{\partial y} - yx\frac{\partial f(x,y)}{\partial y}$$

$$= xf(x,y) + xy\frac{\partial f(x,y)}{\partial y} - yx\frac{\partial f(x,y)}{\partial y} = xf(x,y)$$

Therefore $\left[x\frac{\partial}{\partial y}, y\right] = x$

P17.8) Consider the entangled wave function for two photons,

$$\psi_{12} = \frac{1}{\sqrt{2}}(\psi_1(H)\psi_2(V) + \psi_1(V)\psi_2(H))$$

Assume that the polarization operator $\hat{P}_i$ has the properties $\hat{P}_i\psi_i(H) = -\psi_i(H)$ and $\hat{P}_i\psi_i(V) = +\psi_i(V)$ where $i = 1$ or $i = 2$.

a. Show that ψ_{12} is not an eigenfunction of $\hat{P}_1$ or $\hat{P}_2$.

b. Show that each of the two terms in ψ_{12} is an eigenfunction of the polarization operator $\hat{P}_1$.

c. What is the average value of the polarization P_1 that you will measure on identically prepared systems? It is not necessary to do a calculation to answer this question.

a)

$$\hat{P}_1 \psi_{12} = \frac{1}{\sqrt{2}} \left(-\psi_1(H)\psi_2(V) + \psi_1(V)\psi_2(H) \right)$$

$$\hat{P}_2 \psi_{12} = \frac{1}{\sqrt{2}} \left(\psi_1(H)\psi_2(V) - \psi_1(V)\psi_2(H) \right)$$

In neither case does the operation return the original function multiplied by a constant. Therefore, the function is not an eigenfunction of either operator.

b) Each of the 2 terms of the above expression is the original term multiplied by a constant. Therefore each individual term is an eigenfunction of the operators.

c) A measurement will project the system into the wave function $\psi_1(H)\psi_2(V)$ or $-\psi_1(V)\psi_2(H)$ with equal probability. Therefore it is equally likely to measure the eigenvalue +1 as –1 and the average of the measured values will be zero.

P17.10) Revisit the double-slit experiment of Example Problem 17.2. Using the same geometry and relative uncertainty in the momentum, what electron momentum would give a position uncertainty of 1.00×10^{-9} m? What is the ratio of the wavelength and the slit spacing for this momentum? Would you expect a pronounced diffraction effect for this wavelength?

$$\Delta p = \frac{1}{2} \frac{\hbar}{\Delta x} = \frac{1.055 \times 10^{-34} \text{ J s}}{2.00 \times 10^{-9} \text{ m}} = 5.275 \times 10^{-26} \text{ kg m s}^{-1}$$

$$p = \frac{\Delta p}{0.01} = 5.275 \times 10^{-24} \text{ kg m s}^{-1}$$

$$\lambda = \frac{h}{p} = \frac{6.626 \times 10^{-34} \text{ J s}}{5.275 \times 10^{-24} \text{ kg m s}^{-1}} = 1.26 \times 10^{-10} \text{ m}$$

Because $\dfrac{\lambda}{b} = \dfrac{1.26 \times 10^{-10} \text{ m}}{1.000 \times 10^{-9} \text{ m}} = 0.126$, where b is the slit spacing, the diffraction will hardly be noticeable.

P17.16) Evaluate the commutator $[d/dx, x^2]$ by applying the operators to an arbitrary function $f(x)$.

$$\left[\frac{d}{dx}, x^2\right] f(x) = \frac{d}{dx}\left(x^2 f(x)\right) - x^2 \frac{d}{dx} f(x)$$

$$= 2xf(x) + x^2 \frac{d}{dx} f(x) - x^2 \frac{d}{dx} f(x) = 2xf(x)$$

$$\left[\frac{d}{dx}, x^2\right] = 2x$$

P17.19) Evaluate the commutator $[(d/dx) - x, (d/dx) + x]$ by applying the operators to an arbitrary function $f(x)$.

$$\left[\frac{d}{dx} - x, \frac{d}{dx} + x\right] f(x) = \left(\frac{d}{dx} - x\right)\left(\frac{d}{dx} + x\right) f(x) - \left(\frac{d}{dx} + x\right)\left(\frac{d}{dx} - x\right) f(x)$$

$$= \left(\frac{d}{dx} - x\right)\left(\frac{df(x)}{dx} + xf(x)\right) - \left(\frac{d}{dx} + x\right)\left(\frac{df(x)}{dx} - xf(x)\right)$$

$$= \frac{d^2 f(x)}{dx^2} + f(x) + x\frac{df(x)}{dx} - x\frac{df(x)}{dx} - x^2 f(x) - \frac{d^2 f(x)}{dx^2} + f(x) + x\frac{df(x)}{dx}$$

$$- x\frac{df(x)}{dx} + x^2 f(x)$$

$$= 2f(x) \text{ Therefore,}$$

$$\left[\frac{d}{dx} - x, \frac{d}{dx} + x\right] = 2$$

P17.23) The muzzle velocity of a rifle bullet is about 775 m s^{-1}. If the bullet weighs 28 g, and the uncertainty in its momentum is 0.15%, how accurately can the position of the bullet be measured?

$$p = mv = 28 \times 10^{-3} \text{ kg} \times 775 \text{ m s}^{-1} = 22 \text{ kg m s}^{-1}$$

$$\Delta p = 10^{-3} p = 2.2 \times 10^{-2} \text{ kg m s}^{-1}$$

$$\Delta x = \frac{\hbar}{2\Delta p} = \frac{1.055 \times 10^{-34} \text{ J s}}{2 \times 2.2 \times 10^{-2} \text{ kg m s}^{-1}} = 1.6 \times 10^{-33} \text{ m}$$

Chapter 18: A Quantum Mechanical Model for the Vibration and Rotation of Molecules

P18.1) A gas-phase $^1H^{35}Cl$ molecule, with a bond length of 127.5 pm, rotates in a three-dimensional space.

a. Calculate the zero point energy associated with this rotation.

b. What is the smallest quantum of energy that can be absorbed by this molecule in a rotational excitation?

a) There is no zero point energy because the rotation is not constrained.

b) The smallest energy that can be absorbed is

$$E = \frac{\hbar^2}{2I}J(J+1) = \frac{\hbar^2}{2I}1(1+1)$$

$$= \frac{2\times\left(1.055\times10^{-34} \text{ J s}\right)^2}{2\times\dfrac{1.0078 \text{ amu}\times34.9688 \text{ amu}}{1.0078 \text{ amu} + 34.9688 \text{ amu}}\times1.66\times10^{-27} \text{ kg amu}^{-1}\times\left(127.5\times10^{-12} \text{ m}\right)^2}$$

$$E = 4.21\times10^{-22} \text{ J}$$

P18.3) Using the Boltzmann distribution, calculate n_J/n_0 for $^1H^{19}F$ for $J = 0, 5, 10,$ and 20 at $T = 650$. K. Does n_J/n_0 go through a maximum as J increases? If so, what can you say about the value of J corresponding to the maximum?

$$\frac{n_J}{n_0} = (2J+1)e^{-(E_J-E_0)/kT} = (2J+1)\exp[-E_J/kT]$$

$$I = \mu r_0^2 = \frac{1.0078 \text{ amu} \times 18.9984 \text{ amu}}{1.0078 \text{ amu} + 18.994 \text{ amu}} \times 1.66 \times 10^{-27} \text{ kg amu}^{-1} \times (91.68 \times 10^{-12} \text{ m})^2$$

$$= 1.34 \times 10^{-47} \text{ kg m}^2$$

$$E = \frac{\hbar^2}{2I} J(J+1) = \frac{(1.055 \times 10^{-34} \text{ J s})^2}{2 \times 1.34 \times 10^{-47} \text{ kg m}^2} J(J+1) = 4.17 \times 10^{-22} J(J+1)$$

$$\frac{n_0}{n_0} = 1$$

$$\frac{n_5}{n_0} = (2 \times 5 + 1) \exp\left[-(30 \times 4.17 \times 10^{-22} \text{ J})/1.381 \times 10^{-23} \text{ J K}^{-1} \times 650. \text{ K}\right] = 2.73$$

$$\frac{n_{10}}{n_0} = (2 \times 10 + 1) \exp\left[-(110 \times 4.17 \times 10^{-22} \text{ J})/1.381 \times 10^{-23} \text{ J K}^{-1} \times 650. \text{ K}\right] = 0.127$$

$$\frac{n_{20}}{n_0} = (2 \times 20 + 1) \exp\left[-(420 \times 4.17 \times 10^{-22} \text{ J})/1.381 \times 10^{-23} \text{ J K}^{-1} \times 650. \text{ K}\right] = 1.40 \times 10^{-7}$$

$\frac{n_J}{n_0}$ goes through a maximum because it has a value greater than one for $J = 5$. You can only conclude

that $J_{max} \le 10$.

P18.8) The vibrational frequency for $^{19}F_2$ expressed in wave numbers is 916.64 cm^{-1}. What is the force constant associated with the F-F bond? How much would a classical spring with this force constant be elongated if a mass of 2.50 kg were attached to it? Use the gravitational acceleration on Earth at sea level for this problem.

$$v = c\tilde{v} = \frac{1}{2\pi}\sqrt{\frac{k}{\mu}}$$

so $k = (2\pi c \tilde{v})^2 \mu$

$$k = (2\pi \times 2.998 \times 10^{10} \text{ cm s}^{-1} \times 916.64 \text{ cm}^{-1})^2 \times \frac{18.94 \text{ amu} \times 18.94 \text{ amu}}{2 \times 18.94 \text{ amu}} \times \frac{1.661 \times 10^{-27} \text{ kg}}{\text{amu}}$$

$$k = 470. \text{ N m}^{-1}$$

$$x = \frac{F}{k} = \frac{mg}{k} = \frac{2.50 \text{ kg} \times 9.81 \text{ m s}^{-2}}{470. \text{ N m}^{-1}} = 5.21 \times 10^{-2} \text{ m}$$

P18.9) Calculate E_{rot}/kT for H^{19}F for $J = 0, 5, 10$, and 20 at 298 K. For which of these values of J is $E_{rot}/kT \geq 10$?

$$I = \mu r_0^2 = \frac{1.0078 \times 18.9984}{1.0078 + 18.994} \times 1.66 \times 10^{-27} \text{ kg amu}^{-1} \times \left(91.68 \times 10^{-12} \text{ m}\right)^2 = 1.34 \times 10^{-47} \text{ kg m}^2$$

$$E = \frac{h^2}{2I} J(J+1) = \frac{\left(1.055 \times 10^{-34} \text{ J s}\right)^2}{2 \times 1.34 \times 10^{-47} \text{ kg m}^2} J(J+1) = 4.17 \times 10^{-22} J(J+1)$$

$$E_{J=0} = 0$$

$$E_{J=5} = 30 \times 4.17 \times 10^{-22} \text{ J} = 12.5 \times 10^{-21} \text{ J}$$

$$\frac{E_{J=5}}{kT} = \frac{12.5 \times 10^{-21} \text{ J}}{1.381 \times 10^{-23} \text{ J K}^{-1} \times 298 \text{ K}} = 3.04$$

$$E_{J=10} = 110 \times 4.17 \times 10^{-22} \text{ J} = 4.58 \times 10^{-20} \text{ J}$$

$$\frac{E_{J=10}}{kT} = \frac{4.58 \times 10^{-20} \text{ J}}{1.381 \times 10^{-23} \text{ J K}^{-1} \times 298 \text{ K}} = 11.1$$

$$E_{J=20} = 20 \times 21 \times 4.17 \times 10^{-22} \text{ J} = 1.75 \times 10^{-19} \text{ J}$$

$$\frac{E_{J=20}}{kT} = \frac{17.5 \times 10^{-20} \text{ J}}{1.381 \times 10^{-23} \text{ J K}^{-1} \times 298 \text{ K}} = 42.5$$

$$\frac{E_{rot}}{kT} > 10 \text{ for } J = 10 \text{ and } 20.$$

P18.12) Show by carrying out the appropriate integration that the total energy eigenfunctions for the harmonic oscillator $\psi_0(x) = (\alpha/\pi)^{1/4} e^{-(1/2)\alpha x^2}$ and $\psi_2(x) = (\alpha/4\pi)^{1/4}(2\alpha x^2 - 1)e^{-(1/2)\alpha x^2}$ are orthogonal over the interval $-\infty < x < \infty$ and that $\psi_2(x)$ is normalized over the same interval. In evaluating integrals of this type, $\int_{-\infty}^{\infty} f(x)dx = 0$ if $f(x)$ is an odd function of x and $\int_{-\infty}^{\infty} f(x)dx = 2\int_0^{\infty} f(x)dx$ if $f(x)$ is an even function of x.

We use the standard integrals $\int_0^{\infty} x^{2n} e^{-ax^2} dx = \frac{1 \cdot 3 \cdot 5 \cdots (2n-1)}{2^{n+1} a^n} \sqrt{\frac{\pi}{a}}$ and

$$\int_0^{\infty} e^{-ax^2} dx = \left(\frac{\pi}{4a}\right)^{1/2}$$

$$\int_{-\infty}^{\infty} \psi_2^*(x)\psi_0(x)\,dx = \int_{-\infty}^{\infty}\left(\frac{\alpha}{4\pi}\right)^{1/4}(2\alpha x^2-1)e^{-\frac{1}{2}\alpha x^2}\left(\frac{\alpha}{\pi}\right)^{1/4}e^{-\frac{1}{2}\alpha x^2}\,dx$$

$$=\left(\frac{\alpha^2}{4\pi^2}\right)^{1/4}\int_{-\infty}^{\infty}(2\alpha x^2-1)e^{-\alpha x^2}\,dx = 2\left(\frac{\alpha^2}{4\pi^2}\right)^{1/4}\int_{0}^{\infty}(2\alpha x^2-1)e^{-\alpha x^2}\,dx$$

$$=\left(\frac{\alpha^2}{4\pi^2}\right)^{1/4}\left(2\alpha\frac{1}{4\alpha}\sqrt{\frac{\pi}{\alpha}}-\frac{1}{2}\sqrt{\frac{\pi}{\alpha}}\right)=0$$

$$\int_{-\infty}^{\infty} \psi_2^*(x)\psi_2(x)\,dx = \int_{-\infty}^{\infty}\left(\frac{\alpha}{4\pi}\right)^{1/4}(2\alpha x^2-1)e^{-\frac{1}{2}\alpha x^2}\left(\frac{\alpha}{4\pi}\right)^{1/4}(2\alpha x^2-1)e^{-\frac{1}{2}\alpha x^2}\,dx$$

$$=2\left(\frac{\alpha}{4\pi}\right)^{1/2}\int_{0}^{\infty}(4\alpha^2 x^4-4\alpha x^2+1)e^{-\alpha x^2}\,dx$$

$$=2\left(\frac{\alpha}{4\pi}\right)^{1/2}\left(4\alpha^2\frac{3}{2^3\alpha^2}\sqrt{\frac{\pi}{\alpha}}-4\alpha\frac{1}{2^2\alpha}\sqrt{\frac{\pi}{\alpha}}+\frac{1}{2}\sqrt{\frac{\pi}{\alpha}}\right)=2\left(\frac{\alpha}{4\pi}\right)^{1/2}\sqrt{\frac{\pi}{\alpha}}\left(\frac{3}{2}-1+\frac{1}{2}\right)=1$$

P18.14) Calculate the frequency and wavelength of the radiation absorbed when a quantum harmonic oscillator with a frequency of 5.58×10^{13} s^{-1} makes a transition from the $n = 3$ to the $n = 4$ state.

$$\Delta E = (4+1/2)h\nu - (3+1/2)h\nu = h\nu$$
$$= 6.626\times10^{-34}\text{ J s}\times5.58\times10^{13}\text{ s}^{-1} = 3.70\times10^{-20}\text{ J}$$
$$\nu=\frac{E}{h}=5.58\times10^{13}\text{ s}^{-1}$$
$$\lambda = \frac{c}{\nu} = \frac{2.998\times10^8\text{ m s}^{-1}}{5.58\times10^{13}\text{ s}^{-1}} = 5.37\times10^{-6}\text{ m}$$

P18.16) The vibrational frequency of ^{1}H^{19}F is 1.24×10^{14} s^{-1}. Calculate the force constant of the molecule. How large a mass would be required to stretch a classical spring with this force constant by 1.00 cm? Use the gravitational acceleration on Earth at sea level for this problem.

$$v = \frac{1}{2\pi}\sqrt{\frac{k}{\mu}}; \; k = 4\pi^2 \mu v^2$$

$$k = 4 \times \pi^2 \times \frac{1.008 \text{ amu} \times 18.9984 \text{ amu}}{1.008 \text{ amu} + 18.9984 \text{ amu}} \times \frac{1.661 \times 10^{-27} \text{ kg}}{\text{amu}} \times \left(1.24 \times 10^{14} \text{ s}^{-1}\right)^2$$

$$k = 965 \text{ kg s}^{-2}$$

$$F = kx = mg$$

$$m = \frac{kx}{g} = \frac{965 \text{ kg s}^{-2} \times 10^{-2} \text{ m}}{9.81 \text{ m s}^{-2}} = 0.984 \text{ kg}$$

P18.17) Use $\sqrt{\langle x^2 \rangle}$ as calculated in Problem P18.32 as a measure of the vibrational amplitude for a molecule. What fraction is $\sqrt{\langle x^2 \rangle}$ of the 141.4-pm bond length of the ^{1}H^{81}Br molecule for $n = 0$, 1, and 2? The force constant for the ^{1}H^{81}Br molecule is 412 N m^{-1}.

$$\sqrt{\langle x^2 \rangle} = \left(\frac{\hbar}{2\sqrt{k\mu}}\right)^{\frac{1}{2}} = \left(\frac{1.055 \times 10^{-34} \text{ J s}}{2 \times \sqrt{412 \text{ N m}^{-1} \times \frac{1.0078 \times 80.9163}{1.0078 + 80.9163} \times 1.66 \times 10^{-27} \text{ kg amu}^{-1}}}\right)^{\frac{1}{2}}$$

For $n = 0$, $= 8.00 \times 10^{-12}$ m

$$\frac{\sqrt{\langle x^2 \rangle}}{\text{bond length}} = \frac{8.00 \times 10^{-12} \text{ m}}{141.4 \times 10^{-12} \text{ m}} = 0.0565$$

For $n = 1$

$$\sqrt{\langle x^2 \rangle} = \left(\frac{3\hbar}{2\sqrt{k\mu}}\right)^{\frac{1}{2}}$$

$$= \left(\frac{3 \times 1.055 \times 10^{-34} \text{ J s}}{2 \times \sqrt{412 \text{ N m}^{-1} \times \frac{1.0078 \text{ amu} \times 80.9163 \text{ amu}}{1.0078 \text{ amu} + 80.9163 \text{ amu}} \times 1.66 \times 10^{-27} \text{ kg amu}^{-1}}}\right)^{\frac{1}{2}}$$

$$= 1.38 \times 10^{-11} \text{ m}$$

$$\frac{\sqrt{\langle x^2 \rangle}}{\text{bond length}} = \frac{1.38 \times 10^{-11} \text{ m}}{141.4 \times 10^{-12} \text{ m}} = 0.0979$$

For $n = 2$

$$\sqrt{\langle x^2 \rangle} = \left(\frac{5\hbar}{2\sqrt{k\mu}} \right)^{\frac{1}{2}}$$

$$= \left(\frac{5 \times 1.055 \times 10^{-34} \text{ J s}}{2 \times \sqrt{412 \text{ N m}^{-1} \times \dfrac{1.0078 \text{ amu} \times 80.9163 \text{ amu}}{1.0078 \text{ amu} + 80.9163 \text{ amu}} \times 1.66 \times 10^{-27} \text{ kg amu}^{-1}}} \right)^{\frac{1}{2}}$$

$$= 1.79 \times 10^{-11} \text{ m}$$

$$\frac{\sqrt{\langle x^2 \rangle}}{\text{bond length}} = \frac{1.79 \times 10^{-11} \text{ m}}{141.4 \times 10^{-12} \text{ m}} = 0.126$$

P18.18) A coin with a mass of 5.67 g suspended on a rubber band has a vibrational frequency of 3.00 s^{-1}. Calculate a) the force constant of the rubber band; b) the zero point energy; c) the total vibrational energy if the maximum displacement is 0.500 cm; and d) the vibrational quantum number corresponding to the energy in part c).

$$v = \frac{1}{2\pi}\sqrt{\frac{k}{m}} \qquad k = 4\pi^2 v^2 \mu$$

$$k = 4\pi^2 \times 9.00 \text{ s}^{-2} \times 5.67 \times 10^{-3} \text{ kg} = 2.01 \text{ N m}^{-1}$$

$$E_0 = \frac{1}{2}hv = 6.626 \times 10^{-34} \text{ J s} \times 3.00 \text{ s}^{-1} = 9.94 \times 10^{-34} \text{ J}$$

$$E_n = \frac{1}{2}kx^2_{max} = \frac{1}{2} \times 2.01 \text{ N m}^{-1} \times \left(0.500 \times 10^{-2} \text{ m} \right)^2 = 2.52 \times 10^{-5} \text{ J}$$

$$E_n = \left(n + \frac{1}{2} \right)hv \qquad n = \frac{E_n}{hv} - \frac{1}{2}$$

$$n = \frac{2.52 \times 10^{-5} \text{ J}}{6.626 \times 10^{-34} \text{ J s} \times 3.00 \text{ s}^{-1}} - \frac{1}{2} = 1.27 \times 10^{28}$$

P18.21) Is it possible to simultaneously know the angular orientation of a molecule rotating in a two-dimensional space and its angular momentum? Answer this question by evaluating the commutator $[\phi, -i\hbar(\partial/\partial\phi)]$.

$$\left[\phi, -i\hbar\frac{\partial}{\partial\phi}\right]f(\phi) = -i\hbar\phi\frac{df(\phi)}{d\phi} + i\hbar\frac{d[\phi f(\phi)]}{d\phi}\phi = i\hbar f(\phi)$$

$$\left[\phi, -i\hbar\frac{\partial}{\partial\phi}\right] = i\hbar$$

Because the commutator is not equal to zero, it is not possible to simultaneously know the angular orientation of a molecule rotating in a two-dimensional space and its angular momentum.

P18.23) The force constant for a $H^{35}Cl$ molecule is 516 N m^{-1}.

a. Calculate the zero point vibrational energy for this molecule for a harmonic potential.

b. Calculate the light frequency needed to excite this molecule from the ground state to the first excited state.

a)

$$E_1 = h\sqrt{\frac{k}{\mu}}\left(1+\frac{1}{2}\right) = \frac{3}{2}\times1.055\times10^{-34} \text{ J s}\times\sqrt{\frac{516 \text{ N m}^{-1}}{\frac{1.0078 \text{ amu}\times34.9688 \text{ amu}}{1.0078 \text{ amu} + 34.9688 \text{ amu}}\times1.66\times10^{-27} \text{ kg amu}^{-1}}}$$

$$E_1 = 8.91\times10^{-20} \text{ J}$$

$$E_0 = \hbar\sqrt{\frac{k}{\mu}}\left(\frac{1}{2}\right) = \frac{1}{3}E_1 = 2.97\times10^{-20} \text{ J}$$

b) $v = \dfrac{E_1 - E_0}{h} = \dfrac{8.91\times10^{-20} \text{ J} - 2.97\times10^{-20} \text{ J}}{6.626\times10^{-34} \text{ J s}} = 8.97\times10^{13} \text{ s}^{-1}$

P18.25) A $^1H^{127}I$ molecule, with a bond length of 160.92 pm, absorbed on a surface rotates in two dimensions.

a. Calculate the zero point energy associated with this rotation.

b. What is the smallest quantum of energy that can be absorbed by this molecule in a rotational excitation?

a) There is no zero point energy because the rotation is not constrained.

b) The smallest energy that can be absorbed is

$$E = \frac{\hbar^2 m_l^2}{2I} = \frac{\hbar^2}{2I} = \frac{\left(1.055\times10^{-34} \text{ J s}\right)^2}{2\times\frac{1.0078 \text{ amu}\times126.9045 \text{ amu}}{1.0078 \text{ amu} + 126.9045 \text{ amu}}\times1.66\times10^{-27} \text{ kg amu}^{-1}\times\left(160.92\times10^{-12} \text{ m}\right)^2}$$

$$E = 1.29\times10^{-22} \text{ J}$$

P18.29) Evaluate the average linear momentum of the quantum harmonic oscillator, $\langle p_x \rangle$, for the ground state ($n = 0$) and first two excited states ($n = 1$ and $n = 2$). Use the hint about evaluating integrals in Problem P18.12.

We use the standard integrals $\int_0^{\infty} x^{2n} e^{-ax^2} dx = \dfrac{1 \cdot 3 \cdot 5 \cdots (2n-1)}{2^{n+1} a^n} \sqrt{\dfrac{\pi}{a}}$ and

$$\int_0^{\infty} e^{-ax^2} dx = \left(\frac{\pi}{4a} \right)^{1/2}$$

$$\langle p_x \rangle = \int_{-\infty}^{\infty} \psi_n^*(x) \left(-i\hbar \frac{d}{dx} \right) \psi_n \, dx$$

for $n = 0$, $\langle p_x \rangle = \int_{-\infty}^{\infty} \left(\frac{\alpha}{\pi} \right)^{1/4} e^{-\frac{1}{2}\alpha x^2} \left(-i\hbar \frac{d}{dx} \right) \left(\frac{\alpha}{\pi} \right)^{1/4} e^{-\frac{1}{2}\alpha x^2} \, dx$

$$\langle p_x \rangle = \left(\frac{\alpha}{\pi} \right)^{1/2} (-i\hbar) \int_{-\infty}^{\infty} -\alpha x e^{-\alpha x^2} \, dx$$

Because the integrand is an odd function of x, $\langle p_x \rangle = 0$ for $n = 0$.

for $n = 1$, $\langle p_x \rangle = \int_{-\infty}^{\infty} \left(\frac{4\alpha^3}{\pi} \right)^{1/4} x e^{-\frac{1}{2}\alpha x^2} \left(-i\hbar \frac{d}{dx} \right) \left(\frac{4\alpha^3}{\pi} \right)^{1/4} x e^{-\frac{1}{2}\alpha x^2} \, dx$

$$\langle p_x \rangle = \left(\frac{4\alpha^3}{\pi} \right)^{1/2} (-i\hbar) \int_{-\infty}^{\infty} x \left(1 - \alpha x^2 \right) e^{-\alpha x^2} \, dx$$

Because the integrand is an odd function of x, $\langle p_x \rangle = 0$ for $n = 1$.

for $n = 2$, $\langle p_x \rangle = \int_{-\infty}^{\infty} \left(\frac{\alpha}{4\pi} \right)^{1/4} \left(2\alpha x^2 - 1 \right) e^{-\frac{1}{2}\alpha x^2} \left(-i\hbar \frac{d}{dx} \right) \left(\frac{\alpha}{4\pi} \right)^{1/4} \left(2\alpha x^2 - 1 \right) e^{-\frac{1}{2}\alpha x^2} \, dx$

$$\langle p_x \rangle = \left(\frac{\alpha}{4\pi} \right)^{1/2} (-i\hbar) \int_{-\infty}^{\infty} \left(2\alpha x^2 - 1 \right) e^{-\alpha x^2} \left(-2\alpha^2 x^3 + 4\alpha x + \alpha x \right) dx$$

$$\langle p_x \rangle = \left(\frac{\alpha}{4\pi} \right)^{1/2} (-i\hbar) \int_{-\infty}^{\infty} e^{-\alpha x^2} \left(-4\alpha^3 x^5 + 12\alpha^2 x^3 - 5\alpha x \right) dx$$

Because the integrand is an odd function of x, $\langle p_x \rangle = 0$ for $n = 3$.

The result is general. $\langle p_x \rangle = 0$ for all values of n.

P18.33) For molecular rotation, the symbol J rather than l is used as the quantum number for angular momentum. A $^1H^{19}F$ molecule has the rotational quantum number $J = 10$ and vibrational quantum number $n = 0$.

a. Calculate the rotational and vibrational energy of the molecule. Compare each of these energies with kT at 300. K.

b. Calculate the period for vibration and rotation. How many times does the molecule rotate during one vibrational period?

a)

$$E_{rot} = \frac{J(J+1)\hbar^2}{2\mu r^2} = \frac{10 \times 11 \times \left(1.055 \times 10^{-34} \text{ J s}\right)^2}{2 \times \frac{1.0078 \text{ amu} \times 18.994 \text{ amu}}{1.0078 \text{ amu} + 18.994 \text{ amu}} \times \frac{1.661 \times 10^{-27} \text{ kg}}{\text{amu}} \times \left(91.68 \times 10^{-12} \text{ m}\right)^2}$$

$$E_{rot} = 4.58 \times 10^{-20} \text{ J}$$

$$\frac{E_{rot}}{kT} = \frac{4.58 \times 10^{-20} \text{ J}}{1.381 \times 10^{-23} \text{ J K}^{-1} \times 300. \text{ K}} = 11.1$$

$$E_{vib} = \left(n + \frac{1}{2}\right)\hbar\sqrt{\frac{k}{\mu}} = \frac{1}{2} \times 1.055 \times 10^{-34} \text{ J s} \times \sqrt{\frac{966 \text{ N m}^{-1}}{\frac{1.0078 \text{ amu} \times 18.994 \text{ amu}}{1.0078 \text{ amu} + 18.994 \text{ amu}} \times \frac{1.661 \times 10^{-27} \text{ kg}}{\text{amu}}}}$$

$$E_{vib} = 4.11 \times 10^{-20} \text{ J}$$

$$\frac{E_{vib}}{kT} = \frac{4.11 \times 10^{-20} \text{ J}}{1.381 \times 10^{-23} \text{ J K}^{-1} \times 300. \text{ K}} = 9.93$$

b) Calculate the period for vibration and rotation. How many times does the molecule vibrate during one rotational period?

$$E_{rot} = \frac{1}{2}I\omega^2; \quad \omega = 2\pi\nu = \frac{2\pi}{T_{rot}}$$

$$T_{rot} = \frac{2\pi}{\omega} = \frac{2\pi}{\sqrt{\frac{2E_{rot}}{I}}}$$

$$= \frac{2\pi}{\sqrt{\frac{2 \times 4.58 \times 10^{-20} \text{ J}}{\frac{1.0078 \text{ amu} \times 18.994 \text{ amu}}{1.0078 \text{ amu} + 18.994 \text{ amu}} \times \frac{1.661 \times 10^{-27} \text{ kg}}{\text{amu}} \times \left(91.68 \times 10^{-12} \text{ m}\right)^2}}} = 7.59 \times 10^{-14} \text{ s}$$

$$T_{vib} = \frac{1}{\nu} = 2\pi\sqrt{\frac{\mu}{k}} = 2\pi\sqrt{\frac{\frac{1.0078 \text{ amu} \times 18.994 \text{ amu}}{1.0078 \text{ amu} + 18.994 \text{ amu}} \times \frac{1.661 \times 10^{-27} \text{ kg}}{\text{amu}}}{966 \text{ N m}^{-1}}} = 8.06 \times 10^{-15} \text{ s}$$

It vibrates $\frac{T_{rot}}{T_{vib}} = \frac{7.59 \times 10^{-14} \text{ s}}{8.06 \times 10^{-15} \text{ s}} = 9.41$ times in one rotational period.

Chapter 19: The Vibrational and Rotational Spectroscopy of Diatomic Molecules

P19.2) The infrared spectrum of $^7Li^{35}Cl$ has an intense line at 643 cm^{-1}. Calculate the force constant and period of vibration of this molecule.

$$k = 4\pi^2 v^2 \mu$$

$$= 4\pi^2 \left(2.998 \times 10^{10} \text{ cm s}^{-1} \times 643 \text{ cm}^{-1}\right)^2 \times \frac{34.9688 \text{ amu} \times 7.0160 \text{ amu}}{34.9688 \text{ amu} + 7.0160 \text{ amu}} \times 1.6605 \times 10^{-27} \text{ kg amu}^{-1}$$

$$= 142 \text{ N m}^{-1}$$

$$T = \frac{1}{v} = 5.19 \times 10^{-14} \text{ s}$$

P19.3) Purification of water for drinking using UV light is a viable way to provide potable water in many areas of the world. Experimentally, the decrease in UV light of wavelength 250 nm follows the empirical relation $I/I_0 = e^{-\varepsilon' l}$ where l is the distance that the light passed through the water and ε' is an effective absorption coefficient. $\varepsilon' = 0.070$ cm^{-1} for pure water and 0.30 cm^{-1} for water exiting a waste water treatment plant. What distance corresponds to a decrease in I of 10.% from its incident value for a) pure water and b) waste water?

$$\ln\left[\frac{I(\lambda)}{I_0(\lambda)}\right] = -\varepsilon'(\lambda)l; \quad l = -\frac{\ln\left[\frac{I(\lambda)}{I_0(\lambda)}\right]}{\varepsilon'}$$

$$l = \frac{\ln[0.90]}{0.070 \text{ cm}^{-1}} = 1.5 \text{ cm for pure water}$$

$$l = \frac{\ln[0.90]}{0.30 \text{ cm}^{-1}} = 0.35 \text{ cm for treatment plant water}$$

P19.8) An infrared absorption spectrum of an organic compound is shown in the following figure. Use the characteristic group frequencies listed in Section 19.5 to decide whether this compound is more likely to be ethyl amine, pentanol, or acetone.

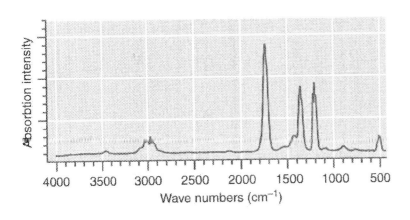

The major peak near 1700 cm^{-1} is the C=O stretch and the peak near 1200 cm^{-1} is a C–C–C stretch. These peaks are consistent with the compound being acetone. Ethyl amine should show a strong peak near 3350 cm^{-1} and pentanol should show a strong peak near 3400 cm^{-1}. Because these peaks are absent, these compounds can be ruled out.

P19.10) Write an expression for the moment of inertia of the acetylene molecule in terms of the bond distances. Does this molecule have a pure rotational spectrum?

The two mutually perpendicular axes of rotation are perpendicular to the molecular axis and go through the center of the molecule. The two moments of inertia are equal.

$$I = \sum_i m_i x_i^2 = 2\left(m_C \left[\frac{x_{C\equiv C}}{2} \right]^2 + m_H \left[\frac{x_{C\equiv C}}{2} + x_{C-H} \right]^2 \right)$$

Because the molecule has no dipole moment, it does not have a pure rotational spectrum.

P19.15) Calculating the motion of individual atoms in the vibrational modes of molecules (called normal modes) is an advanced topic. Given the normal modes shown in the following figure, decide which of the normal modes of CO_2 and H_2O have a nonzero dynamical dipole moment and are therefore infrared active. The motion of the atoms in the second of the two doubly degenerate bend modes for CO_2 is identical to the first, but is perpendicular to the plane of the page.

Copyright © 2006 Pearson Education, Inc., publishing as Benjamin Cummings

All three vibrational modes of water will lead to a change in the dipole moment and are therefore infrared active. The symmetric stretch of carbon dioxide will not lead to a change in the dipole moment and is infrared inactive. The other two modes will lead to a change in the dipole moment and are infrared active.

P19.16) The force constants for F_2 and I_2 are 470. and 172 N m^{-1}, respectively. Calculate the ratio of the vibrational state populations n_1/n_0 and n_2/n_0 at $T = 300.$ and at 1000. K.

The vibrational energy is given by $E_n = \left(n + \dfrac{1}{2}\right)h\nu = \left(n + \dfrac{1}{2}\right)h\sqrt{\dfrac{k}{\mu}}$

For F_2

$$E_0 = \frac{1}{2}\hbar\sqrt{\frac{k}{\mu}} = \frac{1}{2}\times1.055\times10^{-34}\ \text{J s}\times\sqrt{\frac{470.\ \text{N m}^{-1}}{\dfrac{18.994\ \text{amu}\times18.994\ \text{amu}}{18.994\ \text{amu}+18.994\ \text{amu}}\times1.661\times10^{-27}\ \text{kg amu}^{-1}}}$$

$= 9.11\times10^{-21}\ \text{J}$

$$E_1 = \frac{3}{2}\hbar\sqrt{\frac{k}{\mu}} = \frac{3}{2}\times1.055\times10^{-34}\ \text{J s}\times\sqrt{\frac{470.\ \text{N m}^{-1}}{\dfrac{18.994\ \text{amu}\times18.994\ \text{amu}}{18.994\ \text{amu}+18.994\ \text{amu}}\times1.661\times10^{-27}\ \text{kg amu}^{-1}}}$$

$= 2.73\times10^{-20}\ \text{J}$

$$E_2 = \frac{5}{2}\hbar\sqrt{\frac{k}{\mu}} = \frac{5}{2}\times1.055\times10^{-34}\ \text{J s}\times\sqrt{\frac{470.\ \text{N m}^{-1}}{\dfrac{18.994\ \text{amu}\times18.994\ \text{amu}}{18.994\ \text{amu}+18.994\ \text{amu}}\times1.661\times10^{-27}\ \text{kg amu}^{-1}}}$$

$= 4.55\times10^{-20}\ \text{J}$

For I_2

$$E_0 = \frac{1}{2}\hbar\sqrt{\frac{k}{\mu}} = \frac{1}{2} \times 1.055 \times 10^{-34} \text{ J s} \times \sqrt{\frac{172 \text{ N m}^{-1}}{\frac{126.9045 \text{ amu} \times 126.9045 \text{ amu}}{126.9045 \text{ amu} + 126.9045 \text{ amu}} \times 1.661 \times 10^{-27} \text{ kg amu}^{-1}}}$$

$$= 2.13 \times 10^{-21} \text{ J}$$

$$E_1 = \frac{3}{2}\hbar\sqrt{\frac{k}{\mu}} = \frac{3}{2} \times 1.055 \times 10^{-34} \text{ J s} \times \sqrt{\frac{172 \text{ N m}^{-1}}{\frac{126.9045 \text{ amu} \times 126.9045 \text{ amu}}{126.9045 \text{ amu} + 126.9045 \text{ amu}} \times 1.661 \times 10^{-27} \text{ kg amu}^{-1}}}$$

$$= 6.39 \times 10^{-21} \text{ J}$$

$$E_2 = \frac{5}{2}\hbar\sqrt{\frac{k}{\mu}} = \frac{5}{2} \times 1.055 \times 10^{-34} \text{ J s} \times \sqrt{\frac{172 \text{ N m}^{-1}}{\frac{126.9045 \text{ amu} \times 126.9045 \text{ amu}}{126.9045 \text{ amu} + 126.9045 \text{ amu}} \times 1.661 \times 10^{-27} \text{ kg amu}^{-1}}}$$

$$= 1.07 \times 10^{-20} \text{ J}$$

For F_2 at 300. K, $\dfrac{n_1}{n_0} = e^{-\frac{E_1 - E_0}{kT}} = e^{-\frac{(2.73 - 0.911) \times 10^{-20} \text{J}}{1.381 \times 10^{-23} \text{ J K}^{-1} \times 300. \text{ K}}} = 0.0123$

For F_2 at 1000. K, $\dfrac{n_1}{n_0} = e^{-\frac{E_1 - E_0}{kT}} = e^{-\frac{(1.31 - 0.437) \times 10^{-19} \text{J}}{1.381 \times 10^{-23} \text{ J K}^{-1} \times 1000. \text{ K}}} = 0.267$

For F_2 at 300. K, $\dfrac{n_2}{n_0} = e^{-\frac{E_2 - E_0}{kT}} = e^{-\frac{(4.55 - 0.915) \times 10^{-20} \text{J}}{1.381 \times 10^{-23} \text{ J K}^{-1} \times 300. \text{ K}}} = 1.52 \times 10^{-4}$

For F_2 at 1000. K, $\dfrac{n_2}{n_0} = e^{-\frac{E_2 - E_0}{kT}} = e^{-\frac{(4.55 - 0.915) \times 10^{-20} \text{J}}{1.381 \times 10^{-23} \text{ J K}^{-1} \times 1000. \text{ K}}} = 0.0715$

For I_2 at 300. K, $\dfrac{n_1}{n_0} = e^{-\frac{E_1 - E_0}{kT}} = e^{-\frac{(6.39 - 2.13) \times 10^{-21} \text{J}}{1.381 \times 10^{-23} \text{ J K}^{-1} \times 300. \text{ K}}} = 0.357$

For I_2 at 1000. K, $\dfrac{n_1}{n_0} = e^{-\frac{E_1 - E_0}{kT}} = e^{-\frac{(6.39 - 2.13) \times 10^{-21} \text{J}}{1.381 \times 10^{-23} \text{ J K}^{-1} \times 1000. \text{ K}}} = 0.734$

For I_2 at 300. K, $\dfrac{n_2}{n_0} = e^{-\frac{E_2 - E_0}{kT}} = e^{-\frac{(1.07 - 0.213) \times 10^{-20} \text{J}}{1.381 \times 10^{-23} \text{ J K}^{-1} \times 300. \text{ K}}} = 0.127$

For I_2 at 1000. K, $\dfrac{n_2}{n_0} = e^{-\frac{E_2 - E_0}{kT}} = e^{-\frac{(1.07 - 0.213) \times 10^{-20} \text{J}}{1.381 \times 10^{-23} \text{ J K}^{-1} \times 1000. \text{ K}}} = 0.539$

P19.19) Show that the Morse potential approaches the harmonic potential for small values of the vibrational amplitude. (*Hint:* Expand the Morse potential in a Taylor-Maclaurin series.)

$$V(R) = D_e\left[1 - e^{-\alpha(R-R_e)}\right]^2$$

Expanding in a Taylor-Maclaurin series and keeping only the first term,

$$V(R) = D_e\left[1 - \left[e^{-\alpha(R-R_e)}\right]_{R=R_e} - \left[\frac{\partial e^{-\alpha(R-R_e)}}{\partial(R-R_e)}\right]_{R=R_e}(R-R_e)\right]^2$$

$$= D_e\left[1 - 1 - \alpha(R-R_e)\right]^2 = D_e\alpha^2(R-R_e)^2$$

P19.20) The rotational constant for $^{127}\text{I}^{35}\text{Cl}$ determined from microwave spectroscopy is 0.1141619 cm^{-1}. Calculate the bond length in $^{127}\text{I}^{35}\text{Cl}$ to the maximum number of significant figures consistent with this information.

$$B = \frac{h}{8\pi^2\mu r_0^2}; \quad r_0 = \sqrt{\frac{h}{8\pi^2\mu B}}$$

$$r_0 = \sqrt{\frac{6.6260755\times10^{-34}\text{ J s}}{8\pi^2\times\dfrac{126.904473\text{ amu}\times34.9688\text{ amu}}{(126.904473\text{ amu} + 34.9688\text{ amu})}\times1.6605402\times10^{-27}\text{ kg amu}^{-1}\times0.1141619\text{ cm}^{-1}\times2.99792458\times10^{10}\text{ cm s}^{-1}}}$$

$$r_0 = 2.32085\times10^{-10}\text{ m}$$

P19.23) The fundamental vibrational frequencies for $^1\text{H}^{19}\text{F}$ and $^2\text{D}^{19}\text{F}$ are 4138.52 and 2998.25 cm^{-1}, respectively, and D_e for both molecules is 5.86 eV. What is the difference in the bond energy of the two molecules?

$$\left(D_e - \frac{1}{2}hc\tilde{\nu}\right)_{\text{HF}} - \left(D_e - \frac{1}{2}hc\tilde{\nu}\right)_{\text{DF}} = \frac{1}{2}hc(\tilde{\nu}_{\text{HF}} - \tilde{\nu}_{\text{DF}})$$

$$= \frac{1}{2}\times6.626\times10^{-34}\text{ J s}\times3.00\times10^{10}\text{ cm s}^{-1}\times(4138.52\text{ cm}^{-1} - 2998.25\text{ cm}^{-1}) = 1.133\times10^{-20}\text{ J}$$

P19.27) Fill in the missing step in the derivation that led to the calculation of the spectral line shape in Figure 19.22. Starting from

$$a_2(t) = \mu_x^{21}\frac{E_0}{2}\left(\frac{1 - e^{\frac{i}{\hbar}(E_2-E_1+h\nu)t}}{E_2 - E_1 + h\nu} + \frac{1 - e^{-\frac{i}{\hbar}(E_2-E_1-h\nu)t}}{E_2 - E_1 - h\nu}\right)$$

and neglecting the first term in the parentheses, show that

$$a_2^*(t)a_2(t) = E_0^2\left[\mu_x^{21}\right]^2\frac{\sin^2[(E_2-E_1-h\nu)t/2\hbar]}{(E_2-E_1-h\nu)^2}$$

$$a_2^*(t)a_2(t) = \left[\mu_z^{21}\right]^2 E_0^2 \left(\frac{1-e^{+\frac{i}{\hbar}(E_2-E_1-h\nu)t}}{E_2-E_1-h\nu}\right)\left(\frac{1-e^{-\frac{i}{\hbar}(E_2-E_1-h\nu)t}}{E_2-E_1-h\nu}\right)$$

$$= \left[\mu_z^{21}\right]^2 E_0^2 \frac{\left(2-2\cos\left[(E_2-E_1-h\nu)\frac{t}{\hbar}\right]\right)}{(E_2-E_1-h\nu)^2}$$

Using the identity $1-\cos x = 1-\cos\left(\dfrac{x}{2}+\dfrac{x}{2}\right) = \cos^2\dfrac{x}{2}+\sin^2\dfrac{x}{2}-\left(\cos^2\dfrac{x}{2}-\sin^2\dfrac{x}{2}\right) = 2\sin^2\dfrac{x}{2}$

$$a_2^*(t)a_2(t) = \left[\mu_z^{21}\right]^2 \frac{E_0^2}{4}\left(\frac{2-2\cos\left[(E_2-E_1-h\nu)\frac{t}{\hbar}\right]}{(E_2-E_1-h\nu)^2}\right) = \left[\mu_z^{21}\right]^2 E_0^2 \frac{\sin^2\left[(E_2-E_1-h\nu)\frac{t}{2\hbar}\right]}{(E_2-E_1-h\nu)^2}$$

P19.30) A strong absorption band in the infrared region of the electromagnetic spectrum is observed at $\tilde{\nu} = 2649$ cm^{-1} for H^{81}Br. Assuming that the harmonic potential applies, calculate the fundamental frequency ν in units of inverse seconds, the vibrational period in seconds, and the zero point energy for the molecule in joules and electron-volts.

$$\nu = \tilde{\nu}c = 2649 \text{ cm}^{-1} \times 3.00\times10^{10} \text{ cm s}^{-1} = 7.94\times10^{13} \text{ s}^{-1}$$

$$T = \frac{1}{\nu} = \frac{1}{7.94\times10^{13} \text{ s}^{-1}} = 1.26\times10^{-14} \text{ s}$$

$$E = \frac{1}{2}h\nu = \frac{1}{2}\times6.626\times10^{-34} \text{ J s}\times7.94\times10^{13} \text{ s}^{-1} = 2.63\times10^{-20} \text{ J}\times\frac{6.241\times10^{18} \text{ eV}}{\text{J}} = 0.164 \text{ eV}$$

P19.31) The spacing between lines in the pure rotational spectrum of ^{7}Li^{1}H is 4.505×10^{11} s^{-1}. Calculate the bond length of this molecule.

$$B = \frac{v}{2c} = \frac{4.505\times10^{11}\ s^{-1}}{2\times2.998\times10^{10}\ cm\ s^{-1}} = 7.513\ cm^{-1}$$

$$r_0 = \sqrt{\frac{h}{8\pi^2\mu cB}}$$

$$=$$

$$\sqrt{\frac{6.626\times10^{-34}\ J\ s}{8\pi^2\times\dfrac{1.0078\ amu\times7.0160\ amu}{1.0078\ amu + 7.0160\ amu}\times1.6605\times10^{-27}\ kg\ amu^{-1}\times2.998\times10^{10}\ cm\ s^{-1}\times7.513\ cm^{-1}}}$$

$$r_0 = 1.596\times10^{-10}\ m$$

P19.33) Calculate the moment of inertia, the magnitude of the rotational angular momentum, and the energy in the $J = 12$ rotational state for $^1H_2$ in which the bond length of $^1H_2$ is 74.2 pm.

$$I = \mu r_0^2 = \frac{1.007825^2\ amu^2}{2\times1.007825\ amu}\times1.6605402\times10^{-27}\ kg\ amu^{-1}\times\left(74.2\times10^{-12}\ m\right)^2$$

$$= 4.605\times10^{-48}\ kg\ m^2$$

$$|J| = \sqrt{J(J+1)}\hbar = \sqrt{12\times13}\times1.0554\times10^{-34}\ J\ s = 1.3177\times10^{-33}\ J\ s$$

$$= 1.3177\times10^{-33}\ kg\ m\ s^{-2}$$

$$E_J = \frac{J(J+1)\hbar^2}{2I} = \frac{12\times13\times\left(1.0554\times10^{-34}\ J\ s\right)^2}{2\times4.605\times10^{-48}\ kg\ m^2} = 1.89\times10^{-19}\ J$$

P19.38) 50.% of the light incident on a 10.0 mm thick piece of fused silica quartz glass passes through the glass. What percentage of the light will pass through a 20.0 mm thick piece of the same glass?

$$\ln\left[\frac{I(\lambda)}{I_0(\lambda)}\right] = -\varepsilon(\lambda)l;\quad \varepsilon = -\frac{\ln\left[\dfrac{I(\lambda)}{I_0(\lambda)}\right]}{l}$$

$$\varepsilon = -\frac{\ln[0.50]}{1.00\ cm^{-1}} = 0.693\ cm^{-1}$$

$$\ln\left[\frac{I(\lambda)}{I_0(\lambda)}\right] = -0.693\ cm^{-1}\times2.00\ cm = -1.3863$$

$$\frac{I(\lambda)}{I_0(\lambda)} = 0.250$$

25% of the light will pass through the glass

P19.41) An infrared absorption spectrum of an organic compound is shown in the following figure. Use the characteristic group frequencies listed in Section 19.5 to decide whether this compound is more likely to be hexene, hexane, or hexanol.

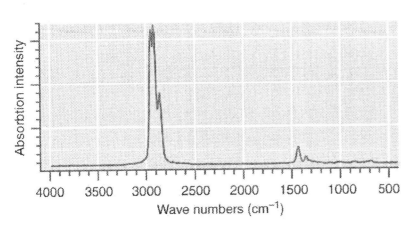

The group of peaks near 2900 cm^{-1} is due to CH_3 and CH_2 stretching vibrations and the small peak near 1400 cm^{-1} is due to a CH_3 umbrella bending mode. These peaks are consistent with the compound being hexane. Hexene should show a strong peak near 1650 cm^{-1} and hexanol should show a strong peak near 3400 cm^{-1}. Because these peaks are absent, these compounds can be ruled out.

Computational Problems

Before solving the computational problems, it is recommended that students work through Tutorials 1–3 under the Help menus in Spartan Student Edition to gain familiarity with the program.

Computational Problem 19.1 Build structures for the gas-phase (a) hydrogen fluoride ($^1H^{19}F$), (b) hydrogen chloride ($^1H^{35}Cl$), (c) carbon monoxide ($^{12}C^{16}O$), and (d) sodium chloride ($^{23}Na^{35}Cl$) molecules. (For Spartan, these are the default isotopic masses.) Calculate the equilibrium geometry and the IR spectrum using the B3LYP method with the 6-311+G** basis set.

a. Compare your result for the vibrational frequency with the experimental value listed in Table 19.3. What is the relative error in the calculation?

b. Calculate the force constant from the vibrational frequency and reduced mass. Determine the relative error using the experimental value in Table 19.3.

c. Calculate the values for the rotational constant, B, using the calculated bond length. Determine the relative error using the experimental value in Table 19.3.

Create each one as a new molecule within the same file using the "New Molecule" command in the File menu.

Your solution to this problem should answer the questions in bold type in Steps 5 and 10.

Step 1: Create a new file, and build structures for hydrogen fluoride ($^1H^{19}F$), hydrogen chloride ($^1H^{35}Cl$), carbon monoxide ($^{12}C^{16}O$), and sodium chloride ($^{23}Na^{35}Cl$, just the diatomic, gas-phase species, not the crystal structure!). All of the isotope values are the defaults, so you don't need to worry about changing them. Create each species as a new molecule within the same file using the "New Molecule" command in the File menu. Remember to minimize your structures using molecular mechanics before starting the next step; good initial geometries help keep calculation times short.

Note that for NaCl and CO, you will have to use the inorganic model kit, which is accessed by clicking the "Inorganic" tab on the model kit. To give the atoms only one valence, click the "-X" button. Verify that CO has a triple bond by going to Display > Properties, then clicking on the C-O bond while in view mode. If it is not listed as a triple bond, change its type to triple. While the displayed bond type is not relevant for the QM calculations, it tells the molecular mechanics minimizer to use the right parameters, and give you a better initial geometry.

Step 2: Go to Setup → Calculations. Set the calculation type to "Equilibrium Geometry" and the method to **B3LYP/6-311+G****. The meaning of the method (Density Functional Theory with B3LYP functional) and basis set (6-311+G**) will be explained later in the course, consider this to be a "black box" for now. "Total Charge" should be neutral, "Spin Multiplicity" should be 1, "Global Calculations" should be on, and the "Calculate IR Spectrum" box should be checked. Click "Submit." Your calculations will start immediately.

Since you are using a large set of basis functions (this will be covered later in the course), and since the computational engine is calculating the normal modes for the molecules as well as the energies, the calculations may take several minutes.

Step 3: Once your calculations complete, go to Display → Spectra. You should see the frequency of the normal mode for each molecule, in wave numbers (cm^{-1}). These can be converted to Hz by multiplying by the speed of light in cm/s. You can also draw the spectrum by clicking "Draw Spectrum." If the peak for one or more of the molecules does not appear, you will need to change the scale of the x-axis of the graph. Make sure the "Properties" window is open, then click on the x-axis of the graph while in view mode, and change the viewable range to 100 to 4000 cm^{-1}. Scroll through the individual molecules in the file using the control in the left lower portion of the window.

Step 4: Record the normal mode frequencies for each molecule. Since you will be manipulating these in the next few steps, a spreadsheet may be helpful. Also, measure and record the bond lengths.

Step 5: On your spreadsheet, calculate the reduced mass μ for each molecule, using values taken from the inside back cover of the textbook for the mass in atomic mass units; these are single-isotope species, not a statistical average of isotopes. Sodium is not included in that table; use a mass of 22.9897 amu for Na. Convert the reduced masses into kilograms.

Step 6: Using the normal modes and the reduced masses, calculate the force constant for each molecule using the following equation derived from the harmonic oscillator. Make sure you are using the correct units! Note that this is just a rearrangement of Equation 18.7 from your textbook. Make a table with the bond length, reduced mass, normal mode, and force constant for each molecule, **leaving one extra line for step #9.**

$$k = 4\pi^2 v^2 \mu$$

Step 7: Using the bond lengths and force constants, **qualitatively** estimate the **relative** strength of the bonds in the four species. Is this congruent with your chemical intuition about how strong these bonds should be?

Step 8: Go to http://webbook.nist.gov, (You can access the webbook through the UW library when you log on as member of the UW community) should have a link called "Constants of Diatomic Molecules." Access that data, and obtain the frequency of the normal mode v_e for each species, **in the $X^1\Sigma$ (ground) state**. Be very careful that you pick the correct state. *Compare these values with your calculated values.*

Step 9: The calculated normal mode for deuterium chloride is at 2129.97 cm^{-1}. Calculate the reduced mass and force constant for DCl, and add to your table. *How do the force constants for HCl and DCl compare with each other?*

Step 10: Your calculation can also yield some useful thermodynamic properties. Go to HCl and open up the output, under the display menu. Find the total energy, which is in another potentially unfamiliar unit, the Hartree. Using the total energy for HCl, and the following values for the total energy of H and Cl, calculate the bond dissociation energy.

Cl: -460.166854

H: -0.5021559

Convert this to kJ/mol using the conversion factor 2625.5 kJ/mol per Hartree. *How does it compare with the experimental bond dissociation energy of 427.71 kJ/mol?*

Next, scroll down to the bottom of the output, where you should find a table containing thermodynamic properties. Find and record the total entropy. Next, go back to NIST Webbook, and on the page for HCl, click on "Gas Phase Thermochemistry Data." Find a measurement of the total entropy. *How does your calculated entropy compare with the experimental data?*

Computational Problem 19.2: Calculate the bond energy in gaseous (a) hydrogen fluoride ($^1H^{19}F$), (b) hydrogen chloride ($^1H^{35}Cl$), (c) carbon monoxide ($^{12}C^{16}O$), and (d) sodium chloride ($^{23}Na^{35}Cl$) molecules by comparing the total energies of the species in the dissociation reactions. Use the B3LYP method with the 6-31G* basis set. Determine the relative error of the calculation using the experimental value in Table 19.3.

Step 1: Create a new file, and build structures for hydrogen fluoride ($^1H^{19}F$), hydrogen chloride ($^1H^{35}Cl$), carbon monoxide ($^{12}C^{16}O$), and sodium chloride ($^{23}Na^{35}Cl$, just the diatomic, gas-phase species, not the crystal structure!). Build structures for all the atoms except H for which you know the exact energy. All of the isotope values are the defaults, so you don't need to worry about changing them. Create each species as a new molecule within the same file using the "New Molecule" command in the File menu. Remember to minimize the molecular structures using molecular mechanics before starting the next step; good initial geometries help keep calculation times short.

Note that for NaCl and CO, you will have to use the inorganic model kit, which is accessed by clicking the "Inorganic" tab on the model kit. To give the atoms only one valence, click the "-X" button. Verify that CO has a triple bond by going to Display > Properties, then clicking on the C-O bond while in view mode. If it is not listed as a triple bond, change its type to triple. While the displayed bond type is not relevant for the QM calculations, it tells the molecular mechanics minimizer to use the right parameters, and give you a better initial geometry.

Step 2: Go to Setup → Calculations. Set the calculation type to "Equilibrium Geometry" and the method to **B3LYP/6-311G**** for the molecules and "Energy" for the atoms except for HCl and NaCl. Use the **6-31G*** basis set for these molecules Cl(g) and Na(g). The meaning of the method (Density Functional Theory with B3LYP functional) and basis set (6-31G*) will be explained later in the course, consider this to be a "black box" for now. "Total Charge" is neutral, and you must enter the appropriate number of unpaired electrons for each species. "Global Calculations" should not be on because the number of unpaired electrons is not the same for all species. Click "Submit." Your calculations will start immediately.

Step 3: When the calculation has finished, obtain the total energy for a species by going to Display →
Output. Scroll through the individual species in the file using the control in the left lower portion of
window. Note that the energies are in atomic units where 1 hartree = 2625.5 kJ mol^{-1} = 27.2114 eV.
Record (copy and paste) the energies in a spreadsheet.

Step 4. Calculate the change in total energy for the reaction AB(g) → A(g) + B(g) using the energies
obtained in your calculations. Determine the relative error of the calculation using the experimental
values in Table 19.3 and find a reference value for the NaCl bond energy (the NaCl molecule and not a
NaCl crystal).

Computational Problem 19.3: Build structures for the gas-phase (a) NF_3, (b) PCl_3, and (c) SO_3
molecules. Calculate the equilibrium geometry and the IR spectrum using the B3LYP method with the
6-31G* basis set. Animate the vibrational normal modes and classify them as symmetrical stretch,
symmetrical deformation, degenerate stretch, and degenerate deformation

Step 1: Create a new file, and build structures for $^{14}N^{19}F_3$, $^{31}P^{35}Cl_3$, and $^{32}S^{16}O_3$. All of the isotope
values are the defaults, so you don't need to worry about changing them. Create each species as a new
molecule within the same file using the "New Molecule" command in the File menu. Use the inorganic
model kit, which is accessed by clicking the "Inorganic" tab on the model kit. Remember to minimize
your structures using molecular mechanics before starting the next step; good initial geometries help
keep calculation times short.

Step 2: Go to Setup → Calculations. Set the calculation type to "Equilibrium Geometry" and the
method to **B3LYP/6-31G***. The meaning of the method (Density Functional Theory with B3LYP
functional) and basis set (6-31G*) will be explained later in the course, consider this to be a "black box"
for now. "Total Charge" should be neutral. Enter the appropriate number of unpaired electrons. "Global
Calculations" should be checked, and **the "Calculate IR Spectrum" box should be checked.** Click
"Submit." Your calculations will start immediately.

Step 3: Once your calculations complete, go to Display → Spectra. You should see the frequency of the
normal modes for each molecule, in wave numbers (cm^{-1}). Note the normal mode frequencies for each
molecule as well as their degeneracies. How many normal modes should each molecule exhibit? If one
or more of the peaks does not appear, you will need to change the scale of the x-axis of the graph. Make
sure the "Properties" window is open, then click on the graph while in view mode, and change the
viewable range to include all normal mode frequencies. Scroll through the individual molecules in the
file using the control in the left lower portion of the window.

Step 4: Make a table showing the normal mode frequencies, the degeneracy (number of modes with the same frequency), and intensity in columns 1–3 for each molecule. How many peaks have sufficient intensity that they are easily observable? If any peaks have zero intensity, explain why. Classify the modes as symmetrical stretch, symmetrical deformation, degenerate stretch, and degenerate deformation. A stretching mode is symmetric if all bonds are being stretched simultaneously, and degenerate if two or more stretching modes have the same frequency. The same classification holds for a deformation, in which the bond angle, rather than the bond length, changes. Compare the frequencies of equivalent normal modes in NF_3 and PCl_3. What factors might lead to the observed difference in frequency?

Computational Problem 19.4: Build structures for the gas-phase (a) F_2CO, (b) Cl_2CO, and (c) O_2NF molecules of the structural form X_2YZ. Calculate the equilibrium geometry and the IR spectrum using the B3LYP method with the 6-311+G**basis set. Animate the vibrational normal modes and classify them as symmetric Y-Z stretch, YX_2 scissors, antisymmetric X-Y stretch, YX_2 rock, and Y-X_2 wag.

Step 1: Create a new file, and build structures for (a) $^{19}F_2^{12}C^{16}O$, (b) $^{35}Cl_2^{12}C^{16}O$, and (c) $^{16}O_2^{14}N^{19}F$. All of the isotope values are the defaults, so you don't need to worry about changing them. Create each species as a new molecule within the same file using the "New Molecule" command in the File menu. Use the inorganic model kit, which is accessed by clicking the "Inorganic" tab on the model kit. Remember to minimize your structures using molecular mechanics before starting the next step; good initial geometries help keep calculation times short.

Step 2: Go to Setup → Calculations. Set the calculation type to "Equilibrium Geometry" and the method to **B3LYP/6-31G***. The meaning of the method (Density Functional Theory with B3LYP functional) and basis set (6-31G*) will be explained later in the course, consider this to be a "black box" for now. "Total Charge" should be neutral. Enter the appropriate number of unpaired electrons. "Global Calculations" should be checked, and **the "Calculate IR Spectrum" box should be checked.** Click "Submit." Your calculations will start immediately.

Step 3: Once your calculations complete, go to Display → Spectra. You should see the frequency of the normal modes for each molecule, in wave numbers (cm^{-1}). Note the normal mode frequencies for each molecule as well as their degeneracies. How many normal modes should each molecule exhibit? If one or more of the peaks does not appear, you will need to change the scale of the x-axis of the graph. Make sure the "Properties" window is open, then click on the graph while in view mode, and change the viewable range to include all normal mode frequencies. Scroll through the individual molecules in the file using the control in the left lower portion of the window.

Step 4: Make a table showing the normal mode frequencies, the degeneracy (number of modes with the same frequency), and intensity in columns 1–3 for each molecule. How many peaks have sufficient intensity that they are easily observable? If any peaks have zero intensity, explain why. Classify the modes as symmetric Y-Z stretch, YX_2 scissors, antisymmetric X-Y stretch, YX_2 rock, and Y-X_2 wag. A stretching mode is symmetric if all bonds are being stretched simultaneously, and degenerate if two or more stretching modes have the same frequency. An out of plane deformation is referred to as a wag. Rocking and scissors modes correspond to their names. Compare the frequencies of equivalent normal modes in F_2CO and Cl_2CO. What factors might lead to the observed difference in frequency? Is the order of the modes symmetric Y-Z stretch, YX_2 scissors, antisymmetric X-Y stretch, YX_2 rock, and Y-X_2 wag the same for all molecules in terms of frequency?

Computational Problem 19.5:
Build structures for the bent gas-phase (a) HOF, (b) ClOO, and (c) HSO molecules of the structural form XYZ. Calculate the equilibrium geometry and the IR spectrum using the B3LYP method with the basis set. Animate the vibrational normal modes and classify them as Y-Z stretch, X-Y stretch, and X-Y-Z bend.

Step 1: Create a new file, and build structures for $^1H^{16}O^{19}F$, (b) $^{35}Cl^{16}O^{16}O$, and (c) $^1H^{32}S^{16}O$. All of the isotope values are the defaults, so you don't need to worry about changing them. Create each species as a new molecule within the same file using the "New Molecule" command in the File menu. Use the inorganic model kit, which is accessed by clicking the "Inorganic" tab on the model kit. Remember to minimize your structures using molecular mechanics before starting the next step; good initial geometries help keep calculation times short.

Step 2: Go to Setup → Calculations. Set the calculation type to "Equilibrium Geometry" and the method to **B3LYP/6-31G***. The meaning of the method (Density Functional Theory with B3LYP functional) and basis set (6-31G*) will be explained later in the course, consider this to be a "black box" for now. "Total Charge" should be neutral. Enter the appropriate number of unpaired electrons. "Global Calculations" should not be checked, and **the "Calculate IR Spectrum" box should be checked.** Click "Submit." Your calculations will start immediately.

Step 3: Once your calculations complete, go to Display → Spectra. You should see the frequency of the normal modes for each molecule, in wave numbers (cm^{-1}). Note the normal mode frequencies for each molecule as well as their degeneracies. How many normal modes should each molecule exhibit? If one or more of the peaks does not appear, you will need to change the scale of the x-axis of the graph. Make sure the "Properties" window is open, then click on the graph while in view mode, and change the viewable range to include all normal mode frequencies. Scroll through the individual molecules in the file using the control in the left lower portion of the window.

Step 4: Make a table showing the normal mode frequencies, the degeneracy (number of modes with the same frequency), and intensity in columns 1–3 for each molecule. How many peaks have sufficient intensity that they are easily observable? If any peaks have zero intensity, explain why. Classify the modes as Y-Z stretch, X-Y stretch, and X-Y-Z bend.

Chapter 20: The Hydrogen Atom

P20.1) Calculate the wave number corresponding to the most and least energetic spectral lines in the Lyman, Balmer, and Paschen series for the hydrogen atom.

$$\tilde{v} = 109{,}678\left(\frac{1}{n_{initial}^2} - \frac{1}{n_{final}^2}\right)\text{cm}^{-1}$$ and the most energetic line corresponds to $n \to \infty$.

$n_{initial}$ is 1, 2, and 3 for the Lyman, Balmer, and Paschen series, respectively.

$\tilde{v} = 109{,}678 \text{ cm}^{-1}$ for the Lyman series,

$$\tilde{v} = 109{,}678\left(\frac{1}{4}\right)\text{cm}^{-1} = 27419.3\,\text{cm}^{-1} \text{ for the Balmer series, and}$$

$$\tilde{v} = 109{,}678\left(\frac{1}{9}\right)\text{cm}^{-1} = 12186.3\,\text{cm}^{-1} \text{ for the Paschen series.}$$

The least energetic transition corresponds to $n_{initial} \to n_{initial} + 1$.

$$\tilde{v} = 109{,}678\left(1 - \frac{1}{4}\right)\text{cm}^{-1} = 82258.5\,\text{cm}^{-1},$$

$$\tilde{v} = 109{,}678\left(\frac{1}{4} - \frac{1}{9}\right)\text{cm}^{-1} = 15233.1\,\text{cm}^{-1}, \text{ and}$$

$$\tilde{v} = 109{,}678\left(\frac{1}{9} - \frac{1}{16}\right)\text{cm}^{-1} = 5331.57\,\text{cm}^{-1} \text{ for the Lyman, Balmer, and Paschen series, respectively.}$$

P20.5) Using the result of Problem P20.13, calculate the probability that the $1s$ electron for H will be found between $r = 0$ and $r = a_0$.

Let $u = r^2$ and $dv = e^{-\frac{r}{\alpha}}dr$

$$\int r^2 e^{-\frac{r}{\alpha}}dr = -\alpha r^2 e^{-\frac{r}{\alpha}} + 2\alpha \int r\, e^{-\frac{r}{\alpha}}dr$$

Integrating by parts again,

$$\int r^2 e^{-\frac{r}{\alpha}}dr = -\alpha r^2 e^{-\frac{r}{\alpha}} + 2\alpha\left(-\alpha r e^{-\frac{r}{\alpha}} + \alpha \int e^{-\frac{r}{\alpha}}dr\right) = -\alpha r^2 e^{-\frac{r}{\alpha}} - 2\alpha\left(\alpha r e^{-\frac{r}{\alpha}} - \alpha^2 e^{-\frac{r}{\alpha}}\right)$$

$$\int r^2 e^{-\frac{r}{\alpha}}dr = e^{-\frac{r}{\alpha}}\left(-\alpha r^2 - 2\alpha^2 r - 2\alpha^3\right)$$

137

$$P = 1 - e^{-2r/a_0} - \frac{2r}{a_0}\left(1 + \frac{r}{a_0}\right)e^{-2r/a_0}$$

for $r = a_0$

P20.6) Using the result of Problem P20.13, calculate the probability of finding the electron in the $1s$ state outside a sphere of radius $0.5a_0$, $3a_0$, and $5a_0$.

The probability of finding the electron inside the sphere of radius r is $1 - e^{-\frac{2r}{a_0}} - \frac{2r}{a_0}\left(1 + \frac{r}{a_0}\right)e^{-\frac{2r}{a_0}}$. The

probability of finding it outside the sphere of radius r is

$$1 - \left[1 - e^{-\frac{2r}{a_0}} - \frac{2r}{a_0}\left(1 + \frac{r}{a_0}\right)e^{-\frac{2r}{a_0}}\right] = e^{-\frac{2r}{a_0}} + \frac{2r}{a_0}\left(1 + \frac{r}{a_0}\right)e^{-\frac{2r}{a_0}}.$$ Evaluating this function at

$0.5a_0$, $3a_0$, and $5a_0$ gives 0.920, 0.0620, and 2.77×10^{-3}, respectively.

P20.12) As the principal quantum number n increases, the electron is more likely to be found far from the nucleus. It can be shown that for H and for ions with only one electron such as He^+,

$$\langle r \rangle_{nl} = \frac{n^2 a_0}{Z}\left[1 + \frac{1}{2}\left(1 - \frac{l(l+1)}{n^2}\right)\right]$$

Calculate the value of n for an s state in the hydrogen atom such that $\langle r \rangle = 1000\, a_0$. Round up to the nearest integer. What is the ionization energy of the H atom in this state in electron-volts? Compare your answer with the ionization energy of the H atom in the ground state.

$$\langle r \rangle_{nl} = \frac{n^2 a_0}{Z}\left[1 + \frac{1}{2}\left(1 - \frac{l(l+1)}{n^2}\right)\right]$$

$$n = \sqrt{\frac{2Z\langle r \rangle_{n0}}{3a_0}} = \sqrt{\frac{2000a_0}{3a_0}} = 25.82 \approx 26$$

$$I = \frac{Z^2 e^2}{8\pi \varepsilon_0 a_0 n^2} = \frac{\frac{Z^2}{n^2} \times (1.602 \times 10^{-19}\text{ C})^2}{8\pi \times 8.854 \times 10^{-12}\text{ J}^{-1}\text{ C}^2\text{ m}^{-1} \times 5.292 \times 10^{-11}\text{ m}} \times \frac{1\text{eV}}{1.602 \times 10^{-19}\text{ J}}$$

$$I = 13.6039\frac{Z^2}{n^2}\text{ eV} = 13.6039\frac{1}{n^2}\text{ eV for the H atom.}$$

For the ground state, $I = 13.6039$ eV and for $n = 26$, $I = 0.0201$ eV.

P20.14) Calculate the expectation value $(r - \langle r \rangle)^2$ if the H atom wave function is $\psi_{100}(r)$.

$$\left\langle (r - \langle r \rangle)^2 \right\rangle = \left\langle r^2 - 2r\langle r \rangle + \langle r \rangle^2 \right\rangle = \left\langle r^2 \right\rangle - 2\langle r \rangle^2 + \langle r \rangle^2 = \left\langle r^2 \right\rangle - \langle r \rangle^2$$

$$\langle r \rangle = \frac{1}{\pi a_0^3} \int_0^{2\pi} d\phi \int_0^\pi \sin\theta \, d\theta \int_0^\infty r^3 e^{-\frac{2r}{a_0}} dr$$

$$\langle r \rangle = \frac{4}{a_0^3} \int_0^\infty r^3 e^{-\frac{2r}{a_0}} dr$$

Using the standard integral $\int_0^\infty r^n e^{-\alpha r} = \dfrac{n!}{\alpha^{n+1}}$

$$\langle r \rangle = \frac{4}{a_0^3} \frac{6a_0^4}{16} = \frac{3}{2} a_0$$

$$\left\langle r^2 \right\rangle = \frac{1}{\pi a_0^3} \int_0^{2\pi} d\phi \int_0^\pi \sin\theta \, d\theta \int_0^\infty r^4 e^{-\frac{2r}{a_0}} dr$$

$$\left\langle r^2 \right\rangle = \frac{4}{a_0^3} \int_0^\infty r^4 e^{-\frac{2r}{a_0}} dr$$

Using the standard integral $\int_0^\infty r^n e^{-\alpha r} = \dfrac{n!}{\alpha^{n+1}}$

$$\left\langle r^2 \right\rangle = \frac{4}{a_0^3} \frac{4!}{(2/a_0)^5} = 3(a_0)^2$$

Therefore, $(r - \langle r \rangle)^2 = \left\langle r^2 \right\rangle - \langle r \rangle^2 = 3(a_0)^2 - \left(\dfrac{3}{2}a_0\right)^2 = \dfrac{3}{4}a_0^2$.

P20.16) The force acting between the electron and the proton in the H atom is given by

$F = -e^2/4\pi\varepsilon_0 r^2$. Calculate the expectation value $\langle F \rangle$ for the $1s$ and $2p_z$ states of the H atom in terms of

e, ε_0, and a_0.

$$\langle F\rangle_{1s}=-\frac{e^2}{4\pi\varepsilon_0}\int\psi^*(\tau)\frac{1}{r^2}\psi(\tau)d\tau$$

$$\langle F\rangle_{1s}=-\frac{e^2}{4\pi\varepsilon_0}\frac{1}{\pi a_0^3}\int_0^{2\pi}d\phi\int_0^{\pi}\sin\theta\,d\theta\int_0^{\infty}\left[e^{-r/a_0}\right]\left(\frac{1}{r^2}\right)\left[e^{-r/a_0}\right]r^2dr$$

$$\langle F\rangle_{1s}=-\frac{e^2}{4\pi\varepsilon_0}\frac{4}{a_0^3}\int_0^{\infty}e^{-2r/a_0}dr=-\frac{e^2}{4\pi\varepsilon_0}\frac{4}{a_0^3}\left[-\frac{a_0}{2}e^{-2r/a_0}\right]_0^{\infty}=-\frac{e^2}{2\pi\varepsilon_0 a_0^2}$$

$$\langle F\rangle_{2pz}=-\frac{e^2}{4\pi\varepsilon_0}\int\psi^*(\tau)\frac{1}{r^2}\psi(\tau)d\tau$$

$$\langle F\rangle_{2pz}=-\frac{e^2}{4\pi\varepsilon_0}\frac{1}{32\pi a_0^3}\int_0^{2\pi}d\phi\int_0^{\pi}\cos^2\theta\sin\theta\,d\theta\int_0^{\infty}\left(\frac{r}{a_0}\right)^2\left[e^{-r/a_0}\right]\left(\frac{1}{r^2}\right)r^2dr$$

$$\langle F\rangle_{2pz}=-\frac{e^2}{4\pi\varepsilon_0}\frac{1}{16a_0^5}\left[\frac{\cos^3\theta}{3}\right]_0^{\pi}\times\int_0^{\infty}r^2e^{-r/a_0}dr$$

$$\langle F\rangle_{2pz}=-\frac{e^2}{4\pi\varepsilon_0}\frac{1}{24a_0^3}\int_0^{\infty}r^2e^{-r/a_0}dr$$

Using the standard integral $\int_0^{\infty}r^ne^{-\alpha r}=\frac{n!}{\alpha^{n+1}}$

$$\langle F\rangle_{2pz}=-\frac{e^2}{4\pi\varepsilon_0}\frac{1}{24a_0^5}\times2a_0^3=-\frac{e^2}{48\pi\varepsilon_0 a_0^2}$$

P20.17) The d orbitals have the nomenclature $d_{z^2},d_{xy},d_{xz},d_{yz}$, and $d_{x^2-y^2}$. Show how the d orbital

$$\psi_{3d_{yz}}(r,\theta,\phi)=\frac{\sqrt{2}}{81\sqrt{\pi}}\left(\frac{1}{a_0}\right)^{3/2}\frac{r^2}{a_0^2}e^{-r/3a_0}\sin\theta\cos\theta\sin\phi$$

can be written in the form $yzF(r)$.

In spherical coordinates, $x=r\sin\theta\cos\phi$, $y=r\sin\theta\sin\phi$, and $z=r\cos\theta$. Therefore,

$$\psi_{3d_{yz}}(r,\theta,\phi)=\frac{\sqrt{2}}{81\sqrt{\pi}}\left(\frac{1}{a_0}\right)^{3/2}\frac{r^2}{a_0^2}e^{-r/3a_0}\sin\theta\cos\theta\sin\phi$$

$$=\frac{\sqrt{2}}{81\sqrt{\pi}}\left(\frac{1}{a_0}\right)^{3/2}\frac{1}{a_0^2}e^{-r/3a_0}(r\cos\theta)(r\sin\theta\sin\phi)=\frac{\sqrt{2}}{81\sqrt{\pi}}\left(\frac{1}{a_0}\right)^{3/2}\frac{1}{a_0^2}e^{-r/3a_0}(yz)$$

P20.18) Calculate the expectation value of the moment of inertia of the H atom in the 2s and $2p_z$ states in terms of μ and a_0.

$$\langle I \rangle = \langle \mu r^2 \rangle = \mu \frac{1}{32\pi a_0^3} \int_0^{2\pi} d\phi \int_0^{\pi} \sin\theta \, d\theta \int_0^{\infty} r^4 \left(2 - \frac{r}{a_0}\right)^2 e^{-r/a_0} dr$$

$$= \mu \frac{1}{8 a_0^3} \int_0^{\infty} \left(4r^4 - \frac{4r^5}{a_0} + \frac{r^6}{a_0^2}\right) e^{-r/a_0} dr = \mu \frac{1}{8 a_0^3} \left(4\int_0^{\infty} r^4 e^{-r/a_0} dr - \frac{4}{a}\int_0^{\infty} r^5 e^{-r/a_0} dr + \frac{1}{a^2}\int_0^{\infty} r^6 e^{-r/a_0} dr\right)$$

Using the standard integral $\int_0^{\infty} r^n e^{-\alpha r} = \frac{n!}{\alpha^{n+1}}$

$$\langle I \rangle = \mu \frac{1}{8 a_0^3}\left(4 \times 4! a_0^5 - \frac{4}{a_0} \times 5! \times a_0^6 + \frac{1}{a_0^2} \times 6! \times a_0^7\right) = 42\mu a_0^2$$

For the $2p_z$ state,

$$\langle I \rangle = \langle \mu r^2 \rangle = \mu \frac{1}{32\pi a_0^3} \int_0^{2\pi} d\phi \int_0^{\pi} \cos^2\theta \sin\theta \, d\theta \int_0^{\infty} r^4 \left(\frac{r}{a_0}\right)^2 e^{-r/a_0} dr$$

$$= \mu \frac{1}{16 a_0^5}\left[\frac{\cos^3\theta}{3}\right]_0^{\pi} \times \int_0^{\infty} r^6 e^{-r/a_0} \, dr = \mu \frac{1}{24 a_0^5} 6! a_0^7 = 30\mu a_0^2$$

P20.19) The energy levels for ions with a single electron such as He^+, Li^{2+}, and Be^{3+} are given by $E_n = -Z^2 e^2 / 8\pi\varepsilon_0 a_0 n^2, n = 1,2,3,4,\ldots$. Calculate the ionization energies of H, He^+, Li^{2+}, and Be^{3+} in their ground states in units of electron-volts (eV).

The ionization potential is the negative of the orbital energy.

$$I = \frac{Z^2 e^2}{8\pi \varepsilon_0 a_0 n^2} = \frac{\frac{Z^2}{n^2} \times \left(1.602\times10^{-19} \text{ C}\right)^2}{8\pi \times 8.854\times10^{-12} \text{ J}^{-1} \text{ C}^2 \text{ m}^{-1} \times 5.292\times10^{-11} \text{ m}} \times \frac{1 \text{ eV}}{1.602\times10^{-19} \text{ J}}$$

$$I = 13.60 \frac{Z^2}{n^2} \text{ eV}$$

$I_H = 13.60 \text{ eV}; \quad I_{He^+} = 54.42 \text{ eV}; \quad I_{Li^{2+}} = 122.4 \text{ eV}; \quad I_{Be^{3+}} = 217.7 \text{ eV}$

P20.22) Locate the radial and angular nodes in the H orbitals $\psi_{3p_x}(r,\theta,\phi)$ and $\psi_{3p_z}(r,\theta,\phi)$.

$$\psi_{3p_x}(r,\theta,\phi) = \frac{1}{81\sqrt{2\pi}}\left(\frac{1}{a_0}\right)^{3/2}\left(6\frac{r}{a_0}-\frac{r^2}{a_0^2}\right)e^{-r/3a_0}\sin\theta\cos\phi \text{ has radial nodes where } 6\frac{r}{a_0}-\frac{r^2}{a_0^2} \text{ has}$$

zeros. This is at $r = 0$, which does not count as a node, and $r = 6a_0$. The angular nodes correspond to

the angles at which the function is zero, or $\theta = 0°$ and $\phi = 90°$, corresponding to the y-z plane.

$$\psi_{3p_z}(r,\theta,\phi) = \frac{1}{81}\left(\frac{2}{\pi}\right)^{1/2}\left(\frac{1}{a_0}\right)^{3/2}\left(6\frac{r}{a_0}-\frac{r^2}{a_0^2}\right)e^{-r/3a_0}\cos\theta \text{ also has a radial node at } r = 6a_0. \text{ The angular}$$

node is for $\theta = 90°$, corresponding to the x-y plane.

P20.25) Show that the total energy eigenfunctions $\psi_{100}(r)$ and $\psi_{200}(r)$ are orthogonal.

$$\iiint \psi_{100}^*(\tau)\psi_{200}(\tau)d\tau = \frac{1}{\sqrt{32}\,\pi a_0^3}\int_0^{2\pi}d\phi\int_0^{\pi}\sin\theta\,d\theta\int_0^{\infty}r^2 e^{-r/a_0}\left(2-\frac{r}{a_0}\right)e^{-r/2a_0}$$

$$= \frac{1}{\sqrt{2}\,a_0^3}\times\left(\int_0^{\infty}2r^2 e^{-3r/2a_0}dr - \int_0^{\infty}\frac{r^3}{a_0}e^{-3r/2a_0}dr\right)$$

$$= \frac{1}{\sqrt{2}\,a_0^3}\times\left(\frac{4}{[3/2a_0]^3}-\frac{6}{a_0[3/2a_0]^4}\right) = \frac{1}{\sqrt{2}\,a_0^3}\times\left(\frac{32a_0^3}{27}-\frac{96a_0^3}{81}\right) = 0$$

Chapter 21: Many-Electron Atoms

P21.1) Is $\psi(1,2) = 1s(1)\alpha(1)1s(2)\beta(2) + 1s(2)\alpha(2)1s(1)\beta(1)$ an eigenfunction of the operator $\hat{S}_z$? If so, what is its eigenvalue M_S?

This problem uses a notation that is not explained in this chapter, but is explained in Chapter 22.

$$\hat{S}_z\left[1s(1)\alpha(1)1s(2)\beta(2) + 1s(2)\alpha(2)1s(1)\beta(1)\right]$$

$$= (\hat{s}_{z1} + \hat{s}_{z2})\left[1s(1)\alpha(1)1s(2)\beta(2) + 1s(2)\alpha(2)1s(1)\beta(1)\right]$$

$$= \frac{\hbar}{2}\left[1s(1)\alpha(1)1s(2)\beta(2) - 1s(2)\alpha(2)1s(1)\beta(1)\right]$$

$$\quad + \frac{\hbar}{2}\left[-1s(1)\alpha(1)1s(2)\beta(2) + 1s(2)\alpha(2)1s(1)\beta(1)\right]$$

$$= \frac{\hbar}{2}(0)$$

The function is an eigenfunction of $\hat{S}_z$ with the eigenvalue $M_S = 0$.

P21.4) In this problem you will prove that the ground-state energy for a system obtained using the variational method is greater than the true energy.

a. The approximate wave function Φ can be expanded in the true (but unknown) eigenfunctions ψ_n of the total energy operator in the form $\Phi = \sum_n c_n \psi_n$. Show that by substituting $\Phi = \sum_n c_n \psi_n$ in the equation

$$E = \frac{\int \Phi^* \hat{H} \Phi \, d\tau}{\int \Phi^* \Phi \, d\tau}$$

you obtain the result

$$E = \frac{\displaystyle\sum_n \sum_m \int (c_n^* \psi_n^*)\hat{H}(c_m \psi_m)\, d\tau}{\displaystyle\sum_n \sum_m \int (c_n^* \psi_n^*)(c_m \psi_m)\, d\tau}$$

b. Because the ψ_n are eigenfunctions of $\hat{H}$, they are orthogonal and $\hat{H}\psi_n = E_n\psi_n$. Show that this information allows us to simplify the expression for D from part (a) to

$$E = \frac{\sum_m E_m c_m^* c_m}{\sum_m c_m^* c_m}$$

c. Arrange the terms in the summation such that the first energy is the true ground-state energy E_0 and the energy increases with the summation index m. Why can you conclude that $E - E_0 \geq 0$?

a) $E = \dfrac{\int \Phi^* \hat{H} \Phi \, d\tau}{\int \Phi^* \Phi \, d\tau} = \dfrac{\int \left(\Phi = \sum_n c_n \psi_n\right)^* \hat{H} \left(\Phi = \sum_m c_m \psi_m\right) d\tau}{\int \left(\Phi = \sum_n c_n \psi_n\right)^* \left(\Phi = \sum_m c_m \psi_m\right) d\tau} = \dfrac{\sum_n \sum_m \int (c_n^* \psi_n^*) \hat{H} (c_m \psi_m) \, d\tau}{\sum_n \sum_m \int (c_n^* \psi_n^*)(c_m \psi_m) \, d\tau}$

b) Simplify the previous expression for E from part (a) to $E = \dfrac{\sum_m E_m c_m^* c_m}{\sum_m c_m^* c_m}$

$$E = \frac{\sum_n \sum_m \int (c_n^* \psi_n^*) \hat{H} (c_m \psi_m) \, d\tau}{\sum_n \sum_m \int (c_n^* \psi_n^*)(c_m \psi_m) \, d\tau} = \frac{\sum_n \sum_m E_m \int (c_n^* \psi_n^*)(c_m \psi_m) \, d\tau}{\sum_n \sum_m \int (c_n^* \psi_n^*)(c_m \psi_m) \, d\tau}$$

$$= \frac{\sum_m E_m c_m^* c_m \int \psi_m^* \psi_m \, d\tau}{\sum_m c_m^* c_m \int \psi_m^* \psi_m \, d\tau} = \frac{\sum_m E_m c_m^* c_m}{\sum_m c_m^* c_m}$$

c)

$$E - E_0 = \frac{\sum_m E_m c_m^* c_m}{\sum_m c_m^* c_m} - \frac{E_0 \sum_m c_m^* c_m}{\sum_m c_m^* c_m} = \frac{\sum_m (E_m - E_0) c_m^* c_m}{\sum_m c_m^* c_m} \geq 0.$$ Both $(E_m - E_0)$ and $c_m^* c_m$ are greater than

zero. Therefore, $E - E_0 \geq 0$.

P21.6) The operator for the square of the total spin of two electrons is $\hat{S}_{total}^2 = (\hat{S}_1 + \hat{S}_2)^2 = \hat{S}_1^2 + \hat{S}_2^2 + 2(\hat{S}_{1x}\hat{S}_{2x} + \hat{S}_{1y}\hat{S}_{2y} + \hat{S}_{1z}\hat{S}_{2z})$. Given that

$$\hat{S}_x\alpha = \frac{\hbar}{2}\beta, \quad \hat{S}_y\alpha = \frac{i\hbar}{2}\beta, \quad \hat{S}_z\alpha = \frac{\hbar}{2}\alpha,$$

$$\hat{S}_x\beta = \frac{\hbar}{2}\alpha, \quad \hat{S}_y\beta = -\frac{i\hbar}{2}\alpha, \quad \hat{S}_z\beta = -\frac{\hbar}{2}\beta,$$

show that $\alpha(1)\,\alpha(2)$ and $\beta(1)\,\beta(2)$ are eigenfunctions of the operator $\hat{S}_{total}^2$. What is the eigenvalue in each case?

This problem uses a notation that is not explained in this chapter, but is explained in Chapter 22.

$\hat{S}_{total}^2\alpha(1)\alpha(2)$

$= \hat{S}_1^2\alpha(1)\alpha(2) + \hat{S}_2^2\alpha(1)\alpha(2) + 2\left(\hat{S}_{1x}\hat{S}_{2x}\alpha(1)\alpha(2) + \hat{S}_{1y}\hat{S}_{2y}\alpha(1)\alpha(2) + \hat{S}_{1z}\hat{S}_{2z}\alpha(1)\alpha(2)\right)$

$= \alpha(2)\hat{S}_1^2\alpha(1) + \alpha(1)\hat{S}_2^2\alpha(2) + 2\left(\hat{S}_{1x}\alpha(1)\hat{S}_{2x}\alpha(2) + \hat{S}_{1y}\alpha(1)\hat{S}_{2y}\alpha(2) + \hat{S}_{1z}\alpha(1)\hat{S}_{2z}\alpha(2)\right)$

$= \frac{3\hbar^2}{4}\alpha(1)\alpha(2) + \frac{3\hbar^2}{4}\alpha(1)\alpha(2) + 2\left(\hat{S}_{1x}\alpha(1)\hat{S}_{2x}\alpha(2) + \hat{S}_{1y}\alpha(1)\hat{S}_{2y}\alpha(2) + \hat{S}_{1z}\alpha(1)\hat{S}_{2z}\alpha(2)\right)$

$= \frac{3\hbar^2}{4}\alpha(1)\alpha(2) + \frac{3\hbar^2}{4}\alpha(1)\alpha(2) + 2\times\frac{\hbar}{2}\left(\hat{S}_{1x}\alpha(1)\beta(2) + i\hat{S}_{1y}\alpha(1)\beta(2) + \hat{S}_{1z}\alpha(1)\alpha(2)\right)$

$= \frac{3\hbar^2}{4}\alpha(1)\alpha(2) + \frac{3\hbar^2}{4}\alpha(1)\alpha(2) + 2\times\left(\frac{\hbar}{2}\right)^2\left(\beta(1)\beta(2) + i^2\beta(1)\beta(2) + \alpha(1)\alpha(2)\right)$

$= \frac{3\hbar^2}{4}\alpha(1)\alpha(2) + \frac{3\hbar^2}{4}\alpha(1)\alpha(2) + \frac{2\hbar^2}{4}\alpha(1)\alpha(2) = 2\hbar^2\alpha(1)\alpha(2)$

The eigenvalue is $2\hbar^2$.

$\hat{S}_{total}^2\beta(1)\beta(2)$

$= \hat{S}_1^2\beta(1)\beta(2) + \hat{S}_2^2\beta(1)\beta(2) + 2\left(\hat{S}_{1x}\hat{S}_{2x}\beta(1)\beta(2) + \hat{S}_{1y}\hat{S}_{2y}\beta(1)\beta(2) + \hat{S}_{1z}\hat{S}_{2z}\beta(1)\beta(2)\right)$

$= \alpha(2)\hat{S}_1^2\beta(1) + \beta(1)\hat{S}_2^2\beta(2) + 2\left(\hat{S}_{1x}\beta(1)\hat{S}_{2x}\beta(2) + \hat{S}_{1y}\beta(1)\hat{S}_{2y}\beta(2) + \hat{S}_{1z}\beta(1)\hat{S}_{2z}\beta(2)\right)$

$= \frac{3\hbar^2}{4}\beta(1)\beta(2) + \frac{3\hbar^2}{4}\beta(1)\beta(2) + 2\left(\hat{S}_{1x}\beta(1)\hat{S}_{2x}\beta(2) + \hat{S}_{1y}\beta(1)\hat{S}_{2y}\beta(2) + \hat{S}_{1z}\beta(1)\hat{S}_{2z}\beta(2)\right)$

$= \frac{3\hbar^2}{4}\alpha(1)\alpha(2) + \frac{3\hbar^2}{4}\beta(1)\beta(2) + 2\times\frac{\hbar}{2}\left(\hat{S}_{1x}\beta(1)\alpha(2) - i\hat{S}_{1y}\beta(1)\alpha(2) - \hat{S}_{1z}\beta(1)\beta(2)\right)$

$= \frac{3\hbar^2}{4}\alpha(1)\alpha(2) + \frac{3\hbar^2}{4}\beta(1)\beta(2) + 2\times\left(\frac{\hbar}{2}\right)^2\left(\alpha(1)\alpha(2) + i^2\alpha(1)\alpha(2) + \beta(1)\beta(2)\right)$

$= \frac{3\hbar^2}{4}\alpha(1)\alpha(2) + \frac{3\hbar^2}{4}\beta(1)\beta(2) + \frac{2\hbar^2}{4}\beta(1)\beta(2) = 2\hbar^2\beta(1)\beta(2)$

The eigenvalue is $2\hbar^2$.

P21.11) Write the Slater determinant for the ground-state configuration of Be.

$$\psi_{Be} = \frac{1}{\sqrt{4!}} \begin{vmatrix} 1s(1)\alpha(1) & 1s(1)\beta(1) & 2s(1)\alpha(1) & 2s(1)\beta(1) \\ 1s(2)\alpha(2) & 1s(2)\beta(2) & 2s(2)\alpha(2) & 2s(2)\beta(2) \\ 1s(3)\alpha(3) & 1s(3)\beta(3) & 2s(3)\alpha(3) & 2s(3)\beta(3) \\ 1s(4)\alpha(4) & 1s(4)\beta(4) & 2s(4)\alpha(4) & 2s(4)\beta(4) \end{vmatrix}$$

Computational problems

Before solving the computational problems, it is recommended that students work through Tutorials 1–3 under the Help menus in Spartan Student Edition to gain familiarity with the program.

Computational Problem 21.1: Calculate the total energy and $1s$ orbital energy for Ne using the Hartree–Fock method and the (a) 3-21G, (b) 6-31G*, and (c) 6-311+G** basis sets. Note the number of basis functions used in the calculations. Calculate the relative error of your result compared with the Hartree–Fock limit of -128.854705 hartree for each basis set. Rank the basis sets in terms of their approach to the Hartree–Fock limit for the total energy.

Step 1: Create a new file, and enter neon using the inorganic palette which is accessed by clicking the "Inorganic" tab on the model kit with a single valence on the atom. Delete the valence. Enter the various basis sets within the same file using the "New Molecule" command in the File menu.

Step 2: Go to Setup → Calculations. Set the calculation type to "Energy" and the method to Hartree–Fock with the appropriate basis set. "Total Charge" should be neutral. Enter the appropriate number of unpaired electrons. "Global Calculations" should be checked. Click "Submit." Your calculations will start immediately.

Step 3: Once your calculations complete, go to Display → Output. Note the values of the total energy for each calculation.

Step 4: Make a table showing the basis set, number of basis functions, total energy, and relative error with respect to the Hartree–Fock limit. Rank the basis sets in terms of their approach to the Hartree–Fock limit for the total energy.

Computational Problem 21.2: Calculate the total energy and $4s$ orbital energy for K using the Hartree–Fock method and the (a) 3-21G and (b) 6-31G* basis sets. Note the number of basis functions used in the calculations. Calculate the percentage deviation from the Hartree–Fock limits which are -16245.7 eV for the total energy and -3.996 eV for the $4s$ orbital energy. Rank the basis sets in terms of their approach to the Hartree–Fock limit for the total energy. What percentage error in the Hartree–Fock limit to the total energy corresponds to a typical reaction enthalpy change of 100 kJ mol^{-1}?

Step 1: Create a new file, and enter K using the inorganic palette which is accessed by clicking the "Inorganic" tab on the model kit with a single valence on the atom. Delete the valence. Enter the various basis sets within the same file using the "New Molecule" command in the File menu.

Step 2: Go to Setup → Calculations. Set the calculation type to "Energy" and the method to Hartree–Fock with the appropriate basis set. "Total Charge" should be neutral. Enter the appropriate number of unpaired electrons. "Global Calculations" should be checked. Click "Submit." Your calculations will start immediately.

Step 3: Once your calculations complete, go to Display → Output. Note the values of the total energy and $4s$ orbital energy for each calculation

Step 4: Make a table showing the basis set, total energy, relative error of the total energy with respect to the Hartree–Fock limit, orbital energy, relative error of the orbital energy with respect to the Hartree–Fock limit. Rank the basis sets in terms of their approach to the Hartree–Fock limit for the total energy.

Computational Problem 21.3: Calculate the ionization energy for (a) Li, (b) F, (c) S, (d) Cl, and (e) Ne using the Hartree–Fock method and the 6-311+G** basis set. Carry out the calculation in two different ways: (a) Use Koopmans' theorem and (b) compare the total energy of the neutral and singly ionized atom. Compare your answers with literature values.

Step 1: Create a new file, and enter Li using the inorganic palette which is accessed by clicking the "Inorganic" tab on the model kit with a single valence on the atom. Delete the valence. Hartree–Fock method and the 6-311+G** basis set. Enter the appropriate number of unpaired electrons. Uncheck the "global calculations" box. Using the "New Molecule" command in the File menu, enter the other atoms and the appropriate cation for each atom. Check the orbitals and energies box. The global calculations box should not be checked. Submit the calculation.

Step 2: Examine the output file and record the energy of the highest occupied AO, the total energy of the neutral atom, the total energy of the cation, the difference in energy of these two previous values, and literature values for the ionization energy as separate columns in a table. Discuss your results by comparing the literature and calculated results.

Computational Problem 21.4: Calculate the electron affinity for (a) Li, (b) F, (c) S, and (d) Cl using the Hartree–Fock method and the 6-311+G** basis set by comparing the total energy of the neutral and singly ionized atom. Compare your answers with literature values.

Step 1: Create a new file, and enter Li using the inorganic palette which is accessed by clicking the "Inorganic" tab on the model kit with a single valence on the atom. Delete the valence. Hartree–Fock method and the 6-311+G** basis set. Enter the appropriate number of unpaired electrons. The global calculations box should not be checked. Using the "New Molecule" command in the File menu, enter the other atoms and the appropriate anion for each atom. Check the orbitals and energies box. Submit the calculation.

Step 2: Examine the output file and record the energy of the highest occupied AO, the total energy of the neutral atom, the total energy of the anion, the difference in energy of these two previous values, and literature values for the ionization energy as separate columns in a table. Discuss your results by comparing the literature and calculated results.

Computational Problem 21.5: Using your results from C21.3 and C21.4, calculate the Mulliken electronegativity for (a) Li, (b) F, (c) S, (d) Cl. Compare your results with literature values.

Computational Problem 21.6: To assess the accuracy of the Hartree–Fock method for calculating energy changes in reactions, calculate the total energy change for the reaction $CH_3OH \rightarrow CH_3 + OH$ by calculating the difference in the total energy of reactants and products using the Hartree–Fock method and the 6-31G* basis set. Compare your result with a calculation using the B3LYP method and the same basis set and with the experimental value of 410. kJ mol^{-1}. As discussed in Chapter 15, the B3LYP method takes electron correlation into account. What percentage error in the Hartree–Fock total energy for CH_3OH would account for the difference between the calculated and experimental value of ΔU?

Chapter 22: Quantum States for Many-Electron Atoms and Atomic Spectroscopy

P22.1) The principal line in the emission spectrum of potassium is violet. On close examination, the line is seen to be a doublet with wavelengths of 393.366 and 396.847 nm. Explain the source of this doublet.

The lower lying state is the level and state $^2S_{1/2}$, and the upper level is ^2P, which contains the states $^2P_{1/2}$ and $^2P_{3/2}$. Both transitions are allowed, and the 589.0 nm wavelength corresponds to the transition between the ground state and $^2P_{3/2}$ because states with higher J lie lower in energy. The 589.6 nm wavelength corresponds to the transition between the ground state and $^2P_{1/2}$.

P22.5) What J values are possible for a ^6H term? Calculate the number of states associated with each level and show that the total number of states is the same as that calculated from $(2S + 1)(2L + 1)$.

$S = 5/2$, $L = 5$.

J lies between $|L + S|$ and $|L - S|$ and can have the values 15/2, 13/2, 11/2, 9/2, 7/2, and 5/2. The number of states is $2J + 1$ or 16, 14, 12, 10, 8, and 6, respectively.

This gives a total number of states of $16 + 14 + 12 + 10 + 8 + 6 = 66$.

$(2S + 1)(2L + 1) = 6 \times 11 = 66$ also.

P22.12) Calculate the wavelengths of the first three lines of the Lyman, Balmer, and Paschen series, and the series limit (the shortest wavelength) for each series.

Lyman Series: $E_n = R_H \left(\dfrac{1}{1^2} - \dfrac{1}{n^2} \right)$

$n = 2 \quad E_2 = R_H \left(1 - \dfrac{1}{4} \right) = \dfrac{3}{4} R_H = 82257.8 \text{ cm}^{-1} \quad \lambda = 121.569 \text{ nm}$

$n = 3 \quad E_2 = R_H \left(1 - \dfrac{1}{9} \right) = \dfrac{8}{9} R_H = 97490.7 \text{ cm}^{-1} \quad \lambda = 102.574 \text{ nm}$

$n = 4 \quad E_2 = R_H \left(1 - \dfrac{1}{16} \right) = \dfrac{15}{16} R_H = 102822 \text{ cm}^{-1} \quad \lambda = 97.2553 \text{ nm}$

$n = \infty \quad E_2 = R_H \left(1 - \dfrac{1}{\infty} \right) = R_H = 109678 \text{ cm}^{-1} \quad \lambda = 91.1768 \text{ nm}$

Balmer Series: $E_n = R_H \left(\dfrac{1}{2^2} - \dfrac{1}{n^2} \right)$

$n = 3 \quad E_2 = R_H \left(\dfrac{1}{4} - \dfrac{1}{9} \right) = \dfrac{5}{36} R_H = 15232.9 \text{ cm}^{-1} \quad \lambda = 656.473 \text{ nm}$

$n = 3 \quad E_2 = R_H \left(\dfrac{1}{4} - \dfrac{1}{16} \right) = \dfrac{3}{16} R_H = 20564.4 \text{ cm}^{-1} \quad \lambda = 486.276 \text{ nm}$

$n = 4 \quad E_2 = R_H \left(\dfrac{1}{4} - \dfrac{1}{25} \right) = \dfrac{21}{100} R_H = 23032.2 \text{ cm}^{-1} \quad \lambda = 434.175 \text{ nm}$

$n = \infty \quad E_2 = R_H \left(\dfrac{1}{4} - \dfrac{1}{\infty} \right) = \dfrac{1}{4} R_H = 27419.3 \text{ cm}^{-1} \quad \lambda = 364.707 \text{ nm}$

Paschen Series: $E_n = R_H \left(\dfrac{1}{3^2} - \dfrac{1}{n^2} \right)$

$n = 4 \quad E_2 = R_H \left(\dfrac{1}{9} - \dfrac{1}{16} \right) = \dfrac{7}{144} R_H = 5331.52 \text{ cm}^{-1} \quad \lambda = 1875.64 \text{ nm}$

$n = 5 \quad E_2 = R_H \left(\dfrac{1}{9} - \dfrac{1}{25} \right) = \dfrac{16}{225} R_H = 7799.25 \text{ cm}^{-1} \quad \lambda = 1282.17 \text{ nm}$

$n = 6 \quad E_2 = R_H \left(\dfrac{1}{9} - \dfrac{1}{36} \right) = \dfrac{1}{12} R_H = 9139.75 \text{ cm}^{-1} \quad \lambda = 1094.12 \text{ nm}$

$n = \infty \quad E_2 = R_H \left(\dfrac{1}{9} - \dfrac{1}{\infty} \right) = \dfrac{1}{9} R_H = 12186 \text{ cm}^{-1} \quad \lambda = 820.591 \text{ nm}$

P22.14) The inelastic mean free path of electrons in a solid, λ, governs the surface sensitivity of techniques such as AES and XPS. The electrons generated below the surface must make their way to the surface without losing energy in order to give elemental and chemical shift information. An empirical expression for elements that give λ as a function of the kinetic energy of the electron generated in AES or XPS is $\lambda = 538 E^{-2} + 0.41 (lE)^{0.5}$. The units of λ are monolayers, E is the kinetic energy of the electron, and l is the monolayer thickness in nanometers. On the basis of this equation, what kinetic energy maximizes the surface sensitivity for a monolayer thickness of 0.3 nm? An equation solver would be helpful in obtaining the answer.

$$\frac{d}{dE} \left(538 E^{-2} + 0.41 (lE)^{0.5} \right) = -\frac{1076}{E^3} + \frac{0.1123}{\sqrt{E}}$$

Setting the derivative equal to zero and solving for E gives two complex roots and $E = 39$ eV.

P22.17) What are the levels that arise from a ^4F term? How many states are there in each level?

$S = 3/2$ and $L = 3$. M_L can range from $-L$ to $+L$ and can have the values $-3, -2, -1, 0, 1, 2$, and 3 in this case. M_S can range from $-S$ to $+S$ and can have the values $-3/2, -1/2, 1/2$, and $3/2$ in this case. J lies between $|L + S|$ and $|L - S|$ and can have the values $9/2, 7/2, 5/2$, and $3/2$ for this case. Because the number of states is $2J+1$, these levels have 10, 8, 6, and 4 states respectively.

P22.23) Use the transition frequencies shown in Example Problem 22.7 to calculate the energy (in joules and electron-volts) of the six levels relative to the $3s\ ^2S_{1/2}$ level. State your answers with the correct number of significant figures.

$$E\left(3p\,^2P_{1/2}\right) = \frac{hc}{\lambda} = \frac{6.626\times10^{-34}\ \text{J s}\times2.998\times10^8\ \text{m s}^{-1}}{589.6\times10^{-9}\ \text{m}} = 3.369\times10^{-19}\ \text{J} = 2.103\ \text{eV}$$

$$E\left(3p\,^2P_{3/2}\right) = \frac{hc}{\lambda} = \frac{6.626\times10^{-34}\ \text{J s}\times2.998\times10^8\ \text{m s}^{-1}}{589.0\times10^{-9}\ \text{m}} = 3.373\times10^{-19}\ \text{J} = 2.105\ \text{eV}$$

$$E\left(4s\,^2S_{1/2}\right) = \frac{hc}{\lambda} = \frac{6.626\times10^{-34}\ \text{J s}\times2.998\times10^8\ \text{m s}^{-1}}{589.6\times10^{-9}\ \text{m}} + \frac{6.626\times10^{-34}\ \text{J s}\times2.998\times10^8\ \text{m s}^{-1}}{1183.3\times10^{-9}\ \text{m}}$$
$$= 5.048\times10^{-19}\text{J} = 3.150\ \text{eV}$$

$$E\left(5s\,^2S_{1/2}\right) = \frac{hc}{\lambda} = \frac{6.626\times10^{-34}\ \text{J s}\times2.998\times10^8\ \text{m s}^{-1}}{589.0\times10^{-9}\ \text{m}} + \frac{6.626\times10^{-34}\ \text{J s}\times2.998\times10^8\text{m s}^{-1}}{616.0\times10^{-9}\ \text{m}}$$
$$= 6.597\times10^{-19}\text{J} = 4.118\ \text{eV}$$

$$E\left(3d\,^2D_{3/2}\right) = \frac{hc}{\lambda} = \frac{6.626\times10^{-34}\ \text{J s}\times2.998\times10^8\ \text{m s}^{-1}}{589.6\times10^{-9}\text{m}} + \frac{6.626\times10^{-34}\ \text{J s}\times2.998\times10^8\ \text{m s}^{-1}}{818.3\times10^{-9}\text{m}}$$
$$= 5.797\times10^{-19}\text{J} = 3.618\ \text{eV}$$

$$E\left(4d\,^2D_{3/2}\right) = \frac{hc}{\lambda} = \frac{6.626\times10^{-34}\ \text{J s}\times2.998\times10^8\ \text{m s}^{-1}}{589.0\times10^{-9}\ \text{m}} + \frac{6.626\times10^{-34}\ \text{J s}\times2.998\times10^8\ \text{m s}^{-1}}{568.2\times10^{-9}\ \text{m}}$$
$$= 6.869\times10^{-19}\text{J} = 4.287\ \text{eV}$$

P22.25) The spectrum of the hydrogen atom reflects the splitting of the $1s^2S$ and $2p^2P$ terms into levels. The energy difference between the levels in each term is much smaller than the difference in energy between the terms. Given this information, how many spectral lines are observed in the $1s^2S \rightarrow 2p^2P$ transition? Are the frequencies of these transitions very similar or quite different?

The ^2S term has a single level, $^2S_{1/2}$. The ^2P term splits into two levels, $^2P_{1/2}$ and $^2P_{3/2}$. Therefore, there will be two closely spaced lines in the spectrum corresponding to the transitions $^2S_{1/2}\rightarrow^2P_{1/2}$ and

$^2S_{1/2} \rightarrow {}^2P_{3/2}$. The energy spacing between the lines will be much smaller than the energy of the transition.

P22.27) What atomic terms are possible for the following electron configurations? Which of the possible terms has the lowest energy?

a) ns^1np^1 b) ns^1nd^1 c) ns^2np^1 d) ns^1np^2

a) ns^1np^1 L can only have the value 1, and S can have the values 0 and 1. The possible terms are ^1P and ^3P. Hund's Rules predict that the ^3P term will have the lower energy.

b) ns^1nd^1 L can only have the value 2, and S can have the values 0 and 1. The possible terms are ^1D and ^3D. Hund's Rules predict that the ^3D term will have the lower energy.

c) ns^2np^1 L can only have the value 1, and S can only have the value 1/2. The only possible term is ^2P.

d) ns^1np^2 A table such as the table in the text for the p^2 configuration will have three columns, one for each of the electrons, for M_L and M_S. Each of the fifteen states for the p^2 configuration can be combined with $m_s = \pm\dfrac{1}{2}$ for the ns electron. This gives a total of 30 states. Working through the table gives ^2D, ^4P, ^2P, and ^2S terms. Hund's Rules predict that the ^4P term will have the lowest energy.

P22.28) Two angular momenta with quantum numbers $j_1 = 3/2$ and $j_2 = 5/2$ are added. What are the possible values of J for the resultant angular momentum states?

$J = |J_1 + J_2|, \ |J_1 + J_2 - 1|, \ |J_1 + J_2 - 2|, ..., \ |J_1 - J_2|$ giving possible J values of 4, 3, 2, and 1.

P22.29) Derive the ground-state term symbols for the following configurations:

a. d^2 b. f^9 c. f^{12}

The method illustrated in Example Problem 22.4 is used for all parts.

a)

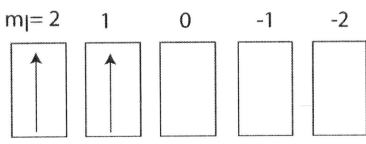

M_{Lmax} = 3 and M_{Smax} = 1. Therefore, the ground state term is ^{3}F.

b)

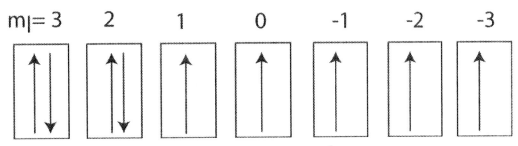

M_{Lmax} = 5 and M_{Smax} = 2.5. Therefore, the ground state term is ^{6}H.

c)

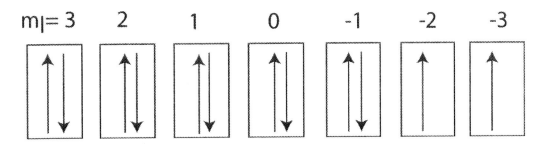

M_{Lmax} = 5 and M_{Smax} = 1. Therefore, the ground state term is ^{3}H.

P22.30) The first ionization potential of ground-state He is 24.6 eV. The wavelength of light associated with the $1s2p$ ^{1}P term is 58.44 nm. What is the ionization energy of the He atom in this excited state?

The photon energy is $E = \dfrac{hc}{\lambda} = \dfrac{6.626 \times 10^{-34} \text{ J s} \times 2.998 \times 10^8 \text{ m s}^{-1}}{58.44 \times 10^{-9} \text{ m}} \times \dfrac{1 \text{ eV}}{1.602 \times 10^{-19} \text{ J}} = 21.2 \text{ eV}$

Therefore, the ionization energy of the He atom in this state is 24.6 eV – 21.2 eV = 3.4eV.

P22.31) In the Na absorption spectrum, the following transitions are observed:

$4p\,^2P \rightarrow 3s\,^2S\ \lambda = 330.26$ nm

$3p\,^2P \rightarrow 3s\,^2S\ \lambda = 589.593$ nm, 588.996 nm

$5s\,^2S \rightarrow 3p\,^2P\ \lambda = 616.073$ nm, 615.421 nm

Calculate the energies of the $4p\,^2P$ and $5s\,^2S$ states with respect to the $3s\,^2S$ ground state.

$$E\left(4p\,^2P\right) = \frac{hc}{\lambda} = \frac{6.626\times10^{-34}\ \text{J s}\times2.998\times10^{8}\ \text{m s}^{-1}}{330.26\times10^{-9}\ \text{m}} = 6.015\times10^{-19}\ \text{J} = 3.754\ \text{eV}$$

By looking at the Grotrian diagram of Example Problem 22.2, it is seen that the $5s\,^2S$ state is accessed by absorption of the photons of wavelength 588.996 nm and 616.073 nm.

$$E\left(5s\,^2S\right) = \frac{hc}{\lambda} = \frac{6.626\times10^{-34}\ \text{J s}\times2.998\times10^{8}\ \text{m s}^{-1}}{588.996\times10^{-9}\ \text{m}} + \frac{6.626\times10^{-34}\ \text{J s}\times2.998\times10^{8}\ \text{m s}^{-1}}{616.073\times10^{-9}\ \text{m}}$$

$$= 6.597\times10^{-19}\ \text{J} = 4.117\ \text{eV}$$

P22.33) List the quantum numbers L and S that are consistent with the following terms:

a. ^4S b. ^4G c. ^3P d. ^2D

a) ^4S: $L = 0$, $2S + 1 = 4$, $S = 3/2$

b) ^4G: $L = 4$, $2S + 1 = 4$, $S = 3/2$

c) ^3P: $L = 1$, $2S + 1 = 3$, $S = 1$

d) ^2D: $L = 2$, $2S + 1 = 2$, $S = \frac{1}{2}$

P22.36) A general way to calculate the number of states that arise from a given configuration is as follows. Calculate the combinations of m_l and m_s for the first electron, and call that number n. The number of combinations used is the number of electrons, which we call m. The number of unused combinations is $n–m$. According to probability theory, the number of distinct permutations that arise from distributing the m electrons among the n combinations is $n!/[m!(n–m)!]$. For example, the number of states arising from a p^2 configuration is $6!/[2!4!]=15$, which is the result obtained in Section 22.2. Using this formula, calculate the number of possible ways to place five electrons in a d subshell. What is the ground-state term for the d^5 configuration and how many states does the term include?

The first electrons can have any combination of 5 m_l and 2 m_s values so that $n = 10$ and $m = 5$. Using

the formula, the calculated number of states is $\dfrac{10!}{5!(10-5)!} = 252$. The number of states in a term is

$(2L+1)(2S+1)$. The terms from Table 22.1 are listed here, with the number of states in each term in

square brackets following the term designation: $^6S[6 \times 1]$, $^4G[4 \times 9]$, $^4F[4 \times 7]$, $^4D[4 \times 5]$, $^4P[4 \times 3]$,

$^2I[2 \times 13]$, $^2H[2 \times 11]$, $^2G(2)[2 \times 2 \times 9]$, $^2F(2)[2 \times 2 \times 7]$, $^2D(3)[3 \times 2 \times 5]$, $^2P[2 \times 3]$, and $^2S[2 \times 1]$. The

number of states in these terms is $6 + 36 + 28 + 20 + 12 + 26 + 22 + 36 + 28 + 30 + 6 + 2 = 252$,

showing consistency.

The ground state is that with the maximum multiplicity which according to Table 21.5 is the ^6S term.

The number of states in the term $= (2L + 1)(2S + 1) = 6$.

Chapter 23: The Chemical Bond in Diatomic Molecules

P23.2) The overlap integral for ψ_g and ψ_u as defined in Section 23.3 is given by

$$S_{ab} = e^{-\zeta R/a_0}\left(1 + \zeta\frac{R}{a_0} + \frac{1}{3}\zeta^2\frac{R^2}{a_0^2}\right)$$

Plot S_{ab} as a function of R/a_0 for $\zeta = 0.8$, 1.0, and 1.2. Estimate the value of R/a_0 for which $S_{ab}=0.4$ for each of these values of ζ.

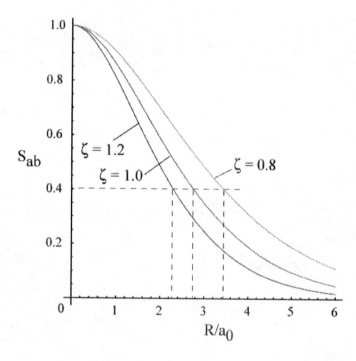

S_{ab} has the value of 0.4 at $R/a_0 = 3.42$, 2.68, and 2.25 for $\zeta = 0.8$, 1.0, and 1.2, respectively.

P23.3) Sketch out a molecular orbital energy diagram for CO and place the electrons in the levels appropriate for the ground state. The AO ionization energies are O2s: 32.3 eV; O2p: 15.8 eV; C2s: 19.4 eV; and C2p: 10.9 eV. The MO energies follow the sequence (from lowest to highest) $1\sigma, 2\sigma, 3\sigma, 4\sigma, 1\pi, 5\sigma, 2\pi, 6\sigma$. Assume that the 1σ and 2σ MOs originate from the $1s$ AOs and that the

3σ and 5σ MOs originate from the $2s$ AOs on C and O. Connect each MO level with the level of the major contributing AO on each atom.

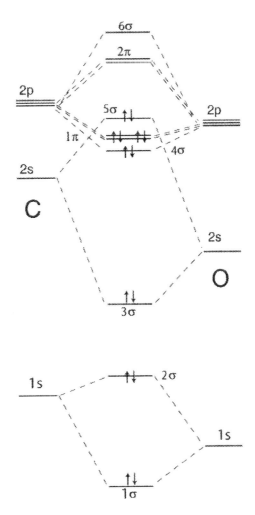

P23.9) Calculate the value for the coefficients of the AOs in Example Problem 23.3 for $S_{12} = 0.6$. How are they different from the values calculated in that problem for $S_{12} = 0.3$? Can you offer an explanation for the changes?

We first obtain the result $H_{12} = -1.75 \, S_{12} \sqrt{H_{11} H_{22}} = -16.7$ eV

Solving for the energies of ε_1 and ε_2 (bonding and antibonding MOs, respectively) gives the values $\varepsilon_1 = -20.9$ eV, and $\varepsilon_2 = +1.94$ eV.

We calculate $\dfrac{c_{12}}{c_{22}}$ by substituting the values for ε_1 and ε_2 in the first of the

Equations (24.9). Both equations give the same result.

$$c_{12}(H_{11}-\varepsilon)+c_{22}(H_{12}-\varepsilon S_{12})=0.$$

For $\varepsilon_2 = 1.94$ eV, $c_{12}(-13.6-1.94)+c_{22}(-16.7-0.6\times1.94)=0$

$$\frac{c_{12}}{c_{22}}=-1.15$$

Using this result in the normalization equation $c_{12}^2+c_{22}^2+2c_{12}c_{22}S_{12}=1$.

$$(-1.15c_{22})^2+c_{22}^2+2(-1.15c_{22})\times0.6=1$$

$$1.3225c_{22}^2+c_{22}^2-1.38c_{22}=0;\ c_{22}=-1.03$$

$$c_{12}=1.18,\ \text{and } \sigma_2=1.18\phi_{H1s}-1.03\phi_{F2p_z}$$

$$c_{12}(H_{11}-\varepsilon)+c_{22}(H_{12}-\varepsilon S_{12})=0.$$

For $\varepsilon_2 = -20.9$ eV, $c_{11}(-13.6+20.9)+c_{21}(-16.7+0.6\times20.9)=0$

$$\frac{c_{11}}{c_{21}}=0.57$$

Using this result in the normalization equation $c_{11}^2+c_{21}^2+2c_{11}c_{21}S_{12}=1$.

$$(0.57c_{21})^2+c_{21}^2+2(0.57c_{21}^2)\times0.6=1$$

$$0.3249c_{21}^2+c_{21}^2+0.684c_{21}^2=0;\ c_{21}=0.71$$

$$c_{11}=0.40,\ \text{and } \sigma_2=0.40\phi_{H1s}+0.71\phi_{F2p_z}$$

The increase in the overlap results in $c_{11}=0.40$ and $c_{21}=0.71$ for $\varepsilon_1=-20.9$ eV and in

$c_{22}=-1.03$ and $c_{12}=1.18$ for $\varepsilon_2=+1.94$ eV. As for $S_{12}=0.3$, the coefficient on the lower lying AO is

greater for the bonding orbital and less for the antibonding orbital. Also as before, the signs of the

coefficients are the same for the bonding orbital and opposite for the antibonding orbital. However, the

magnitude of the coefficients is more nearly equal, due to the greater interaction that arises from a

greater overlap, because H_{12} increases linearly with S_{12}. In the bonding MO, the electron is shared more

equally by the two atoms for the greater overlap.

P23.10) Arrange the following in terms of decreasing bond energy and bond length:

$O_2^+,\ O_2,\ O_2^-,$ and O_2^{2-}.

$$O_2^{2-}: \left(1\sigma_g\right)^2 \left(1\sigma_u^*\right)^2 \left(2\sigma_g\right)^2 \left(2\sigma_u^*\right)^2 \left(3\sigma_g\right)^2 \left(1\pi_u\right)^2 \left(1\pi_u\right)^2 \left(1\pi_g^*\right)^2 \left(1\pi_g^*\right)^2$$

$$\text{Bond Order} = \frac{10-8}{2} = 1$$

$$O_2^{-}: \left(1\sigma_g\right)^2 \left(1\sigma_u^*\right)^2 \left(2\sigma_g\right)^2 \left(2\sigma_u^*\right)^2 \left(3\sigma_g\right)^2 \left(1\pi_u\right)^2 \left(1\pi_u\right)^2 \left(1\pi_g^*\right)^2 \left(1\pi_g^*\right)^1$$

$$\text{Bond Order} = \frac{10-7}{2} = 1.5$$

$$O_2^{-}: \left(1\sigma_g\right)^2 \left(1\sigma_u^*\right)^2 \left(2\sigma_g\right)^2 \left(2\sigma_u^*\right)^2 \left(3\sigma_g\right)^2 \left(1\pi_u\right)^2 \left(1\pi_u\right)^2 \left(1\pi_g^*\right)^1 \left(1\pi_g^*\right)^1$$

$$\text{Bond Order} = \frac{10-6}{2} = 2$$

$$O_2^{+}: \left(1\sigma_g\right)^2 \left(1\sigma_u^*\right)^2 \left(2\sigma_g\right)^2 \left(2\sigma_u^*\right)^2 \left(3\sigma_g\right)^2 \left(1\pi_u\right)^2 \left(1\pi_u\right)^2 \left(1\pi_g^*\right)^1$$

$$\text{Bond Order} = \frac{10-5}{2} = 2.5$$

Bond energy: $O_2^+ > O_2 > O_2^- > O_2^{2-}$

Bond length: $O_2^{-2} > O_2^{-} > O_2 > O_2^{+}$

P23.13) What is the electron configuration corresponding to O_2, O_2^-, and O_2^+? What do you expect the relative order of bond strength to be for these species? Which, if any, have unpaired electrons?

$$O_2: \left(1\sigma_g\right)^2 \left(1\sigma_u^*\right)^2 \left(2\sigma_g\right)^2 \left(2\sigma_u^*\right)^2 \left(3\sigma_g\right)^2 \left(1\pi_u\right)^2 \left(1\pi_u\right)^2 \left(1\pi_g^*\right) \left(1\pi_g^*\right)$$

$$\text{Bond Order} = \frac{10-6}{2} = 2$$

$$O_2^-: \left(1\sigma_g\right)^2 \left(1\sigma_u^*\right)^2 \left(2\sigma_g\right)^2 \left(2\sigma_u^*\right)^2 \left(3\sigma_g\right)^2 \left(1\pi_u\right)^2 \left(1\pi_u\right)^2 \left(1\pi_g^*\right)^2 \left(1\pi_g^*\right)^1$$

$$\text{Bond Order} = \frac{10-7}{2} = 1.5$$

$$O_2^+: \left(1\sigma_g\right)^2 \left(1\sigma_u^*\right)^2 \left(2\sigma_g\right)^2 \left(2\sigma_u^*\right)^2 \left(3\sigma_g\right)^2 \left(1\pi_u\right)^2 \left(1\pi_u\right)^2 \left(1\pi_g^*\right)^1$$

$$\text{Bond Order} = \frac{10-5}{2} = 2.5$$

Because the bond strengths increase with increasing bond order, the relative order of bond strength is

$O_2^+ > O_2 > O_2^-$. O_2 has two unpaired electrons, and the positively and negatively charged species each

have one.

P23.15) Evaluate the energy for the two MOs generated by combining two H1s AOs. Use Equation

(23.23) and carry out the calculation for $S_{12} = 0.1$, 0.2, and 0.6 to mimic the effect of decreasing the

atomic separation in the molecule. Use the parameters $H_{11} = H_{22} = -13.6$ eV and

$H_{12} = -1.75 S_{12} \sqrt{H_{11} H_{22}}$. Explain the trend that you observe in the results.

$$E_1 = \frac{H_{11} + H_{12}}{1 + S_{12}} \text{ and } E_2 = \frac{H_{11} - H_{12}}{1 - S_{12}}$$

S_{12}	H_{12}(eV)	ε_1(eV)	ε_2(eV)
0.1	−2.38	−14.5	−12.5
0.2	−4.76	−15.3	−11.1
0.6	−14.3	−17.4	+1.7

The increase in the overlap mimics a decrease in the bond length. As S_{12} increases, the orbitals overlap

more and their interaction becomes greater. The result is that ε_1 becomes more strongly binding and ε_2

becomes more strongly antibonding.

P23.16) Show that calculating E_u in the manner described by Equation (23.7) gives the result

$E_u = (H_{aa} - H_{ab})/(1 - S_{ab})$.

$$E_u = \frac{\int \psi_u^* \hat{H} \psi_u \, d\tau}{\int \psi_u^* \psi_u \, d\tau}$$

$$= \frac{1}{2(1-S_{ab})} \left(\int \psi_{H1s_a}^* \hat{H} \psi_{H1s_a} \, d\tau + \int \psi_{H1s_b}^* \hat{H} \psi_{H1s_b} \, d\tau - \int \psi_{H1s_b}^* \hat{H} \psi_{H1s_a} \, d\tau - \int \psi_{H1s_a}^* \hat{H} \psi_{H1s_b} \, d\tau \right)$$

$$= \frac{1}{2(1-S_{ab})} \left(H_{aa} + H_{bb} - H_{ba} - H_{ab} \right)$$

$$= \frac{H_{aa} - H_{ab}}{1 + S_{ab}}$$

P23.17) A surface displaying a contour of the total charge density in LiH is shown here. What is the relationship between this surface and the MOs displayed in Problem P23.12? Why does this surface closely resemble one of the MOs?

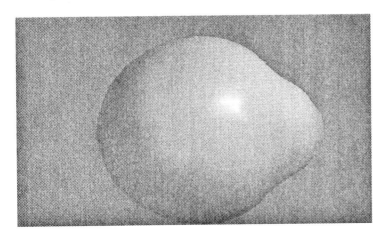

The surface of electron density will be determined by all filled MOs, but the unfilled MOs will not contribute. Therefore, the $3\sigma^*$ MO has no influence in determining the electron density surface. Higher energy MOs are further out from the molecular skeleton than lower lying MOs. Therefore, they have more influence in determining the electron density contour. Therefore, the contour looks like the 2σ rather than the 1σ MO.

P23.23) Calculate the bond order in each of the following species. Which of the species in part a–d do you expect to have the shorter bond length?

a. Li_2 or Li_2^+

c. O_2 or O_2^+

b. C_2 or C_2^+

d. F_2 or F_2^-

Li_2: $\left(1\sigma_g\right)^2 \left(1\sigma_u^*\right)^2 \left(2\sigma_g\right)^2$

Bond Order $= \dfrac{4-2}{2} = 1$

Li_2^+: $\left(1\sigma_g\right)^2 \left(1\sigma_u^*\right)^2 \left(2\sigma_g\right)^1$

Bond Order $= \dfrac{3-2}{2} = 0.5$

C_2: $\left(1\sigma_g\right)^2 \left(1\sigma_u^*\right)^2 \left(2\sigma_g\right)^2 \left(2\sigma_u^*\right)^2 \left(1\pi_u\right)^4$

Bond Order $= \dfrac{8-4}{2} = 2$

C_2^+: $\left(1\sigma_g\right)^2 \left(1\sigma_u^*\right)^2 \left(2\sigma_g\right)^2 \left(2\sigma_u^*\right)^2 \left(1\pi_u\right)^3$

Bond Order $= \dfrac{7-4}{2} = 1.5$

O_2: $\left(1\sigma_g\right)^2 \left(1\sigma_u^*\right)^2 \left(2\sigma_g\right)^2 \left(2\sigma_u^*\right)^2 \left(3\sigma_g\right)^2 \left(1\pi_u\right)^2 \left(1\pi_u\right)^2 \left(1\pi_g^*\right)^1 \left(1\pi_g^*\right)^1$

Bond Order $= \dfrac{10-6}{2} = 2$

O_2^+: $\left(1\sigma_g\right)^2 \left(1\sigma_u^*\right)^2 \left(2\sigma_g\right)^2 \left(2\sigma_u^*\right)^2 \left(3\sigma_g\right)^2 \left(1\pi_u\right)^2 \left(1\pi_u\right)^2 \left(1\pi_g^*\right)^1$

Bond Order $= \dfrac{10-5}{2} = 2.5$

F_2: $\left(1\sigma_g\right)^2 \left(1\sigma_u^*\right)^2 \left(2\sigma_g\right)^2 \left(2\sigma_u^*\right)^2 \left(3\sigma_g\right)^2 \left(1\pi_u\right)^2 \left(1\pi_u\right)^2 \left(1\pi_g^*\right)^2 \left(1\pi_g^*\right)^2$

Bond Order $= \dfrac{10-8}{2} = 1$

F_2^-: $\left(1\sigma_g\right)^2 \left(1\sigma_u^*\right)^2 \left(2\sigma_g\right)^2 \left(2\sigma_u^*\right)^2 \left(3\sigma_g\right)^2 \left(1\pi_u\right)^2 \left(1\pi_u\right)^2 \left(1\pi_g^*\right)^3 \left(1\pi_g^*\right)^2$

Bond Order $= \dfrac{10-9}{2} = 0.5$

Because the bond length is shorter for a greater bond order, the answers are Li_2, C_2, O_2^+, F_2.

Computational problems

Before solving the computational problems, it is recommended that students work through Tutorials 1–3 under the Help menus in Spartan Student Edition to gain familiarity with the program.

Computational Problem 23.1: According to Hund's rules, the ground state of O_2 should be a triplet because the last two electrons are placed in a doubly degenerate set of π MOs. Calculate the energy of the singlet and triplet states of O_2 using the B3LYP method and the 6-31G* basis set. Does the singlet or triplet have the lower energy? Both states will be populated if the energy difference $\Delta E \sim kT$. For which temperature is this the case?

Step 1: Create a new file, and choose O= from the inorganic palette which is accessed by clicking the "Inorganic" tab on the model kit. Add a second O= to the initial atom.

Step 2: Using the calculations menu, enter equilibrium geometry, the B3LYP method, the 6-31G* basis set and the appropriate number of unpaired electrons for the singlet state. Make sure that you check the Infrared spectra box. The global calculations box should not be checked.

Step 3: Repeat step 2 for the triplet state using the mew molecule command. Submit the calculation.

Step 4: Once your calculations complete, go to Display → Output. Note the values of the total energy for the triplet and singlet states. **You cannot conclude that the optimized structure corresponds to an equilibrium structure unless all the vibrational frequencies are real.** Go to display spectra under "Display" to see the vibrational frequencies.

Step 5: Calculate the temperature for which $\Delta E = kT$.

Computational Problem 23.2: If the ground state of oxygen is a diradical, you might think that O_2 would dimerize to form square planar O_4 to achieve a molecule in which all electrons are paired. Optimize the geometry and calculate the energies of triplet O_2 and singlet O_4 using the B3LYP method and the 6-31G* basis set. Do you predict O_4 to be more or less stable than 2 O_2 molecules? Use a nonplanar shape in building your O_4 molecule.

Step 1: In this problem, it is important to start with a nonplanar geometry to avoid biasing the outcome of the calculation. Following these directions will create a "twisted square" initial geometry. Create a new file, and choose **nonplanar** trivalent O from the inorganic palette which is accessed by clicking the "Inorganic" tab on the model kit. Add a second and a third **nonplanar** trivalent O to the initial atom.

Use the "measure angle" button to determine the O-O-O angle. The result displayed in the bottom right corner of the window will be 109.47°. (Don't worry about the added H atoms. They will be removed in the next step.) Replace 109.47° with 95 and press enter. The O-O-O angle is now 95°.

Step 2: Hit the delete button and the H atoms will be removed. Add a fourth **nonplanar** trivalent O to the structure. Change the newly created O-O-O bond angle to 85° as in step 1. Using the dihedral angle button, measure the dihedral angle of the four O atoms. Change this value to 10°. Again use the delete button to remove the H atoms. Use the "make bond" button to complete the ring by clicking on the two closest valences. Use the delete button to remove the four dangling valences. You now have a "twisted square" starting geometry. Do not minimize the energy before continuing the calculation because the minimization will lead to a planar structure.

Step 3: Using the calculations menu, enter equilibrium geometry, the B3LYP method, the 6-31G* basis set and the appropriate number of unpaired electrons. Make sure that you check the Infrared spectra box. The global calculations box should not be checked. Submit the calculation.

Step 4: Once your calculations complete, go to Display → Output. Note the values of the total energy. **You cannot conclude that the optimized structure corresponds to an equilibrium structure unless all the vibrational frequencies are real.** Go to display spectra under "Display" to see the vibrational frequencies. If there is an equilibrium structure for O_4, describe it. Is the geometry optimized molecule planar or nonplanar?

Step 5: Compare the energy of O_4 to 2 molecules of O_2.

Computational Problem 23.3: O_6 might be more stable than O_4 because the bond angle is larger, leading to less steric strain. Optimize the geometry and compare the energy of O_6 with 1.5 times the energy of O_4 using the B3LYP method and the 6-31G* basis set. Is O_6 more stable than O_4? Use a nonplanar shape in building your O_6 molecule. Is the geometry optimized molecule planar or nonplanar?

Step 1: In this problem, it is important to start with a nonplanar geometry to avoid biasing the outcome of the calculation. Following these directions will create a "boat" initial geometry. Create a new file, and choose **nonplanar** trivalent O from the inorganic palette which is accessed by clicking the "Inorganic" tab on the model kit. Add a second and then 3, 4, and 5 **nonplanar** trivalent O to the initial atom. Add to the appropriate valence to generate the "boat" structure.

Step 2: Use the "make bond" button to complete the ring by clicking on the two closest valences. Use the delete button to remove the six dangling valences. You now have a "boat" starting geometry. Minimize the energy.

Step 3: Using the calculations menu, enter equilibrium geometry, the B3LYP method, the 6-31G* basis set and the appropriate number of unpaired electrons. Make sure that you check the Infrared spectra box. The global calculations box should not be checked. Submit the calculation.

Step 4: Once your calculations complete, go to Display → Output. Note the values of the total energy. **You cannot conclude that the optimized structure corresponds to an equilibrium structure unless all the vibrational frequencies are real.** Go to display spectra under "Display" to see the vibrational frequencies. If there is an equilibrium structure for O_6, describe it. Is the geometry optimized molecule planar or nonplanar?

Step 5: Compare the energy of O_6 to 3 molecules of O_2. Compare the energy of O_6 to 1.5 times the energy of O_4. Is O_6 more stable than O_2? Is O_6 more stable than O_4?

Computational Problem 23.4: In a LiF crystal, both the Li and F are singly ionized species. Optimize the geometry and calculate the charge on Li and F in a single LiFl molecule using the B3LYP method and the 6-31G* basis set. Are the atoms singly ionized? Compare the value of the bond length with the distance between Li^+ and F^- ions in the crystalline solid.

Step 1: Create a new file, and build the LiF molecule using the inorganic palette which is accessed by clicking the "Inorganic" tab on the model kit.

Step 2: Using the calculations menu, enter equilibrium geometry, the B3LYP method, the 6-31G* basis set and the appropriate number of unpaired electrons. Submit the calculation.

Step 3: Once your calculations complete, go to Display → Output. Note the values of the positions of Li and F and the charges on each atom. Are the atoms fully ionized? Is the LiF spacing equal to that in an LiF crystal?

Computational Problem 23.5: Does LiF dissociate into neutral atoms or into Li^+ and F^-? Answer this question by comparing the energy difference between reactants and products for the reactions $LiF(g) \rightarrow Li(g) + F(g)$ and $LiF(g) \rightarrow Li^+(g) + F^-(g)$ using the B3LYP method and the 6-31G* basis set.

Step 1: Create a new file, and build the LiF molecule using the inorganic palette which is accessed by clicking the "Inorganic" tab on the model kit. Using the "new molecule" command, add Li, F, Li^+, and F^- sequentially.

Step 2: Using the calculations menu, enter equilibrium geometry, the B3LYP method, the 6-31G* basis set and the appropriate number of unpaired electrons. Make sure the global calculations box is not checked. Submit the calculation.

Step 3: Once your calculations complete, go to Display → Output. Note the values of the energies of each species? Calculate the energy change for the reactions

LiF(g) → Li(g) + F(g) and LiF(g) → Li$^+$(g) + F$^-$(g). Decide whether dissociation into atoms or ions will occur.

Computational Problem 23.6: Calculate Hartree–Fock MO energy values for HF using the MP2 method and the 6-31G* basis set. Make a molecular energy diagram to scale omitting the lowest energy MO. Why can you neglect this MO? Characterize the other MOs as bonding, antibonding, or nonbonding.

Step 1: Create a new file, and build the HF molecule using the inorganic palette which is accessed by clicking the "Inorganic" tab on the model kit.

Step 2: Using the calculations menu, enter equilibrium geometry, the MP2 method, the 6-31G* basis set and the appropriate number of unpaired electrons. Check the "orbitals and energies" box. Submit the calculation.

Step 3: Once your calculations complete, go to Display → Surfaces. By using the "add" button, add the LUMO, HOMO, and HOMO-x surfaces where x takes on integer values. What is the maximum value of x that you should enter to obtain surfaces for all occupied MOs?

Step 4: Look at each MO by checking the appropriate box. By clicking on the MO and going to the properties menu, you can make the MO transparent to better see the atom positions. Classify each MO as bonding, antibonding, and nonbonding. Explain how you reached your conclusions.

Computational Problem 23.7: a) Based on the molecular orbital energy diagram in Problem C23.6, would you expect triplet neutral HF in which an electron is promoted from the 1π to the $4\sigma^*$ MO to be more or less stable than singlet HF?

Because an electron is promoted from a nonbonding to an antibonding MO, the triplet is less stable.

b) Calculate the equilibrium bond length and total energy for singlet and triplet HF using the MP2 method and the 6-31G* basis set. Using the frequency as a criterion, are both stable molecules? Compare the bond lengths and vibrational frequencies.

Singlet: 0.934A, 4036 cm^{-1}, no imaginary frequencies, so molecule is stable

Triplet: Minimum energy not reached. Molecule is unstable and will dissociate into atoms.

c) Calculate the bond energy of singlet and triplet HF by comparing the total energies of the molecules with the total energy of F and H. Are your results consistent with the bond lengths and vibrational frequencies obtained in part b)?

Computational Problem 23.8: Computational chemistry allows you to carry out calculations for hypothetical molecules that do not exist in order to see trends in molecular properties. Calculate the charge on the atoms in singlet HF and in triplet HF for which the bond length is fixed at 10% greater than the bond length for singlet HF. Are the trends that you see consistent with those predicted by Figure 23.4? Explain your answer.

Step 1: Create a new file, and build the HF molecule using the inorganic palette which is accessed by clicking the "Inorganic" tab on the model kit.

Step 2: Using the calculations menu, enter equilibrium geometry, the MP2 method, the 6-311+G** basis set and the appropriate number of unpaired electrons for the ground state of HF. Check the "orbitals and energies" box. The global calculations box should not be checked. Click "OK."

Step 3: Build another HF molecule in the same file using the "new molecule" command. For this triplet state molecule, you will constrain the H-F distance. Under the "Geometry menu," select "Constrain distance." Click on the "measure bond length" and on H and af using the shift key. Click on the open pink lock in the lower right corner of the window and change the H-F distance from the equilibrium value for singlet LiF of 0.934 A to a value that is 10% larger. Using the calculations menu, enter equilibrium geometry, the MP2 method, the 6-311+G** basis set and the appropriate number of unpaired electrons for the triplet state. Check the "orbitals and energies" box. Click "OK."

Step 4: Repeat step 3 for H-F distances that are 20 and 30% larger than 0.934A.

Step 5: Using the "New Molecule" command and the same basis set and method, set up calculations for singlet HF and for the total energy of the F atom.

Step 6: Submit your calculation.

Step 7: Make a table for the 3H-F distances in which the columns are the total energy for the molecule, ΔE for the reaction HF $\rightarrow$ H + F, and the charge on the F atom. The rows list singlet HF and triplet HF for each of the H-F distances.

Step 8: Review Section 21.5 and discuss your results for the difference in the charge on the F atom for the singlet and triplet state. Explain the trend that you observe in the charge on the F atom with the H-F distance.

Chapter 24: Molecular Structure and Energy Levels for Polyatomic Molecules

P24.2) Predict whether LiH_2^+ and NH_2^- should be linear or bent based on the Walsh correlation diagram in Figure 24.10. Explain your answers.

The LiH_2^+ molecular ion has two valence electrons. These fill only the $1a_1$ MO. The correlation diagram shows that the energy of this MO is lowered if the molecule is bent. The molecule NH_2^- has eight valence electrons, as does H_2O, and is bent for the same reason.

P24.5) Use the method described in Example Problem 24.3 to show that the *sp*-hybrid orbitals

$\psi_a = 1/\sqrt{2}\,(-\phi_{2s} + \phi_{2p_z})$ and $\psi_b = 1/\sqrt{2}\,(-\phi_{2s} - \phi_{2p_z})$ are oriented 180° apart.

We differentiate ψ_a with respect to θ and set the derivative equal to zero.

$$\frac{d\psi_a}{d\theta} = \frac{1}{\sqrt{2}}\frac{d}{d\theta}\left[-\frac{1}{\sqrt{32\pi}}\left(\frac{\zeta}{a_0}\right)^{3/2}\left(2-\frac{r}{a_0}\right)e^{-r/a_0} + \frac{1}{\sqrt{32\pi}}\left(\frac{\zeta}{a_0}\right)^{3/2}\frac{r}{a_0}e^{-\zeta r/a_0}\cos\theta \right]$$

$$= -\frac{1}{\sqrt{32\pi}}\left(\frac{\zeta}{a_0}\right)^{3/2}\frac{r}{a_0}e^{-\zeta r/a_0}\sin\theta = 0$$

This equation is satisfied for $\theta = 0$ and $\theta = 180°$. The second derivative is used to establish which of these corresponds to a maximum.

$$\frac{d^2\psi_a}{d\theta^2} = -\frac{1}{\sqrt{32\pi}}\left(\frac{\zeta}{a_0}\right)^{3/2}\left(\frac{r}{a_0}\right)e^{-\zeta r/a_0}\cos\theta$$

$$= -\frac{1}{\sqrt{32\pi}}\left(\frac{\zeta}{a_0}\right)^{3/2}\left(\frac{r}{a_0}\right)e^{-\zeta r/a_0} \text{ at } \theta = 0$$

$$= \frac{1}{\sqrt{32\pi}}\left(\frac{\zeta}{a_0}\right)^{3/2}\left(\frac{r}{a_0}\right)e^{-\zeta r/a_0} \text{ at } \theta = 180°$$

At a maximum, $\dfrac{d^2\psi_a}{d\theta^2} < 0$ and at a minimum, $\dfrac{d^2\psi_a}{d\theta^2} > 0$. Therefore, the maximum is at $0°$, and a

minimum at $180°$. Applying the same procedure to ψ_b shows that the maximum is at $0°$. Therefore, ψ_a

and ψ_b point in opposite directions separated by $180°$.

P24.11) Determine the AO coefficients for the lowest energy Hückel π MO for butadiene.

The energy is given by $\varepsilon = \alpha + 1.62\beta$ and $\psi_\pi = c_1\psi^a_{2p_z} + c_2\psi^b_{2p_z} + c_3\psi^c_{2p_z} + \psi^d_{2p_z}$

Thus, the secular equations are

1): $c_1(H_{aa} - \varepsilon S_{aa}) + c_2(H_{ab} - \varepsilon S_{ab}) + c_3(H_{ac} - \varepsilon S_{ac}) + c_4(H_{ad} - \varepsilon S_{ad}) = 0$

2): $c_1(H_{ba} - \varepsilon S_{ba}) + c_2(H_{bb} - \varepsilon S_{bb}) + c_3(H_{bc} - \varepsilon S_{bc}) + c_4(H_{bd} - \varepsilon S_{bd}) = 0$

3): $c_1(H_{ca} - \varepsilon S_{ca}) + c_2(H_{cb} - \varepsilon S_{cb}) + c_3(H_{cc} - \varepsilon S_{cc}) + c_4(H_{cd} - \varepsilon S_{cd}) = 0$

4): $c_1(H_{da} - \varepsilon S_{da}) + c_2(H_{db} - \varepsilon S_{db}) + c_3(H_{dc} - \varepsilon S_{dc}) + c_4(H_{dd} - \varepsilon S_{dd}) = 0$

The overlap is given by $S_{jk} = \delta_{jk}$, and we assign values to the H_{jk} as follows:

$$H_{jk} = \begin{cases} \alpha \text{ if } j = k \\ \beta \text{ if } j \text{ and } k \text{ differ by 1, and} \\ 0 \text{ otherwise} \end{cases}$$

Substituting in the four secular equations gives the following relations:

1) $c_1\left[\alpha - (\alpha + 1.62\beta)\right] + c_2[\beta] = 0$

    We conclude that $c_2 = 1.62\, c_1$

2) $c_1\beta + c_2\left[\alpha - (\alpha + 1.62\beta)\right] + c_3\beta = 0$

    We conclude that $c_1 + c_3 = 1.62\, c_2$ or $c_3 = \left(1.62^2 - 1\right)c_1$

3) $c_2\beta + c_3\left[\alpha - (\alpha + 1.62\beta)\right] + c_4\beta = 0$

We conclude that $c_2 + c_4 = 1.62\,c_3$ or $c_4 = \dfrac{(1.62^2 - 1)}{1.62}\,c_1$

The 4th equation needed to solve this system is the normalization condition, defined by

$$1 = \int (\psi_\pi)^* \psi_\pi d\tau = \int \left(c_1\psi_{2p_z}^a + c_2\psi_{2p_z}^b + c_3\psi_{2p_z}^c + c_4\psi_{2p_z}^d \right)^2 d\tau$$

$$= c_1^2 \int \left(\psi_{2p_z}^a\right)^2 d\tau + c_2^2 \int \left(\psi_{2p_z}^b\right)^2 d\tau + c_3^2 \int \left(\psi_{2p_z}^c\right)^2 d\tau + c_4^2 \int \left(\psi_{2p_z}^d\right)^2 d\tau$$

$$+ 2c_1 c_2 \int \psi_{2p_z}^a \psi_{2p_z}^b d\tau + 2c_1 c_3 \int \psi_{2p_z}^a \psi_{2p_z}^c d\tau + 2c_1 c_4 \int \psi_{2p_z}^a \psi_{2p_z}^d d\tau$$

$$+ 2c_2 c_3 \int \psi_{2p_z}^b \psi_{2p_z}^c d\tau + 2c_2 c_4 \int \psi_{2p_z}^b \psi_{2p_z}^d d\tau$$

$$= c_1^2 + c_2^2 + c_3^2 + c_4^2 + 2c_1 c_2 S_{ab} + 2c_1 c_3 S_{ac} + 2c_1 c_4 S_{ad} + 2c_2 c_3 S_{bc} + 2c_2 c_4 S_{bd}$$

$$= c_1^2 + c_2^2 + c_3^2 + c_4^2$$

Substituting 1–3 into the normalization equation yields

$$c_1^2 + 1.62^2 c_1^2 + \left(1.62^2 - 1\right)^2 c_1^2 + \left[\frac{(1.62^2 - 1)}{1.62}\right]^2 c_1^2 = 1$$

or

$$c_1^2 = \frac{1}{1 + 1.62^2 + \left(1.62^2 - 1\right)^2 + \left[\dfrac{1.62^2 - 1}{1.62}\right]^2}$$

and

$c_1 = 0.3715$

$c_2 = 1.62 \cdot 0.3715 = 0.602$

$c_3 = \left(1.62^2 - 1\right)0.3715 = 0.602$

$c_4 = \left(1.62 - \dfrac{1}{1.62}\right)0.3715 = 0.3715$

$\psi_\pi = 0.3715\psi_{2p_z}^a + 0.602\psi_{2p_z}^b + 0.602\psi_{2p_z}^c + 0.3715\psi_{2p_z}^d$

P24.13) Use the geometrical construction shown in Example Problem 24.10 to derive the π electron MO levels for the cyclopentadienyl radical. What is the total π energy of the molecule? How many unpaired electrons will the molecule have?

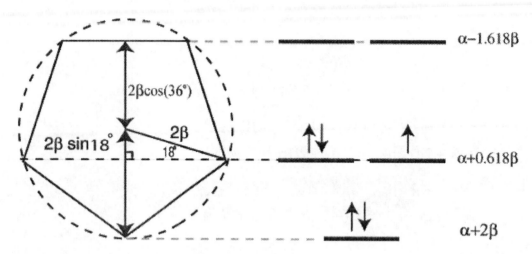

$$2\beta \sin\left(18^\circ\right) = 0.618$$
$$2\beta \cos\left(36^\circ\right) = 1.618$$

There is one unpaired electron. The π energy is

$$E_\pi = 2\left[\alpha + 2\beta\right] + 3\left(\alpha + 0.618\beta\right)$$
$$E_\pi = 5\alpha + 5.85\beta$$

P24.21) Use the VSEPR method to predict the structures of the following:

a. PF_3 b. CO_2 c. BrF_5 d. SO_3^{2-}

a) PF_3 has 3 ligands and a lone pair. VSEPR predicts that the structure is pyramidal.

b) CO_2 has 2 double-bonded ligands and no lone pairs. VSEPR predicts that the structure is linear.

c) BrF_5 has 5 ligands and a lone pair. VSEPR predicts that the structure is square pyramidal.

d) SO_3^{2-} has 3 ligands and a lone pair. VSEPR predicts that the structure is pyramidal.

Computational problems

Before solving the computational problems, it is recommended that students work through Tutorials 1–3 under the Help menus in Spartan Student Edition to gain familiarity with the program.

Computational Problem 24.1: Calculate the bond angles in NH_3 and in NF_3 using the density functional method with the B3LYP functional and the 6-31G* basis set. Compare your result with literature values. Do your results agree with the predictions of the VSEPR model and Bent's rule?

Step 1: Create a new file, and build the NH_3 molecule using **nonplanar trivalent N** inorganic palette which is accessed by clicking the "Inorganic" tab on the model kit.

Step 2: Using the calculations menu, enter equilibrium geometry, the B3LYP method, the 6-31G* basis set and the appropriate number of unpaired electrons for the ground state of NH_3. Check the "infrared spectra" box. Click "OK."

Step 3: Using the "New Molecule" command and the same basis set and method, set up calculations for NF_3. Submit the calculation.

Step 4: Look at the vibrational frequencies by going to "display" and then "spectra." The result of the calculation is an equilibrium structure only if all frequencies are real.

Step 5: Using the measure angle button, record the H-N-H and F-N-F bond angles and compare your results with the predictions of the VSEPR model, Bent's rule, and with literature values.

Computational Problem 24.2: Calculate the bond angles in H_2O and in H_2S using the density functional method with the B3LYP functional and the 6-31G* basis set. Compare your result with literature values. Do your results agree with the predictions of the VSEPR model and Bent's rule?

Step 1: Create a new file, and build the H_2O molecule using **nonlinear divalent O** in inorganic palette which is accessed by clicking the "Inorganic" tab on the model kit.

Step 2: Using the calculations menu, enter equilibrium geometry, the B3LYP method, the 6-31G* basis set and the appropriate number of unpaired electrons for the ground state of H_2O. Check the "infrared spectra" box. Click "OK."

Step 3: Using the "New Molecule" command and the same basis set and method, set up calculations for H_2S. Submit the calculation.

Step 4: Look at the vibrational frequencies by going to "display" and then "spectra." The result of the calculation is an equilibrium structure only if all frequencies are real.

Step 5: Using the measure angle button, record the H-O-H and H-S-H bond angles and compare your results with the predictions of the VSEPR model, Bent's rule, and with literature values.

Computational Problem 24.3: Calculate the bond angle in ClO_2 using the density functional method with the B3LYP functional and the 6-31G* basis set. Compare your result with literature values. Does your result agree with the predictions of the VSEPR model?

Step 1: Create a new file, and build the ClO_2 molecule using **nonlinear divalent Cl** in the inorganic palette which is accessed by clicking the "Inorganic" tab on the model kit.

Step 2: Using the calculations menu, enter equilibrium geometry, the B3LYP method, the 6-31G* basis set and the appropriate number of unpaired electrons for the ground state of ClO_2. Check the "infrared spectra" box. Click "OK."

Step 3: Look at the vibrational frequencies by going to "display" and then "spectra." The result of the calculation is an equilibrium structure only if all frequencies are real.

Step 5: Using the measure angle button, record the O-Cl-O bond angle and compare your results with the predictions of the VSEPR model, Bent's rule, and with literature values.

Computational Problem 24.4: SiF_4 has four ligands and one lone pair on the central S atom. Which of the following structures do you expect to be the equilibrium form based on a calculation using the density functional method with the B3LYP functional and the 6-31G* basis set? In the figure below, from top to bottom, the structures are a trigonal bipyramid, a see-saw structure, and a square planar structure.

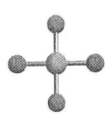

Step 1: Create a new file, and build trigonal bipyramidal SiF_4 using the appropriate S template in the inorganic palette which is accessed by clicking the "Inorganic" tab on the model kit.

Step 2: Using the calculations menu, enter equilibrium geometry, the B3LYP method, the 6-31G* basis set and the appropriate number of unpaired electrons for the ground state of ClO_2. Check the "infrared spectra" box. Check the "global calculations" box. Click "OK"

Step 3: Repeat step 1 for see-saw SiF_4 and square planar SiF_4.

Step 5: Using your results for the total energy of the different isomers, determine the equilibrium structure. Measure the 2 different F-S-F bond angle in your equilibrium structure and compare them with the experimentally determined values of 101.6° and 173°. Justify this structure by making a Lewis diagram for the molecule. Is this the structure that would have been predicted using the VSEPR model?

Computational Problem 24.5: Calculate the bond angles in singlet BeH_2, doublet NH_2, and doublet BH_2 using the Hartree–Fock method and the 6-31G* basis set. Explain your results using the Walsh diagram of Figure 24.11.

Step 1: Create a new file, and build the BeH_2 molecule using **nonlinear divalent Be** from the inorganic palette which is accessed by clicking the "Inorganic" tab on the model kit.

Step 2: Using the calculations menu, enter the equilibrium geometry, the Hartree–Fock method, the 6-31G* basis set and the appropriate number of unpaired electrons for the ground state of BeH_2. Check the "infrared spectra" box. The global calculations box should not be checked. Click "OK."

Step 3: Using the "New Molecule" command and the same basis set and method, set up calculations for doublet NH_2, and doublet BH_2. Submit the calculation.

Step 4: Look at the vibrational frequencies by going to "display" and then "spectra." The result of the calculation is an equilibrium structure only if all frequencies are real.

Step 5: Using the measure angle button, record the H-Be-H, H-B-H, and H-N-H bond angles and compare your results with the prediction of the Walsh diagram of Figure 24.11.

Computational Problem 24.6: Calculate the bond angle in singlet LiH_2^+ using the Hartree–Fock method and the 6-31G* basis set. Can you explain your results using the Walsh diagram of Figure 24.11? (Hint: Determine the calculated bond lengths in the molecular ion.)

Step 1: Create a new file, and build the LiH_2^+ molecular ion using **nonlinear divalent Be** from the inorganic palette which is accessed by clicking the "Inorganic" tab on the model kit.

Step 2: Using the calculations menu, enter equilibrium geometry, the Hartree-Fock method, the 6-31G* basis set and the appropriate number of unpaired electrons for the ground state of LiH_2^+. Check the "infrared spectra" box. Click "OK."

Step 3: Using the "New Molecule" command and the same basis set and method, set up calculations for the ground state of H_2, Submit the calculation.

Step 4: Look at the vibrational frequencies by going to "display" and then "spectra." The result of the calculation is an equilibrium structure only if all frequencies are real. Is LiH_2^+ a stable molecular ion?

Step 5: Compare the H-H distances in H_2, and the H-H vibrational stretching frequency. On the basis of these results, can you explain your results using the Walsh diagram of Figure 24.11?

Computational Problem 24.7: Calculate the bond angle in singlet and triplet CH_2 and doublet CH_2^+ using the Hartree–Fock method and the 6-31G* basis set. Can you explain your results using the Walsh diagram of Figure 24.11?

Step 1: Create a new file, and build the CH_2 molecule using **nonlinear divalent C** from the inorganic palette which is accessed by clicking the "Inorganic" tab on the model kit.

Step 2: Using the calculations menu, enter equilibrium geometry, the Hartree–Fock method, the 6-31G* basis set and the appropriate number of unpaired electrons for the ground state of CH_2. Check the "infrared spectra" box. The global calculations box should not be checked. Click "OK."

Step 3: Using the "New Molecule" command and the same basis set and method, set up calculations for triplet CH_2 and doublet CH_2^+. Submit the calculation.

Step 4: Look at the vibrational frequencies by going to "display" and then "spectra." The result of the calculation is an equilibrium structure only if all frequencies are real. Are all molecules stable?

Step 5: Compare the bond angles in the three molecules. Can you explain your results using the Walsh diagram of Figure 24.11?

Computational Problem 24.8: Calculate the bond angle in singlet NH_2^+, doublet NH_2, and singlet NH_2^- using the Hartree–Fock method and the 6-31G* basis set. Can you explain your results using the Walsh diagram of Figure 24.11?

Step 1: Create a new file, and build the NH_2^+ molecule using **nonlinear divalent N** from the inorganic palette which is accessed by clicking the "Inorganic" tab on the model kit.

Step 2: Using the calculations menu, enter equilibrium geometry, the Hartree–Fock method, the 6-31G* basis set and the appropriate number of unpaired electrons for the ground state of NH_2^+. Check the "infrared spectra" box. The global calculations box should not be checked. Click "OK."

Step 3: Using the "New Molecule" command and the same basis set and method, set up calculations for doublet NH_2, and singlet NH_2^-. Submit the calculation.

Step 4: Look at the vibrational frequencies by going to "display" and then "spectra." The result of the calculation is an equilibrium structure only if all frequencies are real. Are all molecules stable?

Step 5: Compare the bond angles in the three molecules. Can you explain your results using the Walsh diagram of Figure 24.11?

Computational Problem 24.9: How essential is coplanarity to conjugation? Answer this question by calculating the total energy of 1, 3 butadiene using the Hartree–Fock method and the 6-31G* basis set for dihedral angles of 0, 45, and 90 degrees.

Step 1: Create a new file, and build the 1,3-butadiene using sp^2 C from the organic palette which is accessed by clicking the "organic" tab on the model kit.

Step 2: Using the calculations menu, enter equilibrium geometry, the Hartree–Fock method, the 6-31G* basis set and the appropriate number of unpaired electrons for the ground state of 1,3-butadiene. Check the "infrared spectra" box. Click "OK."

Step 3: Using the "New Molecule" command and the same basis set and method, build 1,3-butadiene again. Click on "measure dihedral angle" button and then the 4 carbon atoms. Set this dihedral angle to 135° and hit Enter. You will now constrain this angle. From the "geometry" menu, choose "constrain dihedral" and click on the 4 carbon atoms. Click on the pink lock at the bottom corner of the window and enter 135°. Click minimize.

Step 4: Repeat the procedure of the previous step for a dihedral angle of 90°.

Step 5: Repeat the procedure of the step 3, this time constraining the H-C=C-H dihedral angle where the Cs refer to C1 and C2 to 135°. Click on "measure dihedral angle" button and then the H-C-C-H atoms using the trans H. Submit the calculation.

Step 6: Calculate the energy difference in kJ mol^{-1} between planar 1,3-butadiene and the three twisted structures. Which costs more energy, to rotate by 45° around the C1-C2 or the C2-C3 bond? Why?

Computational Problem 24.10: Calculate the equilibrium structures for singlet and triplet formaldehyde using the density functional method with the B3LYP functional and the 6-311+G** basis set. Choose (a) planar and (b) pyramidal starting geometries. Calculate vibrational frequencies for both starting geometries. Are any of the frequencies imaginary? Explain your results.

Step 1: Create a new file, and build the planar formaldehyde using sp^2 C from the organic palette which is accessed by clicking the "organic" tab on the model kit.

Step 2: Using the calculations menu, enter equilibrium geometry, the B3LYP method, the 6-311+G** basis set and the appropriate number of unpaired electrons for singlet formaldehyde. Check the "infrared spectra" box. The global calculations box should not be checked. Click "OK."

Step 3: Using the "New Molecule" command, sp^2 C from the organic palette and the same basis set and method, build planar triplet formaldehyde. Check the "infrared spectra" box. The global calculations box should not be checked. Click "OK."

Step 4: Using the "New Molecule" command, tetrahedral C from the organic palette and the same basis set and method, build nonplanar singlet and triplet formaldehyde. Delete one of the C valences. Check the "infrared spectra" box. The global calculations box should not be checked. Click "submit."

Step 5: Compare the geometry and total energy of the calculated singlet and triplet structures from the two starting geometries. Are they identical? If they are different, decide which structure is the true equilibrium structure, and explain why the two starting geometries gave different results.

Computational Problem 24.11: Calculate the equilibrium structure for Cl_2O using the density functional method with the B3LYP functional and the 6-31G* basis set. Obtain an infrared spectrum and activate the normal modes. What are the frequencies corresponding to the symmetric stretch, the asymmetric stretch, and the bending modes?

Step 1: Create a new file, and build Cl-O-Cl using nonlinear divalent O from the inorganic palette which is accessed by clicking the "inorganic" tab on the model kit.

Step 2: Using the calculations menu, enter equilibrium geometry, the B3LYP method, the 6-31G* basis set and the appropriate number of unpaired electrons for the molecule. Check the "infrared spectra" box. The global calculations box can be checked. Click "OK."

Step 3: Using the "New Molecule" command, build O-Cl-Cl using nonlinear divalent O from the inorganic palette which is accessed by clicking the "inorganic" tab on the model kit. Check the "infrared spectra" box. The global calculations box can be checked. Click "OK."

Step 4: Which of the two isomers has the lower energy? Compare the bond angle with what would be predicted by the VSEPR model.

Step 5: Obtain an infrared spectrum and activate the normal modes. Determine the frequencies corresponding to the symmetric stretch, the asymmetric stretch, and the bending modes. Compare these frequencies with experimental values from the chemical literature.

Computational Problem 24.12: Calculate the equilibrium structures for PF_3 using the density functional method with the B3LYP functional and the 6-31G* basis set. Obtain an infrared spectrum and activate the normal modes. What are the frequencies corresponding to the symmetric stretch, the symmetric deformation, the degenerate stretch, and the degenerate deformation modes?

Step 1: Create a new file, and build PF_3 using nonplanar trivalent P from the inorganic palette which is accessed by clicking the "inorganic" tab on the model kit.

Step 2: Using the calculations menu, enter equilibrium geometry, the B3LYP method, the 6-31G* basis set and the appropriate number of unpaired electrons for the molecule. Check the "infrared spectra" box. The global calculations box can be checked. Click "OK."

Step 3: Obtain an infrared spectrum and activate the normal modes. Determine the frequencies corresponding to the symmetric stretch, the symmetric deformation, the degenerate stretch, and the degenerate deformation modes. Compare these frequencies with experimental values from the chemical literature.

Computational Problem 24.13: Calculate the equilibrium structures for C_2H_2 using the density functional method with the B3LYP functional and the 6-31G* basis set. Obtain an infrared spectrum and activate the normal modes. What are the frequencies corresponding to the symmetric stretch, the antisymmetric stretch, the stretch, and the two bending modes?

Step 1: Create a new file, and build H-C≡C-H using *sp* C from the inorganic palette which is accessed by clicking the "organic" tab on the model kit.

Step 2: To avoid biasing the outcome of the calculation, make the molecule bent. Click on the measure angle button, and then on the three atoms. Set the angle to 150°. Hit the enter key. Do not minimize because the process will make the molecule linear.

Step 3: Using the calculations menu, enter equilibrium geometry, the B3LYP method, the 6-31G* basis set and the appropriate number of unpaired electrons for the molecule. Check the "infrared spectra" box. The global calculations box can be checked. Click "OK."

Step 4: Obtain an infrared spectrum and activate the normal modes. Determine the frequencies corresponding to the antisymmetric bending, symmetric bending, C≡C stretch, antisymmetric C-H stretch, and symmetric C-H stretch. Compare these frequencies with experimental values from the chemical literature.

Computational Problem 24.14: Calculate the structure of $N \equiv C - Cl$ using the density functional method with the B3LYP functional and the 6-31G* basis set. Which is more electronegative, the Cl or the cyanide group? What result of the calculation did you use to answer this question?

Step 1: Create a new file, and build linear $N \equiv C - Cl$ using *sp* C and N from the organic palette which is accessed by clicking the "organic" tab on the model kit.

Step 2: To avoid biasing the outcome of the calculation, make the molecule bent. Click on the measure angle button, and then on the three atoms. Set the angle to 150°. Hit the enter key. Do not minimize because the process will make the molecule linear.

Step 3: Using the calculations menu, enter equilibrium geometry, the B3LYP method, the 6-31G* basis set and the appropriate number of unpaired electrons for the molecule. Check the "infrared spectra" box. Click "submit."

Step 4: Using the output, decide if the Cl or the cyanide group is more electronegative.

Chapter 27: Molecular Symmetry

P27.2) Use the 3 × 3 matrices for the C_{2v} group in Equation (27.2) to verify the group multiplication table for the following successive operations:

a. $\hat{\sigma}_v \hat{\sigma}_v'$ b. $\hat{\sigma}_v \hat{C}_2$ c. $\hat{C}_2 \hat{C}_2$

$$\hat{E} = \begin{pmatrix} 1 & 0 & 0 \\ 0 & 1 & 0 \\ 0 & 0 & 1 \end{pmatrix} \quad \hat{C}_2 = \begin{pmatrix} -1 & 0 & 0 \\ 0 & -1 & 0 \\ 0 & 0 & 1 \end{pmatrix}$$

$$\hat{\sigma}_v = \begin{pmatrix} 1 & 0 & 0 \\ 0 & -1 & 0 \\ 0 & 0 & 1 \end{pmatrix} \quad \hat{\sigma}_v' = \begin{pmatrix} -1 & 0 & 0 \\ 0 & 1 & 0 \\ 0 & 0 & 1 \end{pmatrix}$$

a)

$$\hat{\sigma}_v \hat{\sigma}_v' = \begin{pmatrix} 1 & 0 & 0 \\ 0 & -1 & 0 \\ 0 & 0 & 1 \end{pmatrix} \begin{pmatrix} -1 & 0 & 0 \\ 0 & 1 & 0 \\ 0 & 0 & 1 \end{pmatrix} = \begin{pmatrix} -1 & 0 & 0 \\ 0 & -1 & 0 \\ 0 & 0 & 1 \end{pmatrix}$$

$$\hat{\sigma}_v \hat{\sigma}_v' = \hat{C}_2$$

b)

$$\hat{\sigma}_v \hat{C}_2 = \begin{pmatrix} 1 & 0 & 0 \\ 0 & -1 & 0 \\ 0 & 0 & 1 \end{pmatrix} \begin{pmatrix} -1 & 0 & 0 \\ 0 & -1 & 0 \\ 0 & 0 & 1 \end{pmatrix} = \begin{pmatrix} -1 & 0 & 0 \\ 0 & 1 & 0 \\ 0 & 0 & 1 \end{pmatrix}$$

$$\hat{\sigma}_v \hat{C}_2 = \hat{\sigma}_v'$$

c)

$$\hat{C}_2 \hat{C}_2 = \begin{pmatrix} -1 & 0 & 0 \\ 0 & -1 & 0 \\ 0 & 0 & 1 \end{pmatrix} \begin{pmatrix} -1 & 0 & 0 \\ 0 & -1 & 0 \\ 0 & 0 & 1 \end{pmatrix} = \begin{pmatrix} 1 & 0 & 0 \\ 0 & 1 & 0 \\ 0 & 0 & 1 \end{pmatrix}$$

$$\hat{C}_2 \hat{C}_2 = \hat{E}$$

P27.4) The D_3 group has the following classes: E, $2C_3$, and $3C_2$. How many irreducible representations does this group have and what is the dimensionality of each?

The D_3 group has three classes and six elements ($1E$, $2C_3$, $3C_2$).

There are three irreducible representations because there are three classes. Their dimensions are determined by the equation

$$d_1^2 + d_2^2 + d_3^2 = 6$$

This can only be satisfied by $d_1 = 2$, $d_2 = 1$, $d_3 = 1$. Thus

3 irreducible representations; two 1-dimensional and one 2-dimensional.

P27.7) XeF$_4$ belongs to the D_{4h} point group with the following symmetry elements: E, C_4, C_4^2, C_2, C_2', C_2'', i, S_4, S_4^2, σ, $2\sigma'$, and $2\sigma''$. Make a drawing similar to Figure 27.1 showing these elements. The 4-fold axis is perpendicular to the plane of the molecule.

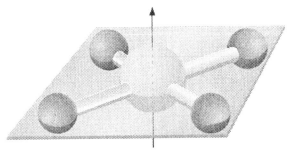

There are three 2-fold axes, as shown below. Two pass through the opposed F atoms, and the third bisects the F-Xe-F angle.

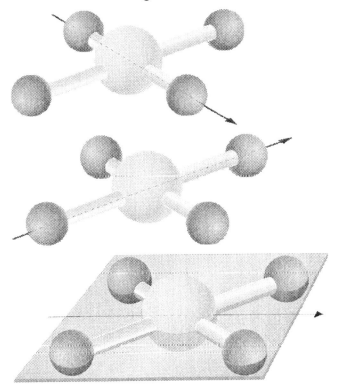

The mirror plane denoted σ lies in the plane of the molecule.

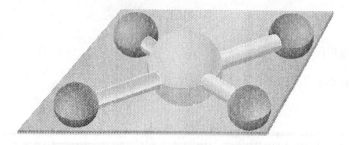

The two mirror planes denoted σ' are mutually perpendicular, and are perpendicular to σ. The Xe atom and two opposed F atoms lie at the intersection of σ' with σ.

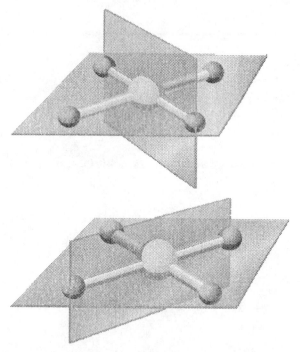

The two mutually perpendicular mirror planes σ'' contain the Xe, but no F atoms.

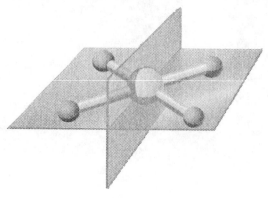

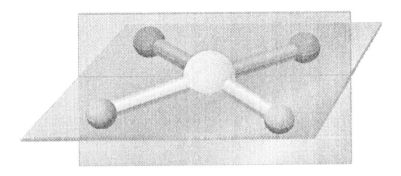

The Xe atom is the invasion center. This element is not depicted.

P27.13) Decompose the following reducible representation into irreducible representations of the C_{2v} group:

E	C_2	σ_v	σ_v'
4	0	0	0

We set up the following table and apply Equation (27.22). The reducible representation is in the first row.

	E	C_2	σ_v	σ_v'
Red	4	0	0	0
A_1	1	1	1	1
A_2	1	1	−1	−1
B_1	1	−1	1	−1
B_2	1	−1	−1	1

$$n_{A_1} = \frac{1}{4}\left[\left(1\times4\times1\right)+\left(1\times0\times1\right)+\left(1\times0\times1\right)+\left(1\times0\times1\right)\right] = 1$$

$$n_{A_2} = \frac{1}{4}\left[\left(1\times4\times1\right)+\left(1\times0\times1\right)+\left(1\times0\times(-1)\right)+\left(1\times0\times(-1)\right)\right] = 1$$

$$n_{B_1} = \frac{1}{4}\left[\left(1\times4\times1\right)+\left(1\times0\times(-1)\right)+\left(1\times0\times1\right)+\left(1\times0\times(-1)\right)\right] = 1$$

$$n_{B_2} = \frac{1}{4}\left[\left(1\times4\times1\right)+\left(1\times0\times(-1)\right)+\left(1\times0\times(-1)\right)+\left(1\times0\times1\right)\right] = 1$$

Thus: $\Gamma_{red} = A_1 + A_2 + B_1 + B_2$

P27.14) Show that z is a basis for the A_1 representation and that R_z is a basis for the A_2 representation of the C_{3v} group.

All of the operations of the C_{3v} group are described by the matrix equation

$$\begin{pmatrix} x' \\ y' \\ z' \end{pmatrix} - \begin{pmatrix} \pm 1 & 0 & 0 \\ 0 & \pm 1 & 0 \\ 0 & 0 & 1 \end{pmatrix} \begin{pmatrix} x \\ y \\ z \end{pmatrix} - \begin{pmatrix} \pm x \\ \pm y \\ z \end{pmatrix}$$

The signs of at least one of x and y are changed by the operations other than E, but it is always the case that $z' = z$. Therefore, the character of each operation on z is 1, which makes z a basis for the A_1 representation.

For a rotation about the z axis, the character is +1 if the direction of rotation is unchanged, and −1 if the direction of the rotation is changed. Because the identity operator does not change the direction of the rotation, the character is +1. The same is true of the $\hat{C}_2$ operator. However, reflection if both the σ_v and σ'_v planes changes the direction of rotation because the rotation axis lies in the plane. Therefore the character for these operations is −1. The sequence of these characters correspond to the A_2 representation.

P27.15) Use the logic diagram of Figure 27.2 to determine the point group for PCl_5. Indicate your decision-making process as was done in the text for NH_3.

1) linear?	No	
2) C_n axis?	Yes	C_3 axis $\Rightarrow z$ axis
3) more than 1 C_n axis	Yes	
3) more than 1 C_n axis, $n > 2$?	No	
4) σ plane?	Yes	
5) $\sigma \perp$ to C_3?	Yes	

We conclude that the point group is D_{3h}.

P27.20) Show that the presence of a C_2 axis and a mirror plane perpendicular to the rotation axis imply the presence of a center of inversion.

Under a 180° rotation around the z axis, $x, y, z \rightarrow -x, -y, z$. The effect of the mirror plane perpendicular

to the rotation axis is $-x, -y, z \rightarrow -x, -y, -z$, which is the same as the inversion operation.

Chapter 29: Probability

P29.1) Suppose that you draw a card from a standard deck of 52 cards. What is the probability of drawing:
a) an ace of any suit?
b) the ace of spades?
c) How would your answers to parts (a) and (b) change if you were allowed to draw three times, replacing the card drawn back into the deck after each draw?

a) In a deck of 52 cards there are four aces, therefore:

$$P_E = \frac{E}{N} = \frac{4}{52}$$

b) There is only one card that corresponds to the event of interest, therefore:

$$P_E = \frac{E}{N} = \frac{1}{52}$$

c) By replacing the card, the probability for each drawing is independent, and the total probability is the sum of probabilities for each event. Therefore, the probability will be $3P_E$ with P_E as given above.

P29.3) A pair of standard dice are rolled. What is the probability of observing the following:
a) The sum of the dice is equal to 7.
b) The sum of the dice is equal to 9.
c) The sum of the dice is less than or equal to 7.

a) We are interested in the outcome where the sum of two dice is equal to 7. If any side of a die has an equal probability of being observed, then the probability of any number appearing is 1/6.

$$P_{sum=7} = \left[2 \times \left(P_1 \times P_6\right)\right] + \left[2 \times \left(P_2 \times P_5\right)\right] + \left[2 \times \left(P_3 \times P_4\right)\right]$$

$$= 3\left[\frac{2}{36}\right] = \frac{1}{6}$$

b) Using the nomenclature developed above:

$$P_{sum=9} = \left[2 \times \left(P_3 \times P_6\right)\right] + \left[2 \times \left(P_4 \times P_5\right)\right]$$

$$= 2\left[\frac{2}{36}\right] = \frac{1}{9}$$

c) Now, one has to sum all the probabilities that correspond to the event of interest:

$$P_{sum \leq 7} = \left[\left(P_1 \times P_1\right)\right] + \left[2 \times \left(P_1 \times P_2\right)\right] + \left[2 \times \left(P_1 \times P_3\right)\right] + \left[2 \times \left(P_1 \times P_4\right)\right] + \left[2 \times \left(P_1 \times P_5\right)\right] + \left[2 \times \left(P_1 \times P_6\right)\right]$$

$$+ \left[\left(P_2 \times P_2\right)\right] + \left[2 \times \left(P_2 \times P_3\right)\right] + \left[2 \times \left(P_2 \times P_4\right)\right] + \left[2 \times \left(P_2 \times P_5\right)\right] + \left[\left(P_3 \times P_3\right)\right] + \left[2 \times \left(P_3 \times P_4\right)\right]$$

$$= 21\left[\frac{1}{36}\right] = \frac{21}{36}$$

P29.5) Atomic chlorine has two naturally occurring isotopes, ^{35}Cl and ^{37}Cl. If the molar abundance of these isotopes is 75.4% and 24.6%, respectively, what fraction of a mole of molecular chlorine (Cl_2) will have one of each isotope? What fraction will contain just the ^{35}Cl isotope?

The probabilities for observing a given isotopic composition of Cl_2 is equal to the product of probabilities for each isotope:

$$^{35}Cl^{35}Cl \quad P = 0.754 \times 0.754 = 0.569$$
$$^{37}Cl^{37}Cl \quad P = 0.246 \times 0.246 = 0.061$$
$$^{35}Cl^{37}Cl \quad P = 0.754 \times 0.246 = 0.186$$
$$^{37}Cl^{35}Cl \quad P = 0.186$$

Using the above probabilities, the fraction of a mole that will contain one of each isotope ($^{35}Cl^{37}Cl$ and $^{37}Cl^{35}Cl$) is $0.186 + 0.186 = 0.372$. The fraction of Cl_2 that will contain just the ^{35}Cl isotope is 0.569.

P29.9) Determine the numerical values for the following:
a) The number of configurations employing all objects in a six-object set
b) The number of configurations employing 4 objects from a six-object set
c) The number of configurations employing no objects from a six-object set
d) $C(50,10)$

a) $C(n,j) = C(6,6) = \left(\dfrac{n!}{j!(n-j)!} \right) = \dfrac{6!}{6!0!} = 1$

b) $C(n,j) = C(6,4) = \left(\dfrac{n!}{j!(n-j)!} \right) = \dfrac{6!}{4!2!} = 15$

c) $C(n,j) = C(6,0) = \left(\dfrac{n!}{j!(n-j)!} \right) = \dfrac{6!}{0!6!} = 1$

d) $P(n,j) = P(50,10) = \left(\dfrac{n!}{j!(n-j)!} \right) = \dfrac{50!}{10!40!} \cong 1.03 \times 10^{10}$

P29.11) Four bases (A, C, T, and G) appear in DNA. Assume that the appearance of each base in a DNA sequence is random.
a) What is the probability of observing the sequence AAGACATGCA?
b) What is the probability of finding the sequence GGGGGAAAAA?
c) How do your answers to parts (a) and (b) change if the probability of observing A is twice that of the probabilities used in parts (a) and (b) of this question when the preceding base is G?

a) There are four choices for each base, and the probability of observing any base is equal. Therefore, for a decamer the number of possible sequences is:

$$N_{total} = (4)^{10} \cong 1.05 \times 10^6$$

Since there is only one sequence that corresponds to the event of interest:

$$P_E = \frac{E}{N} \cong \frac{1}{1.05 \times 10^6} \cong 9.52 \times 10^{-7}$$

b) Identical to part (a).

c) In this case, the probability of observing a base at a given location is dependent on which base is present. If G appears in the sequence, then the probability of observing A is 1/2 while the probability of observing any other base is 1/6 (watch the normalization!). Therefore, the probability of observing the sequence in part (a) is:

$$P = \left(\frac{1}{4}\right)^3 \left(\frac{1}{2}\right)\left(\frac{1}{4}\right)^4 \left(\frac{1}{6}\right)\left(\frac{1}{4}\right) \cong 1.27 \times 10^{-6}$$

and the probability for the sequence in part (b) is:

$$P = \left(\frac{1}{4}\right)\left(\frac{1}{6}\right)^4 \left(\frac{1}{2}\right)\left(\frac{1}{4}\right)^4 \cong 3.77 \times 10^{-7}$$

P29.12) The natural abundance of ^{13}C is roughly 1%, and the abundance of deuterium (^2H or D) is 0.015%. Determine the probability of finding the following in a mole of acetylene:
   a)   H-^{13}C-^{13}C-H
   b)   D-^{12}C-^{12}C-D
   c)   H-^{13}C-^{12}C-D

The probability of observing a each of these species is equal to the product of probabilities for observing the individual isotopes:

a. $P = P(H) \times P(^{13}C) \times P(^{13}C) \times P(H) = P(H)^2 \times P(^{13}C)^2$
    $= (0.99985)^2 \times (0.01)^2 = 9.997 \times 10^{-5} \approx 1 \times 10^{-4}$

b. $P = P(D) \times P(^{12}C) \times P(^{12}C) \times P(D) = P(D)^2 \times P(^{12}C)^2$
    $= (0.00015)^2 \times (0.99)^2 = 2.2 \times 10^{-8}$

c. $P = 2(P(H) \times P(^{13}C) \times P(^{12}C) \times P(D)) =$
    $= 2((0.99985) \times (0.01) \times (0.099) \times (0.00015)) = 2.970 \times 10^{-6} \approx 3 \times 10^{-6}$

The factor of two in part c arises from the fact that there are two configurations corresponding to the same isotopic composition.

P29.14) The Washington State Lottery consists of drawing five balls numbered 1 to 43, and a single ball numbered 1 to 23 from a separate machine.
a) What is the probability of hitting the jackpot in which the values for all six balls are correctly predicted?
b) What is the probability of predicting just the first five balls correctly?
c) What is the probability of predicting the first five balls in the exact order they are picked?

a) The total probability is the product of probabilities for the five-ball outcome and the one-ball outcome. The five-ball outcome is derived by considering the configurations possible using five objects from a set of 43 total objects:

$$P_{fiveball} = [C(43,5)]^{-1} = \left(\frac{43!}{5!38!}\right)^{-1} \cong 1.04 \times 10^{-6}$$

The one-ball outcome is associated with the configurations possible using a single object from a set of 23 objects:

$$P_{oneball} = [C(23,1)]^{-1} = \left(\frac{23!}{1!22!}\right)^{-1} \cong 4.35 \times 10^{-2}$$

The total probability is the product of the above probabilities:
$$P_{total} = P_{fiveball} \times P_{oneball} \cong 4.52 \times 10^{-8}$$

b) The probability is that for the five-ball case determined above.

c) This case corresponds to a specific permutation of all permutations possible using five objects from a set of 43 objects:

$$P = [P(43,5)]^{-1} = \left(\frac{43!}{38!}\right)^{-1} \cong 8.66 \times 10^{-9}$$

P29.17) Imagine an experiment in which you flip a coin four times. Furthermore, the coin is balanced fairly such that the probability of landing heads or tails is equivalent. After tossing the coin 10 times, what is the probability of observing
a) no heads?
b) two heads?
c) five heads?
d) eight heads?

a) The quantity of interest is the probability of observing a given number of successful trials (j) in a series of n trials in which the probability of observing a successful trial, P_E, is equal to 1/2:

$$P(j) = C(n,j)(P_E)^j (1 - P_E)^{n-j} = C(n,j)\left(\frac{1}{2}\right)^n$$

Substituting in for the specific case of $j = 0$ and $n = 10$ yields:

$$P(0) = C(10,0)\left(\frac{1}{2}\right)^{10} \cong 9.77 \times 10^{-4}$$

b) In this case, $j = 2$ and $n = 10$:

$$P(2) = C(10,2)\left(\frac{1}{2}\right)^{10} = \left(\frac{10!}{2!8!}\right)\left(\frac{1}{2}\right)^{10} \cong 0.044$$

c) In this case, $j = 5$ and $n = 10$:

$$P(5) = C(10,5)\left(\frac{1}{2}\right)^{10} = \left(\frac{10!}{5!5!}\right)\left(\frac{1}{2}\right)^{10} \cong 0.246$$

d) In this case, $j = 8$ and $n = 10$:

$$P(8) = C(10,8)\left(\frac{1}{2}\right)^{10} = \left(\frac{10!}{8!2!}\right)\left(\frac{1}{2}\right)^{10} \cong 0.044$$

P29.22) Radioactive decay can be thought of as an exercise in probability theory. Imagine that you have a collection of radioactive nuclei at some initial time (N_0) and are interested in how many nuclei will still remain at a later time (N). For first-order radioactive decay, $N/N_0 = e^{-kt}$. In this expression, k is known as the decay constant and t is time.

a) What is the variable of interest in describing the probability distribution?
b) At what time will the probability of nuclei undergoing radioactive decay be 0.50?

a) The variable (k) defines the width of the distribution of population versus time.

b)

$$\frac{N}{N_0} = 0.5 = e^{-kt}$$

$$\ln(0.5) = -kt$$

$$\frac{-\ln(0.5)}{k} = \frac{\ln(2)}{k} = t$$

P29.23) First order decay processes as described in the previous problem can also be applied to a variety of atomic and molecular processes. For example, in aqueous solution the decay of singlet molecular oxygen ($O_2(^1\Delta_g)$) to the ground state triplet configuration proceeds according to:

$$\frac{\left[O_2\left(^1\Delta_g\right)\right]}{\left[O_2\left(^1\Delta_g\right)\right]_0} = e^{-\left(2.4 \times 10^5 \text{ s}^{-1}\right)t}$$

In the above expression, $[O_2(^1\Delta_g)]$ is the concentration of singlet oxygen at a given time, and the subscript "0" indicates that this is the concentration of singlet oxygen present at the beginning of the decay process ($t = 0$).

a. How long does one have to wait until 90% of the singlet oxygen has decayed?
b. How much singlet oxygen remains after $t = (2.4 \times 10^5 \text{ s}^{-1})^{-1}$?

a. After 90% of the singlet oxygen has decayed the ratio of concentrations is equal to 0.1. Substitution in this value for the concentration ratio and solving for t yields:

$$0.1 = e^{-\left(2.4 \times 10^5 \text{ s}^{-1}\right)t}$$

$$t = \frac{\ln(0.1)}{-2.4 \times 10^5 \text{ s}^{-1}} = 9.6 \times 10^{-6} \text{ s}$$

b.   Using the value for t provided, the equation can be readily solved for the ratio of singlet oxygen concentrations:

$$\frac{\left[O_2\left(^1\Delta_g\right)\right]}{\left[O_2\left(^1\Delta_g\right)\right]_0} = e^{-\left(2.4\times10^5\text{ s}^{-1}\right)t} = e^{-\left(2.4\times10^5\text{ s}^{-1}\right)\times\left(2.4\times10^5\text{ s}^{-1}\right)^{-1}}$$

$$\frac{\left[O_2\left(^1\Delta_g\right)\right]}{\left[O_2\left(^1\Delta_g\right)\right]_0} = e^{-1} = 0.37$$

P29.24) In Chapter 13, we will encounter the energy distribution $P(\varepsilon) = Ae^{-\varepsilon/kT}$, where $P(\varepsilon)$ is the probability of a molecule occupying a given energy state, ε is the energy of the state, k is a constant equal to 1.38×10^{-23} J K^{-1}, and T is temperature. Imagine that there are three energy states at 0, 100, and 500 J mol^{-1}.

a) Determine the normalization constant for this distribution.
b) What is the probability of occupying the highest energy state at 298 K?
c) What is the average energy at 298 K?
d) Which state makes the largest contribution to the average energy?

a) Since the energies are given in units of J mol^{-1}, dividing by Avogadro's number will convert this energy to a per particle unit. Alternatively, Avogadro's number can be

included with Boltzmann's constant resulting in $k \times N_a = R$, where $R = 8.314$ J mol^{-1} K^{-1}. Using this relationship:

$$P_{500} = e^{-500 \text{ J mol}^{-1}/\left(8.314 \text{ J mol}^{-1}\text{ K}^{-1}\right)\left(298 \text{ K}\right)} = 0.817$$

$$P_{100} = e^{-100 \text{ J mol}^{-1}/\left(8.314 \text{ J mol}^{-1}\text{ K}^{-1}\right)\left(298 \text{ K}\right)} = 0.960$$

$$P_0 = e^{-0 \text{ J mol}^{-1}/\left(8.314 \text{ J mol}^{-1}\text{ K}^{-1}\right)\left(298 \text{ K}\right)} = 1$$

Using these probabilities, the normalization constant becomes:

$$A = \frac{1}{\sum\limits_{i=1}^{3} P_i} = \frac{1}{P_0 + P_{100} + P_{500}} \cong 0.360$$

b)   With normalization, the normalized probabilities are given by the product of the probabilities determined in part (a) of this question and the normalization constant:

$$P_{500,norm} = A \times P_{500} = 0.360 \times 0.817 = 0.294$$

$$P_{100,norm} = A \times P_{100} = 0.360 \times 0.960 = 0.345$$

$$P_{0,norm} = A \times P_{100} = 0.360$$

c) The average energy is given by:

$$\langle E \rangle = \sum_{i=1}^{3} E_i P_{i,norm} = \left(0 \text{ J mol}^{-1}\right)(0.360) + \left(100 \text{ J mol}^{-1}\right)(0.345) + \left(500 \text{ J mol}^{-1}\right)(0.294) = 182 \text{ J mol}^{-1}$$

d) Inspection of part (c) of this question illustrates that the highest-energy state makes the largest contribution to the average energy.

P29.26) Consider the following probability distribution corresponding to a particle located between point $x = 0$ and $x = a$:

$$P(x)dx = C \sin^2 \left[\frac{\pi x}{a}\right] dx$$

a) Determine the normalization constant, C.
b) Determine $\langle x \rangle$.
c) Determine $\langle x^2 \rangle$.
d) Determine the variance.

a)

$$1 = C \int_0^a \sin^2 \left(\frac{\pi x}{a}\right) dx = C \left(\frac{a}{2}\right)$$

$$C = \frac{2}{a}$$

b)

$$\langle x \rangle = \int_0^a (x)\left(\frac{2}{a}\sin^2\left(\frac{\pi x}{a}\right)\right) dx = \frac{2}{a}\int_0^a x \sin^2\left(\frac{\pi x}{a}\right) dx$$

$$= \frac{2}{a}\left(\frac{a^2}{4}\right) = \frac{a}{2}$$

c)

$$\langle x^2 \rangle = \int_0^a (x^2)\left(\frac{2}{a}\sin^2\left(\frac{\pi x}{a}\right)\right) dx = \frac{2}{a}\int_0^a x^2 \sin^2\left(\frac{\pi x}{a}\right) dx$$

$$= \left(\frac{2}{a}\right)\left[\left(\frac{a^3}{6}\right) - \left(\frac{a^3}{4\pi^2}\right)\right] = a^2\left(\frac{1}{3} - \frac{1}{2\pi^2}\right)$$

d)

$$\sigma^2 = \langle x^2 \rangle - \langle x \rangle^2 = a^2\left(\frac{1}{12} - \frac{1}{2\pi^2}\right)$$

Chapter 30: The Boltzmann Distribution

P30.2)

a) Realizing that the most probable outcome from a series of N coin tosses is $N/2$ heads and $N/2$ tails, what is the expression for W_{max} corresponding to this outcome?

b) Given your answer for part (a), derive the following relationship between the weight for an outcome other than the most probable and W_{max}:

$$\log\left(\frac{W}{W_{max}}\right) = -H\log\left(\frac{H}{N/2}\right) - T\log\left(\frac{T}{N/2}\right)$$

c) We can define the deviation of a given outcome from the most probable outcome using a "deviation index," $\alpha = \dfrac{H-T}{N}$. Show that the number of heads or tails can be expressed as $H = \dfrac{N}{2}(1+\alpha)$ and $T = \dfrac{N}{2}(1-\alpha)$.

d) Finally, demonstrate that $\dfrac{W}{W_{max}} = e^{-N\alpha^2}$.

a)

$$W = \frac{N!}{H!\,T!} = \frac{N!}{\left(N/2\right)!\left(N/2\right)!} = \frac{N!}{\left[\left(N/2\right)!\right]^2}$$

b)

$$\ln\left(\frac{W}{W_{max}}\right) = \ln W - \ln W_{max} = \ln\left(\frac{N!}{H!T!}\right) - \ln\left(\frac{N!}{\left[\left(N/2\right)!\right]^2}\right)$$

$$= \ln(N!) - \ln(H!) - \ln(T!) - \ln(N!) + 2\ln\left(\left(N/2\right)!\right)$$

$$= -\ln(H!) - \ln(T!) + 2\ln\left(\left(N/2\right)!\right)$$

$$= -H\ln H + H - T\ln T + T + N\ln\left(N/2\right) - N$$

$$= -H\ln H - T\ln T + N\ln\left(N/2\right)$$

$$= -H\ln H - T\ln T + (H+T)\ln\left(N/2\right)$$

$$= -H\ln\left(\frac{H}{N/2}\right) - T\ln\left(\frac{T}{N/2}\right)$$

194

c) Substituting the definition of part (a) into the expressions for H and T:

$$H = \frac{N}{2}(1+\alpha) = \frac{N}{2}\left(1 + \frac{H-T}{N}\right) = \frac{N}{2} + \frac{H-T}{2} = \frac{H+T}{2} + \frac{H-T}{2} = H$$

$$T = \frac{N}{2}(1-\alpha) = \frac{N}{2}\left(1 - \frac{H-T}{N}\right) = \frac{N}{2} - \frac{H-T}{2} = \frac{H+T}{2} - \frac{H-T}{2} = T$$

d) Substituting in the result of part (c) into the final equation of part (b):

$$\ln\left(\frac{W}{W_{max}}\right) = -\frac{N}{2}(1+\alpha)\ln(1+\alpha) - \frac{N}{2}(1-\alpha)\ln(1-\alpha)$$

If $|\alpha| \ll 1$, then $\ln(1 \pm \alpha) = \pm\,\alpha$, therefore:

$$\ln\left(\frac{W}{W_{max}}\right) = -\frac{N}{2}(1+\alpha)\ln(1+\alpha) - \frac{N}{2}(1-\alpha)\ln(1-\alpha)$$

$$= -\frac{N}{2}(1+\alpha)(\alpha) - \frac{N}{2}(1-\alpha)(-\alpha) = -N\alpha^2$$

$$\frac{W}{W_{max}} = e^{-N\alpha^2}$$

P30.4) Determine the weight associated with the following card hands:

a) Having any five cards

b) Having five cards of the same suit (known as a "flush")

c)

    a) The problem can be solved by recognizing that there are 52 total cards ($N = 52$), with 5 cards in the hand ($a_1 = 5$), and 47 out of the hand ($a_0 = 47$):

$$W = \frac{N!}{a_1! a_0!} = \frac{52!}{5!47!} \cong 2.60 \times 10^6$$

    b) For an individual suit, there are 13 total cards ($N = 13$), 5 of which must be in the hand ($a_1 = 5$) while the other 8 remain in the deck ($a_0 = 47$). Finally, there are four total suits:

$$W = 4\left(\frac{N!}{a_1!a_0!}\right) = 4\left(\frac{13!}{5!8!}\right) = 5148$$

P30. 7) Barometric pressure can be understood using the Boltzmann distribution. The potential energy associated with being a given height above the Earth's surface is mgh, where m is the mass of the particle of interest, g is the acceleration due to gravity, and h is height. Using this definition of the potential energy, derive the following expression for pressure:

$$P = P_o \, e^{-mgh/kT}$$

Assuming that the temperature remains at 298 K, what would you expect the relative pressures of N_2 and O_2 to be at the tropopause, the boundary between the troposphere and stratosphere roughly 11 km above the Earth's surface? At the Earth's surface, the composition of air is roughly 78% N_2, 21% O_2, and the remaining 1% is other gases.

At the Earth's surface, $h = 0$ meters and the total pressure is 1 atm. Using the mole fractions of N_2 and O_2, the partial pressures at the Earth's surface are 0.78 and 0.21 atm, respectively. Given this information, the pressure of N_2 at 11 km is given by:

$$P_{11\text{ km}} = P_{0\text{ km}} e^{-mgh/kT} = (0.78 \text{ atm}) e^{-\left(0.028 \text{ kg mol}^{-1} \times N_A^{-1}\right)\left(9.8 \text{ m s}^{-1}\right)\left(1.1 \times 10^4 \text{ m}\right) \Big/ \left(1.38 \times 10^{-23} \text{ J K}^{-1}\right)(298 \text{ K})} = 0.23$$

Performing the identical calculation for O_2 yields:

$$P_{11\text{ km}} = P_{0\text{ km}} e^{-mgh/kT} = (0.21 \text{ atm}) e^{-\left(0.032 \text{ kg mol}^{-1} \times N_A^{-1}\right)\left(9.8 \text{ m s}^{-1}\right)\left(1.1 \times 10^4 \text{ m}\right) \Big/ \left(1.38 \times 10^{-23} \text{ J K}^{-1}\right)(298 \text{ K})} = 0.052$$

P30.9) Consider the following energy-level diagrams, modified from Problem P30.8 by the addition of another excited state with energy of 600. cm^{-1}:

a) At what temperature will the probability of occupying the second energy level be 0.15 for the states depicted in part (a) of the figure?

b) Perform the corresponding calculation for the states depicted in part (b) of the figure.

a)

$$0.15 = p_1 = \frac{e^{-\beta\varepsilon_1}}{q} = \frac{e^{-\beta(300. \text{ cm}^{-1})}}{1 + e^{-\beta(300. \text{ cm}^{-1})} + e^{-\beta(600. \text{ cm}^{-1})}}$$

$$0.15 + 0.15\left(e^{-\beta(300. \text{ cm}^{-1})}\right) + 0.15\left(e^{-\beta(600. \text{ cm}^{-1})}\right) = e^{-\beta(300. \text{ cm}^{-1})}$$

$$0.15 - 0.85\left(e^{-\beta(300. \text{ cm}^{-1})}\right) + 0.15\left(e^{-\beta(600. \text{ cm}^{-1})}\right) = 0$$

The last expression is a quadratic equation with $x = \exp(-\beta\,(300.\ \text{cm}^{-1}))$. This equation has two roots equal to 0.183 and 5.48. Only the 0.183 root will provide temperature greater than zero, therefore:

$$0.183 = e^{-\beta(300.\ \text{cm}^{-1})}$$

$$1.70 = \frac{300.\ \text{cm}^{-1}}{(0.695\ \text{cm}^{-1}\ \text{K}^{-1})(T)}$$

$$T = 254\ \text{K}$$

b)

$$0.15 = p_1 = \frac{2e^{-\beta\varepsilon_1}}{q} = \frac{2e^{-\beta(300.\ \text{cm}^{-1})}}{1+2e^{-\beta(300.\ \text{cm}^{-1})}+e^{-\beta(600.\ \text{cm}^{-1})}}$$

$$0.15 + 0.30\left(e^{-\beta(300.\ \text{cm}^{-1})}\right) + 0.15\left(e^{-\beta(600.\ \text{cm}^{-1})}\right) = 2e^{-\beta(300.\ \text{cm}^{-1})}$$

$$0.15 - 1.70\left(e^{-\beta(300.\ \text{cm}^{-1})}\right) + 0.15\left(e^{-\beta(600.\ \text{cm}^{-1})}\right) = 0$$

The last expression is a quadratic equation with $x = \exp(-\beta\,(300.\ \text{cm}^{-1}))$. This equation has two roots equal to 0.090 and 11.24. Only the 0.090 root will provide temperature greater than zero, therefore:

$$0.090 = e^{-\beta(300.\ \text{cm}^{-1})}$$

$$2.41 = \frac{300.\ \text{cm}^{-1}}{(0.695\ \text{cm}^{-1}\ \text{K}^{-1})(T)}$$

$$T = 179\ \text{K}$$

P30.11) A set of 13 particles occupies states with energies of 0, 100., and 200. cm^{-1}. Calculate the total energy and number of microstates for the following configurations of energy:

a) $a_0 = 8$, $a_1 = 5$, and $a_2 = 0$
b) $a_0 = 9$, $a_1 = 3$, and $a_2 = 1$
c) $a_0 = 10$, $a_1 = 1$, and $a_2 = 2$

Do any of these configurations correspond to the Boltzmann distribution?

The total energy is equal to the sum of energy associated with each level times the number of particles in that level. For the occupation numbers in configuration (a):

$$E = \sum_n \varepsilon_n a_n = \varepsilon_0 a_0 + \varepsilon_1 a_1 + \varepsilon_2 a_2$$

$$= (0\ \text{cm}^{-1})(8) + (100.\ \text{cm}^{-1})(5) + (200.\ \text{cm}^{-1})(0) = 500.\ \text{cm}^{-1}$$

Repeating the calculation for the occupation numbers in (b) and (c) yields the same energy of 500. cm^{-1}. The number of microstates associated with each distribution is given by the weight:

$$W_a = \frac{N!}{\prod_n a_n!} = \frac{N!}{a_0!a_1!a_2!} = \frac{13!}{(8!)(5!)(0!)} = 1287$$

$$W_b = \frac{13!}{(9!)(3!)(1!)} = 2860$$

$$W_c = \frac{13!}{(10!)(1!)(2!)} = 858$$

The ratio of any two occupation numbers for a set of non-degenerate energy levels is given by:

$$\frac{a_i}{a_j} = e^{-\beta(\varepsilon_i - \varepsilon_j)} = e^{-\left(\frac{\varepsilon_i - \varepsilon_j}{k}\right)\frac{1}{T}}$$

The above expression suggests that the ratio of occupation numbers can be used to determine the temperature. For set (b), comparing the occupation numbers for level 2 and level 0 results in:

$$\frac{a_2}{a_0} = e^{-\left(\frac{\varepsilon_2 - \varepsilon_0}{k}\right)\frac{1}{T}}$$

$$\frac{1}{9} = e^{-\left(\frac{200.\ cm^{-1} - 0\ cm^{-1}}{0.695\ cm^{-1}\ K^{-1}}\right)\frac{1}{T}}$$

$$T = 131\ K$$

Repeating the same calculation for level 1 and level 0:

$$\frac{a_1}{a_0} = e^{-\left(\frac{\varepsilon_2 - \varepsilon_0}{k}\right)\frac{1}{T}}$$

$$\frac{3}{9} = e^{-\left(\frac{100.\ cm^{-1} - 0\ cm^{-1}}{0.695\ cm^{-1}\ K^{-1}}\right)\frac{1}{T}}$$

$$T = 131\ K$$

The distribution of energy in (b) is in accord with the Boltzmann distribution. Performing a similar calculation for (a) and (c) will demonstrate that the temperatures are not equivalent, that these are not in accord with the Boltzmann distribution.

P30.13) For two non-degenerate energy levels separated by an amount of energy $\varepsilon/k = 500.$ K, at what temperature will the population in the higher-energy state be ½ that of the lower-energy state? What temperature is required to make the populations equal?

a)   The populations are directly related to the probability of occupying the energy levels, and the ratio of energy-level probabilities is related to the energy difference between these levels as follows:

$$\frac{p_1}{p_0} = \frac{1}{2} = e^{-\left(\frac{\varepsilon_1 - \varepsilon_0}{kT}\right)} = e^{-\left(\frac{500.\ K}{T}\right)}$$

$$\ln 2 = \frac{500.\ K}{T}$$

$$T = 721\ K$$

b)   For equal energy-level probabilities the ratio of probabilities will equal 1. This is only achieved when $T = \infty$:

$$\frac{p_1}{p_0} = 1 = e^{-\left(\frac{\varepsilon_1 - \varepsilon_0}{kT}\right)} = e^{-\left(\frac{500.\ K}{T}\right)}$$

$$\ln 1 = 0 = \frac{500\ K}{T}$$

$$T = \infty$$

P30.16) The ^{13}C nucleus is a spin 1/2 particle as is a proton. However, the energy splitting for a given field strength is roughly 1/4 of that for a proton. Using a 1.45-T magnet as in Example Problem 30.6, what is the ratio of populations in the excited and ground spin states for ^{13}C at 298 K?

Using the information provided in the example problem, the separation in energy is given by:

$$\Delta E = \frac{1}{4}\left(2.82 \times 10^{-26}\ J\ T^{-1}\right) B = \frac{1}{4}\left(2.82 \times 10^{-26}\ J\ T^{-1}\right)(1.45\ T) = 1.02 \times 10^{-26}\ J$$

Using this separation in energy, the ratio in spin-state occupation numbers is:

$$\frac{a_+}{a_-} = e^{-\left(\frac{\varepsilon_+ - \varepsilon_-}{kT}\right)} = e^{-\left(\frac{\Delta E}{kT}\right)}$$

$$\frac{a_+}{a_-} = e^{-\left(\frac{1.02 \times 10^{-26}\ J}{\left(1.38 \times 10^{-23}\ J\ K^{-1}\right)(298\ K)}\right)}$$

$$\frac{a_+}{a_-} = 0.999998$$

P30.18) The vibrational frequency of I_2 is 208 cm^{-1}. At what temperature will the population in the first excited state be half that of the ground state?

$$\frac{a_1}{a_0} = \frac{1}{2} = e^{-\beta(\varepsilon_1 - \varepsilon_0)} = e^{-\beta(208 \text{ cm}^{-1})}$$

$$0.5 = e^{\left(\frac{-208 \text{ cm}^{-1}}{(0.695 \text{ cm}^{-1} \text{ K})(T)}\right)}$$

$$0.693 = \frac{208 \text{ cm}^{-1}}{(0.695 \text{ cm}^{-1} \text{ K})(T)}$$

$$T = 432 \text{ K}$$

P30.23) The lowest two electronic energy levels of the molecule NO are illustrated here. Determine the probability of occupying one of the higher energy states at 100., 500., and 2000. K.

Both the lower and higher energy states are two-fold degenerate, with an energy spacing of 121.1 cm^{-1}. At 100. K, the partition function is:

$$q = \sum_n e^{-\beta\varepsilon_n} = 2 + e^{-\beta(121.1 \text{ cm}^{-1})} = 2 + e^{-121.1 \text{ cm}^{-1}/(0.695 \text{ cm}^{-1} \text{ K}^{-1})(100. \text{ K})} = 2.35$$

With the partition function evaluated, the probability of occupying the excited energy level is readily determined:

$$p_1 = \frac{g_1 e^{-\beta\varepsilon_1}}{q} = \frac{2e^{-121.1 \text{ cm}^{-1}/(0.695 \text{ cm}^{-1} \text{ K}^{-1})(100. \text{ K})}}{2.35} = 0.149$$

At 500. K:

$$q = 2 + e^{-121.1 \text{ cm}^{-1}/(0.695 \text{ cm}^{-1} \text{ K}^{-1})(500. \text{ K})} = 3.41$$

$$p_1 = \frac{g_1 e^{-\beta\varepsilon_1}}{q} = \frac{2e^{-121.1 \text{ cm}^{-1}/(0.695 \text{ cm}^{-1} \text{ K}^{-1})(500. \text{ K})}}{3.41} = 0.414$$

Finally, at 2000. K:

$$q = 2 + e^{-121.1 \text{ cm}^{-1}/(0.695 \text{ cm}^{-1} \text{ K}^{-1})(2000. \text{ K})} = 3.83$$

$$p_1 = \frac{g_1 e^{-\beta\varepsilon_1}}{q} = \frac{2e^{-121.1 \text{ cm}^{-1}/(0.695 \text{ cm}^{-1} \text{ K}^{-1})(2000. \text{ K})}}{3.83} = 0.479$$

Since there are two states per energy level, the probability of occupying an individual excited state is 1/2 of the above probabilities.

Chapter 31: Ensemble and Molecular Partition Functions

P31.2) Evaluate the translational partition function for $^{35}Cl_2$ confined to a volume of 1.00 L at 298 K. How does your answer change if the gas is $^{37}Cl_2$? (*Hint:* Can you reduce the ratio of translational partition functions to an expression involving mass only?)

The translational partition function for $^{35}Cl_2$ is calculated as follows:

$$q_T\left(^{35}Cl_2\right) = \frac{V}{\Lambda^3}$$

$$\Lambda = \left(\frac{h^2}{2\pi mkT}\right)^{1/2} = \left(\frac{\left(6.626\times10^{-34}\ J\ s\right)^2}{2\pi\left(\dfrac{0.0700\ kg\ mol^{-1}}{N_A}\right)\left(1.38\times10^{-23}\ J\ K^{-1}\right)\left(298\ K\right)}\right)^{1/2} = 1.21\times10^{-11}\ m$$

$$q_T\left(^{35}Cl_2\right) = \frac{V}{\left(1.21\times10^{-11}\ m\right)^3} = \frac{\left(1000\ cm^3\right)\left(10^{-6}\ m^3\ cm^{-3}\right)}{\left(1.21\times10^{-11}\ m\right)^3} = 5.66\times10^{29}$$

Taking the ratio of translational partition functions for $^{35}Cl_2$ and $^{37}Cl_2$ and cancelling out common terms yields:

$$\frac{q_T\left(^{37}Cl_2\right)}{q_T\left(^{35}Cl_2\right)} = \frac{\dfrac{V}{\Lambda^3\left(^{37}Cl_2\right)}}{\dfrac{V}{\Lambda^3\left(^{35}Cl_2\right)}} = \frac{\Lambda^3\left(^{35}Cl_2\right)}{\Lambda^3\left(^{37}Cl_2\right)} = \left(\frac{m\left(^{37}Cl_2\right)}{m\left(^{35}Cl_2\right)}\right)^{3/2} = \left(\frac{74.0\ g\ mol^{-1}}{70.0\ g\ mol^{-1}}\right)^{3/2} = 1.087$$

$$q_T\left(^{37}Cl_2\right) = 1.087\left(q_T\left(^{35}Cl_2\right)\right)$$

P31.4) Evaluate the translational partition function for Ar confined to a volume of 1000. cm^3 at 298 K. At what temperature will the translational partition function of Ne be identical to that of Ar at 298 K confined to the same volume?

$$q_T\left(Ar\right) = \frac{V}{\Lambda^3}$$

$$\Lambda = \left(\frac{h^2}{2\pi mkT}\right)^{1/2} = \left(\frac{\left(6.626\times10^{-34}\ J\ s\right)^2}{2\pi\left(\dfrac{0.0399\ kg\ mol^{-1}}{N_A}\right)\left(1.38\times10^{-23}\ J\ K^{-1}\right)\left(298\ K\right)}\right)^{1/2} = 1.60\times10^{-11}\ m$$

$$q_T\left(Ar\right) = \frac{V}{\left(1.60\times10^{-11}\ m\right)^3} = \frac{\left(1000.\ cm^3\right)\left(10^{-6}\ m^3\ cm^{-3}\right)}{\left(1.60\times10^{-11}\ m\right)^3} = 2.44\times10^{29}$$

If the gases are confined to the same volume, then the partition functions will be equal when the thermal wavelengths are equal:

$$\Lambda(Ne) = \Lambda(Ar) = 1.60 \times 10^{-11} \text{ m}$$

$$\left(\frac{h^2}{2\pi mkT}\right)^{1/2} = 1.60 \times 10^{-11} \text{ m}$$

$$T = \frac{h^2}{2\pi mk\left(1.60 \times 10^{-11} \text{ m}\right)^2} - \frac{\left(6.626 \times 10^{-34} \text{ J s}\right)^2}{2\pi\left(\frac{0.0202 \text{ kg mol}^{-1}}{N_A}\right)\left(1.38 \times 10^{-23} \text{ J K}^{-1}\right)\left(1.60 \times 10^{-11} \text{ m}\right)^2}$$

$$T = 590 \text{ K}$$

P31.7) For N_2 at 77.3 K, 1 atm, in a 1-cm^3 container, calculate the translational partition function and the ratio of this partition function to the number of N_2 molecules present under these conditions.

$$q_T(N_2) = \frac{V}{\Lambda^3}$$

$$\Lambda = \left(\frac{h^2}{2\pi mkT}\right)^{1/2} = \left(\frac{\left(6.626 \times 10^{-34} \text{ J s}\right)^2}{2\pi\left(\frac{0.028 \text{ kg mol}^{-1}}{N_A}\right)\left(1.38 \times 10^{-23} \text{ J K}^{-1}\right)(77.3 \text{ K})}\right)^{1/2} = 3.75 \times 10^{-11} \text{ m}$$

$$q_T(N_2) = \frac{V}{\left(3.75 \times 10^{-11} \text{ m}\right)^3} = \frac{\left(1 \text{ cm}^3\right)\left(10^{-6} \text{ m}^3 \text{ cm}^{-3}\right)}{\left(3.75 \times 10^{-11} \text{ m}\right)^3} = 1.90 \times 10^{25}$$

Next, the number of molecules (N) present at this temperature is determined using the ideal gas law:

$$n = \frac{PV}{RT} = \frac{(1 \text{ atm})\left(1.00 \times 10^{-3} \text{ L}\right)}{\left(0.0821 \text{ L atm mol}^{-1} \text{ K}^{-1}\right)(77.3 \text{ K})} = 1.58 \times 10^{-4} \text{ mol}$$

$$N = n \times N_A = 9.49 \times 10^{19} \text{ molecules}$$

With N, the ratio is readily determined:

$$\frac{q_T}{N} = \frac{1.90 \times 10^{25}}{9.49 \times 10^{19}} = 2.00 \times 10^5$$

P31.11) Consider *para*-H_2 ($B = 60.853$ cm^{-1}) for which only even-J levels are available. Evaluate the rotational partition function for this species at 50. K. Perform this same calculation for HD ($B = 45.655$ cm^{-1}).

For para-H_2, only even J levels are allowed; therefore, the rotational partition function is:

$$q_R = \sum_{J=0,2,4,6,\dots} (2J+1)e^{-\beta hcBJ(J+1)} = 1+5e^{-\frac{(6.626\times10^{-34}\text{ J s})(3.00\times10^{10}\text{ m s}^{-1})(60.853\text{ cm}^{-1})(6)}{(1.38\times10^{-23}\text{ J K}^{-1})(50.\text{ K})}}$$

$$+9e^{-\frac{(6.626\times10^{-34}\text{ J s})(3.00\times10^{10}\text{ m s}^{-1})(60.853\text{ cm}^{-1})(20)}{(1.38\times10^{-23}\text{ J K}^{-1})(50.\text{ K})}} + \dots$$

$$= 1+1.35\times10^{-4}+\dots$$

$$\cong 1.00$$

Performing this same calculation for HD where both even and odd J states are allowed:

$$q_R = \sum_{J=0,1,2,3,\dots} (2J+1)e^{-\beta hcBJ(J+1)} = 1+3e^{-\frac{(6.626\times10^{-34}\text{ J s})(3.00\times10^{10}\text{ m s}^{-1})(45.655\text{ cm}^{-1})(2)}{(1.38\times10^{-23}\text{ J K}^{-1})(50.\text{ K})}}$$

$$+5e^{-\frac{(6.626\times10^{-34}\text{ J s})(3.00\times10^{10}\text{ m s}^{-1})(45.655\text{ cm}^{-1})(6)}{(1.38\times10^{-23}\text{ J K}^{-1})(50.\text{ K})}} + \dots$$

$$= 1+0.216+3.74\times10^{-4}+\dots$$

$$\cong 1.22$$

P31.12) Calculate the rotational partition function for the interhalogen compound $F^{35}Cl$ ($B = 0.516$ cm^{-1}) at 298 K.

$$q_R = \frac{1}{\sigma\beta B} = \frac{kT}{B} = \frac{(0.695\text{ cm}^{-1}\text{ K}^{-1})(298\text{ K})}{(0.516\text{ cm}^{-1})} = 401$$

P31.15) Calculate the rotational partition function for SO_2 at 298 K where $B_A = 2.03$ cm^{-1}, $B_B = 0.344$ cm^{-1}, and $B_C = 0.293$ cm^{-1}.

$$q_R = \frac{\sqrt{\pi}}{\sigma}\left(\frac{1}{\beta B_A}\right)^{1/2}\left(\frac{1}{\beta B_B}\right)^{1/2}\left(\frac{1}{\beta B_C}\right)^{1/2}$$

$$= \frac{\sqrt{\pi}}{2}\left(\frac{(0.695\text{ cm}^{-1})(298\text{ K})}{2.03\text{ cm}^{-1}}\right)^{1/2}\left(\frac{(0.695\text{ cm}^{-1})(298\text{ K})}{0.344\text{ cm}^{-1}}\right)^{1/2}\left(\frac{(0.695\text{ cm}^{-1})(298\text{ K})}{0.293\text{ cm}^{-1}}\right)^{1/2}$$

$$\cong 5832$$

P31.18) What R-branch transition in the ro-vibrational spectrum of IF ($B = 0.280$ cm^{-1}) is expected to be the most intense at 298 K?

This problem can be solved using the expression relating temperature to J for the maximum transition in the ro-vibrational spectrum and solving for J:

$$T = \frac{(2J+1)^2 hcB}{2k} \rightarrow (2J+1)^2 = \frac{2kT}{hcB}$$

$$(2J+1)^2 = \frac{2(1.38 \times 10^{-23} \text{ J K}^{-1})(298 \text{ K})}{(6.626 \times 10^{-34} \text{ J s})(3.00 \times 10^{10} \text{ cm s}^{-1})(0.280 \text{ cm}^{-1})}$$

$$(2J+1)^2 = 1450$$

$$2J+1 = 38$$

$$J = 18.5$$

Rounding this answer for J up to 19 results in the 19–20 transition as being predicted to be most intense.

P31.20)

a) Calculate the percent population of the fist 10 rotational energy levels for HBr (B = 8.46 cm^{-1}) at 298 K.

b) Repeat this calculation for HF assuming that the bond length of this molecule is identical to that of HBr.

Since $T \gg \Theta_R$, the high-temperature limit is valid. In this limit, the probability of occupying a specific rotational state (p_J) is:

$$p_J = \frac{(2J+1)e^{-\beta hcBJ(J+1)}}{q} = \frac{(2J+1)e^{-\beta hcBJ(J+1)}}{\left(\dfrac{1}{\sigma\beta hcB}\right)}$$

Evaluating the above expression for $J = 0$:

$$p_J = \frac{(2J+1)e^{-\beta hcBJ(J+1)}}{\left(\dfrac{1}{\sigma\beta hcB}\right)} = \frac{1}{\left(\dfrac{1}{\sigma\beta hcB}\right)}$$

$$= \frac{\sigma hcB}{kT} = \frac{(1)(6.626 \times 10^{-34} \text{ J s})(3.00 \times 10^{10} \text{ cm s}^{-1})(8.46 \text{ cm}^{-1})}{(1.38 \times 10^{-23} \text{ J K}^{-1})(298 \text{ K})} = 0.0409$$

Performing similar calculations for $J = 1$ to 9:

J	p_J	J	p_J
0	0.0409	5	0.132
1	0.113	6	0.0955
2	0.160	7	0.0622
3	0.175	8	0.0367
4	0.167	9	0.0196

a) The rotational constant of HF must be determined before the corresponding level probabilities can be evaluated. The ratio of rotational constants for HBr versus HF yields:

$$\frac{B_{HBr}}{B_{HF}} = \frac{\left(\dfrac{h}{8\pi^2 c I_{HBr}}\right)}{\left(\dfrac{h}{8\pi^2 c I_{HF}}\right)} = \frac{I_{HF}}{I_{HBr}} = \frac{\mu_{HF} r^2}{\mu_{HBr} r^2} = \frac{\mu_{HF}}{\mu_{HBr}} = \frac{\dfrac{m_H m_F}{m_H + m_F}}{\dfrac{m_H m_{Br}}{m_H + m_{Br}}} = 0.962$$

$$B_{HF} = \frac{B_{HBr}}{0.962} = 8.80 \text{ cm}^{-1}$$

With this rotational constant, the p_J values for $J = 0$ to 9 are:

J	p_J	J	p_J
0	0.0425	5	0.131
1	0.117	6	0.0927
2	0.165	7	0.0590
3	0.179	8	0.0338
4	0.163	9	0.0176

P31.24) Evaluate the vibrational partition function for H_2O at 2000. K where the vibrational frequencies are 1615, 3694, and 3802 cm^{-1}.

The total vibrational partition function is the product of partition functions for each vibrational degree of freedom:

$$q_{V,total} = (q_{V,1})(q_{V,2})(q_{V,3})$$

$$= \left(\frac{1}{1 - e^{\beta hc \bar{v}_1}}\right)\left(\frac{1}{1 - e^{\beta hc \bar{v}_2}}\right)\left(\frac{1}{1 - e^{\beta hc \bar{v}_3}}\right)$$

$$= \left(\frac{1}{1 - e^{-\frac{(6.626 \times 10^{-34} \text{ J s})(3.00 \times 10^{10} \text{ cm s}^{-1})(1615 \text{ cm}^{-1})}{(1.38 \times 10^{-23} \text{ J K}^{-1})(2000. \text{ K})}}}\right)\left(\frac{1}{1 - e^{-\frac{(6.626 \times 10^{-34} \text{ J s})(3.00 \times 10^{10} \text{ cm s}^{-1})(3694 \text{ cm}^{-1})}{(1.38 \times 10^{-23} \text{ J K}^{-1})(2000. \text{ K})}}}\right)$$

$$\times \left(\frac{1}{1 - e^{-\frac{(6.626 \times 10^{-34} \text{ J s})(3.00 \times 10^{10} \text{ cm s}^{-1})(3802 \text{ cm}^{-1})}{(1.38 \times 10^{-23} \text{ J K}^{-1})(2000. \text{ K})}}}\right)$$

$$= 1.67$$

P31.26) Evaluate the vibrational partition function for NH_3 at 1000. K for which the vibrational frequencies are 950., 1627.5 (doubly degenerate), 3335, and 3414 cm^{-1} (doubly degenerate). Are there any modes that you can disregard in this calculation? Why or why not?

The total vibrational partition function is the product of partition functions for each vibrational degree of freedom, with the partition function for the mode with degeneracy raised to the power equal to the degeneracy:

$$q_{V,total} = (q_{V,1})(q_{V,2})^2(q_{V,3})(q_{V,4})^2$$

$$= \left(\frac{1}{1-e^{\beta hc\tilde{v}_1}}\right)\left(\frac{1}{1-e^{\beta hc\tilde{v}_2}}\right)^2\left(\frac{1}{1-e^{\beta hc\tilde{v}_3}}\right)\left(\frac{1}{1-e^{\beta hc\tilde{v}_4}}\right)^2$$

$$= \left(\frac{1}{1-e^{\frac{(6.626\times10^{-34}\,\text{J s})(3.00\times10^{10}\,\text{cm s}^{-1})(950.\,\text{cm}^{-1})}{(1.38\times10^{-23}\,\text{J K}^{-1})(1000.\,\text{K})}}}\right)\left(\frac{1}{1-e^{\frac{(6.626\times10^{-34}\,\text{J s})(3.00\times10^{10}\,\text{cm s}^{-1})(1627.5\,\text{cm}^{-1})}{(1.38\times10^{-23}\,\text{J K}^{-1})(1000.\,\text{K})}}}\right)^2$$

$$\times \left(\frac{1}{1-e^{\frac{(6.626\times10^{-34}\,\text{J s})(3.00\times10^{10}\,\text{cm s}^{-1})(3335\,\text{cm}^{-1})}{(1.38\times10^{-23}\,\text{J K}^{-1})(1000.\,\text{K})}}}\right)\left(\frac{1}{1-e^{\frac{(6.626\times10^{-34}\,\text{J s})(3.00\times10^{10}\,\text{cm s}^{-1})(3414\,\text{cm}^{-1})}{(1.38\times10^{-23}\,\text{J K}^{-1})(1000.\,\text{K})}}}\right)^2$$

$$= (1.34)(1.11)^2(1.01)(1.01)^2$$

$$= 1.70$$

Notice that the two highest-frequency vibrational degrees of freedom have partition functions that are near unity; therefore, their contribution to the total partition function is modest, and could be ignored to a reasonable approximation in evaluating the total partition function.

$$\%error = \frac{q_{anharm} - q_{harm}}{q_{anharm}} \times 100\% = 0.03\%$$

P31.32) Consider a particle free to translate in one dimension. The classical Hamiltonian is $H = \frac{p^2}{2m}$.

a) Determine $q_{classical}$ for this system. To what quantum system should you compare it in order to determine the equivalence of the classical and quantum statistical mechanical treatments?
b) Derive $q_{classical}$ for a system with translational motion in three dimensions for which:
$$H = \left(p_x^2 + p_y^2 + p_z^2\right)/2m.$$

a) The particle in a one-dimensional box model is the appropriate quantum-mechanical model for comparison. Integrating

$$q_{class} = \frac{1}{h}\int_{-\infty}^{\infty}\int_0^L e^{\frac{-\beta p^2}{2m}}\,dx\,dp = \frac{2L}{h}\int_0^{\infty} e^{\frac{-\beta p^2}{2m}}\,dp = \frac{2L}{h}\left(\frac{1}{2}\sqrt{\frac{\pi}{\beta/2m}}\right) = L\left(\frac{h^2}{2\pi mkT}\right)^{-1/2} = \frac{L}{\Lambda}$$

b)

$$q_{class} = \frac{1}{h^3} \int_{-\infty}^{\infty}\int_{-\infty}^{\infty}\int_{-\infty}^{\infty}\int_{0}^{L_x}\int_{0}^{L_y}\int_{0}^{L_z} e^{\frac{-\beta(p_x^2+p_y^2+p_z^2)}{2m}} dx\,dy\,dz\,dp_x\,dp_y\,dp_z$$

$$= \frac{8(L_x L_y L_z)}{h^3} \int_{0}^{\infty}\int_{0}^{\infty}\int_{0}^{\infty} e^{\frac{-\beta(p_x^2+p_y^2+p_z^2)}{2m}} dp_x\,dp_y\,dp_z$$

$$= \frac{8V}{h^3}\left(\frac{1}{2}\sqrt{\frac{\pi}{\beta/2m}}\right)^3 = V\left(\frac{h^2}{2\pi mkT}\right)^{-3/2} = \frac{L}{\Lambda^3}$$

P31.34)
a) Evaluate the electronic partition function for atomic Si at 298 K given the following energy levels:

Level (n)	Energy (cm^{-1})	Degeneracy
0	0	1
1	77.1	3
2	223.2	5
3	6298	5

b) At what temperature will the $n = 3$ energy level contribute 0.1 to the electronic partition function?

a)

$$q_E = \sum_n g_n e^{-\beta\varepsilon_n} = 1e^{-0} + 3e^{-\beta(77.1 \text{ cm}^{-1})} + 5e^{-\beta(223.2 \text{ cm}^{-1})} + 5e^{-\beta(6298 \text{ cm}^{-1})}$$

$$= 1 + 3e^{-\frac{77.1 \text{ cm}^{-1}}{(0.695 \text{ cm}^{-1} \text{ K}^{-1})(298 \text{ K})}} + 5e^{-\frac{223.2 \text{ cm}^{-1}}{(0.695 \text{ cm}^{-1} \text{ K}^{-1})(298 \text{ K})}} + 3e^{-\frac{6298 \text{ cm}^{-1}}{(0.695 \text{ cm}^{-1} \text{ K}^{-1})(298 \text{ K})}}$$

$$= 1 + 3(0.689) + 5(0.340) + 5(6.22 \times 10^{-14})$$

$$q_E = 4.77$$

b)   Focusing on the contribution to q_E from the $n = 3$ level:

$$0.1 = g_3 e^{-\beta\varepsilon_3} = 5e^{-\frac{6298 \text{ cm}^{-1}}{(0.695 \text{ cm}^{-1} \text{ K}^{-1})(T)}}$$

$$3.91 = \frac{6298 \text{ cm}^{-1}}{(0.695 \text{ cm}^{-1} \text{ K}^{-1})(T)}$$

$$T = \frac{6298 \text{ cm}^{-1}}{(0.695 \text{ cm}^{-1} \text{ K}^{-1})(3.91)}$$

$$T = 2318 \text{ K}$$

P31.37) Determine the total molecular partition function for I_2, confined to a volume of 1000. cm^3 at 298 K. Other information you will find useful: $B = 0.0374$ cm^{-1}, $\tilde{v} = 208$ cm^{-1}, and the ground electronic state is non-degenerate.

Since q_{Total} is the product of partition functions for each energetic degree of freedom (translational, rotational, vibrational, and electronic), it is more straightforward to calculate the partition function for each of these degrees of freedom separately, and then take the product of these functions:

$$q_T = \frac{V}{\Lambda^3}$$

$$\Lambda = \left(\frac{h^2}{2\pi mkT}\right)^{1/2} = \left(\frac{\left(6.626\times10^{-34}\text{ J s}\right)^2}{2\pi\left(\dfrac{0.254\text{ kg mol}^{-1}}{N_A}\right)\left(1.38\times10^{-23}\text{ J K}^{-1}\right)\left(298\text{ K}\right)}\right)^{1/2} = 6.35\times10^{-12}\text{ m}$$

$$q_T = \frac{V}{\left(6.35\times10^{-12}\text{ m}\right)^3} = \frac{\left(1000.\text{ cm}^3\right)\left(10^{-6}\text{ m}^3\text{ cm}^{-3}\right)}{\left(6.35\times10^{-12}\text{ m}\right)^3} = 3.91\times10^{30}$$

$$q_R = \left(\frac{1}{\sigma\beta hcB}\right) = \left(\frac{\left(1.38\times10^{-23}\text{ J K}^{-1}\right)\left(298\text{ K}\right)}{\left(2\right)\left(6.626\times10^{-34}\text{ J s}\right)\left(3.00\times10^{10}\text{ cm s}^{-1}\right)\left(0.0374\text{ cm}^{-1}\right)}\right)$$

$$= 2.77\times10^3$$

$$q_V = \frac{1}{1-e^{-\beta hc\tilde{v}}} = \frac{1}{1-e^{\frac{\left(6.626\times10^{-34}\text{ J s}\right)\left(3.00\times10^{10}\text{ cm s}^{-1}\right)\left(208\text{ cm}^{-1}\right)}{\left(1.38\times10^{-23}\text{ J K}^{-1}\right)\left(298\text{ K}\right)}}}$$

$$= 1.58$$

$$q_E = 1$$

$$q_{total} = q_T q_R q_V q_E = \left(3.91\times10^{30}\right)\left(2.77\times10^3\right)\left(1.58\right)\left(1\right) = 1.71\times10^{34}$$

Computational Problems

Before solving the computational problems, it is recommended that students work through Tutorials 1–3 under the Help menus in Spartan Student Edition to gain familiarity with the program.

Computational Problem 31.1: Using the Hartree–Fock 6-31G* basis set, determine the vibrational frequencies for F_2O and Br_2O and calculate the vibrational partition function for these species at 500 K.

Procedure

Step 1: Create a new file and build F_2O.
Step 2: Go to "Setup > Calculations." Set the calculation type to equilibrium geometry and the method to **Hartree-Fock 6-31G***.

Step 3: In the Calculations dialog window, make sure "**Equilibrium Geometry**" is selected under the Calculate pull-down menu, check the box next to "**IR.**"
Step 4: Click the "**Submit**" button in the Calculations dialog window. You will be asked to provide a file name (F_2O would work just fine), and the calculation will begin.

Step 5: When the calculation is completed a dialog box will appear. Click "ok" in this box. Go to "**Display > Spectra**" and the three vibrational frequencies for F_2O will appear. Record those frequencies.
Step 6: Repeat the above steps for Br_2O.

Computational Problem 31.2: Halogenated methanes are of interest as greenhouse gases. Perform a Hartree–Fock 6-31G* calculation of CFH_3 including determination of the IR spectrum. Does this molecule possess any strong transitions in the infrared? Calculate the vibrational partition function for the vibrational degree of freedom with the strongest predicted IR-absorption intensity.

Procedure

Step 1: Create a new file and build CFH_3.
Step 2: Go to "**Setup > Calculations.**" Set the calculation type to equilibrium geometry and the method to **Hartree–Fock 6-31G***.
Step 3: In the Calculations dialog window, make sure "**Equilibrium Geometry**" is selected under the Calculate pull-down menu, check the box next to "**IR**".
Step 4: Click the "**Submit**" button in the Calculations dialog window. Once you provide a file name the calculation will begin.
Step 5: When the calculation is completed a dialog box will appear. Click "ok" in this box. Go to "**Display > Spectra**" and a dialog box displaying the vibrational frequencies for CFH_3 will appear with corresponding IR intensities.
Step 6: To view the IR spectrum click the "**Draw IR Spectrum**" box. Record the frequency of the most intense IR transition.

Computational Problem 31.3: Nitryl chloride ($ClNO_2$) is an important compound in atmospheric chemistry. It serves as a reservoir compound for Cl, and is produced in the lower troposphere through the reaction of N_2O_5 with Cl^--containing aerosols. Using the Hartree–Fock 3-21G basis set, determine the value of the vibrational partition function at 260. K. Does your answer change appreciably if you use the 6-31G* basis set?

Procedure

Step 1: Create a new file and build $ClNO_2$ (Hint: you may want to draw the Lewis dot structure first to figure out the bonding).
Step 2: Go to "**Setup > Calculations.**" Set the calculation type to equilibrium geometry and the method to **Hartree–Fock 3-21G**.
Step 3: In the Calculations dialog window, make sure "**Equilibrium Geometry**" is selected under the Calculate pull-down menu, check the box next to "**IR**".
Step 4: Click the "**Submit**" button in the Calculations dialog window. Once you provide a file name the calculation will begin.
Step 5: When the calculation is completed a dialog box will appear. Click "ok" in this box. Go to "**Display > Spectra**" and a dialog box displaying the vibrational frequencies will appear.

Computational Problem 31.4: The rotational constant for 1,3-butadiyne (C_4H_2) is 4391.19 MHz. Performing a Hartree–Fock 6-31G* calculation, minimize the geometry of this species and compare the effective bond length determined using the rotational constant to the actual geometry of this species.

Procedure

Step 1: Create a new file and build 1,3-butadiyne (Hint: all of the carbons are *sp* hybridized corresponding to two triple bonds).
Step 2: Go to "Setup > Calculations." Set the calculation type to equilibrium geometry and the method to **Hartree–Fock 6-31G***.
Step 3: In the Calculations dialog window, make sure "**Equilibrium Geometry**" is selected under the Calculate pull-down menu.
Step 4: Click the "**Submit**" button in the Calculations dialog window. Once you provide a file name the calculation will begin.
Step 5: When the calculation is completed a dialog box will appear. Click "ok" in this box.
Step 6: On the toolbar select the "**<?>**" button to select the distance tool. Click on the two terminal hydrogens to determine the bond length.

Computational Problem 31.5: Using Hartree–Fock with the 3-21 basis set, determine the value of the vibrational partition function for CF_2Cl_2 and CCl_4 at 298 K. What do you notice about the vibrational frequencies for the higher-symmetry species?

Procedure

Step 1: Create a new file and build CF_2Cl_2.
Step 2: Go to "Setup > Calculations." Set the calculation type to equilibrium geometry and the method to **Hartree–Fock 3-21G**.
Step 3: In the Calculations dialog window, make sure "**Equilibrium Geometry**" is selected under the Calculate pull-down menu, check the box next to "**IR**."
Step 4: Click the "**Submit**" button in the Calculations dialog window. You will be asked to provide a file name, and the calculation will begin.
Step 5: When the calculation is completed a dialog box will appear. Click "ok" in this box. Go to "**Display > Spectra**" and three vibrational frequencies will appear. Record those frequencies.
Step 6: Repeat the above steps for CCl_4.

Computational Problem 31.6: The high-temperature limit was shown to be of limited applicability to vibrational degrees of freedom. As will be shown in the next chapter, this has a profound consequence when exploring the role of vibrations in statistical thermodynamics. For example, a coarse rule of thumb is that one usually does not need to worry about the contribution of CH, NH, and OH stretch vibrations. Perform a Hartree–Fock 6-31G* calculation on 1,3-cyclohexadiene, and calculate the IR intensities. Looking at the atomic displacements that make up the normal modes, identify those modes that are predominately of C-H stretch character. In what frequency range are these modes located? Identify the lowest-energy C-H stretch, and the lowest-frequency mode in general. Compare the value of the partition function for these two vibrational degrees of freedom at 298 K. How does this comparison support the "rule of thumb"?

Procedure

Step 1: Create a new file and build 1,3-cyclohexadiene.
Step 2: Go to "**Setup > Calculations.**" Set the calculation type to equilibrium geometry and the method to **Hartree-Fock 6-31G***.
Step 3: In the Calculations dialog window, make sure "**Equilibrium Geometry**" is selected under the Calculate pull-down menu, check the box next to "IR."
Step 4: Click the "**Submit**" button in the Calculations dialog window. Once you provide a file name the calculation will begin.
Step 5: When the calculation is completed a dialog box will appear. Click "ok" in this box. Go to "**Display > Spectra**" and a dialog box displaying the vibrational frequencies will appear. Record the lowest vibrational frequency.
Step 6: To view the IR spectrum click the "**Draw IR Spectrum**" box. Record the frequency of the most intense IR transition.
Step 7: In the Spectra dialog box, select one of the vibrational modes and view the corresponding atomic displacements depicted. View all of the vibrational modes to determine which modes have substantial C-H stretch character, and identify the mode with the lowest frequency.

Computational Problem 31.7: You are interested in designing a conducting polyene-based polymer. The design calls for maximizing the conjugation length, but as the conjugation length increases, the HOMO-LUMO energy gap corresponding to the lowest-energy electronic transition decreases. If the gap becomes too small, then thermal excitation can result in population of the excited state. If thermal excitation is significant, it will degrade the performance of the polymer.

  a. With a tolerance of 2% population in the first electronic excited state at 373 K, what is the smallest electronic energy gap that can be tolerated?

$$\frac{p_1}{p_0} = e^{-(\varepsilon_1 - \varepsilon_0)/kT}$$

$$\frac{0.02}{0.98} = e^{-(\varepsilon_1 - \varepsilon_0)/kT}$$

$$\ln(49) = \frac{(\varepsilon_1 - \varepsilon_0)}{\left(1.38 \times 10^{-23} \text{ J K}^{-1}\right)(373 \text{ K})}$$

$$2.00 \times 10^{-20} \text{ J} = 0.125 \text{ eV} = (\varepsilon_1 - \varepsilon_0)$$

  b. Using the Hartree–Fock 3-21G basis set, calculate the HOMO-LUMO energy gap for 1,3,5-hexatriene, 1,3,5,7-octatetraene, and 1,3,5,7,9-decapentaene. Using a plot of energy gap versus number of double bonds, determine the longest polyene structure that can be achieved while maintaining the 2% excited-state population tolerance assuming that the energy gap varies linearly with conjugation length. (Note: This method provides a very rough estimate of the actual energy gap.)

Procedure

Step 1: Create a new file and build 1,3,5-hexatriene.

Step 2: Go to "Setup > Calculations." Set the calculation type to equilibrium geometry and the method to **Hartree–Fock 3-21G**.

Step 3: In the Calculations dialog window, make sure "**Equilibrium Geometry**" is selected under the Calculate pull-down menu, check the box next to "**IR.**"

Step 4: Click the "**Submit**" button in the Calculations dialog window. Once you provide a file name the calculation will begin.

Step 5: When the calculation is finished, go to "**Display > Output**" to view the output file for the calculation. Scroll down until you locate the orbital energies. Identify the HOMO and LUMO, and calculate the energy difference between these two molecular orbitals.

Step 6: Repeat the above steps for the two longer polyenes.

Step 7: Construct a graph of energy gap versus number of double bonds, and determine which polyene is predicted to have an energy gap as close to 0.125 eV as possible.

c. (Advanced) In part b it was assumed that the HOMO-LUMO energy gap decreases linearly with an increase in conjugation length. Check this assumption by calculating this energy gap for the next two polyenes. Does the linear correlation continue?

Chapter 32: Statistical Thermodynamics

P32.2) Consider two separate ensembles of particles characterized by the energy-level diagram provided in the text. Derive expressions for the internal energy for each ensemble. At 298 K, which ensemble is expected to have the greatest internal energy?

$$U = -\left(\frac{\partial \ln Q}{\partial \beta}\right)_V = -N\left(\frac{\partial \ln q}{\partial \beta}\right)_V = \frac{-N}{q}\left(\frac{\partial q}{\partial \beta}\right)_V$$

$$q_A = \sum_n g_n e^{-\beta \varepsilon_n} = 1 + e^{-\beta(300.\ \text{cm}^{-1})} + e^{-\beta(600.\ \text{cm}^{-1})}$$

$$U_A = \frac{-N}{1 + e^{-\beta(300.\ \text{cm}^{-1})} + e^{-\beta(600.\ \text{cm}^{-1})}}\left(\frac{\partial}{\partial \beta}\left(1 + e^{-\beta(300.\ \text{cm}^{-1})} + e^{-\beta(600.\ \text{cm}^{-1})}\right)\right)$$

$$= \frac{-N}{1 + e^{-\beta(300.\ \text{cm}^{-1})} + e^{-\beta(600.\ \text{cm}^{-1})}}\left(-\left(\left(300.\ \text{cm}^{-1}\right)e^{-\beta(300\ \text{cm}^{-1})} + \left(600.\ \text{cm}^{-1}\right)e^{-\beta(600.\ \text{cm}^{-1})}\right)\right)$$

$$= \frac{N\left(\left(300.\ \text{cm}^{-1}\right)e^{-\beta(300.\ \text{cm}^{-1})} + \left(600.\ \text{cm}^{-1}\right)e^{-\beta(600.\ \text{cm}^{-1})}\right)}{1 + e^{-\beta(300.\ \text{cm}^{-1})} + e^{-\beta(600.\ \text{cm}^{-1})}}$$

$$q_B = \sum_n g_n e^{-\beta \varepsilon_n} = 1 + 2e^{-\beta(300.\ \text{cm}^{-1})} + e^{-\beta(600.\ \text{cm}^{-1})}$$

$$U_B = \frac{-N}{1 + 2e^{-\beta(300.\ \text{cm}^{-1})} + e^{-\beta(600.\ \text{cm}^{-1})}}\left(\frac{\partial}{\partial \beta}\left(1 + 2e^{-\beta(300.\ \text{cm}^{-1})} + e^{-\beta(600.\ \text{cm}^{-1})}\right)\right)$$

$$= \frac{-N}{1 + 2e^{-\beta(300.\ \text{cm}^{-1})} + e^{-\beta(600.\ \text{cm}^{-1})}}\left(-\left(\left(600.\ \text{cm}^{-1}\right)e^{-\beta(300.\ \text{cm}^{-1})} + \left(600.\ \text{cm}^{-1}\right)e^{-\beta(600.\ \text{cm}^{-1})}\right)\right)$$

$$= \frac{N\left(600.\ \text{cm}^{-1}\right)\left(e^{-\beta(300.\ \text{cm}^{-1})} + e^{-\beta(600.\ \text{cm}^{-1})}\right)}{1 + 2e^{-\beta(300.\ \text{cm}^{-1})} + e^{-\beta(600.\ \text{cm}^{-1})}}$$

Evaluating the expressions for U_A and U_B at 298 K:

$$U_A = \frac{N\left((300.\ \text{cm}^{-1})e^{-\frac{(300.\ \text{cm}^{-1})}{(0.695\ \text{cm}^{-1}\ \text{K}^{-1})(298\ \text{K})}} + (600.\ \text{cm}^{-1})e^{-\frac{(600.\ \text{cm}^{-1})}{(0.695\ \text{cm}^{-1}\ \text{K}^{-1})(298\ \text{K})}}\right)}{1 + e^{-\frac{(300\ \text{cm}^{-1})}{(0.695\ \text{cm}^{-1}\ \text{K}^{-1})(298\ \text{K})}} + e^{-\frac{(600\ \text{cm}^{-1})}{(0.695\ \text{cm}^{-1}\ \text{K}^{-1})(298\ \text{K})}}}$$

$$= N(80.3\ \text{cm}^{-1})$$

$$U_B = \frac{N(600.\ \text{cm}^{-1})\left(e^{-\frac{(300.\ \text{cm}^{-1})}{(0.695\ \text{cm}^{-1}\ \text{K}^{-1})(298\ \text{K})}} + e^{-\frac{(600.\ \text{cm}^{-1})}{(0.695\ \text{cm}^{-1}\ \text{K}^{-1})(298\ \text{K})}}\right)}{1 + 2e^{-\frac{(300.\ \text{cm}^{-1})}{(0.695\ \text{cm}^{-1}\ \text{K}^{-1})(298\ \text{K})}} + e^{-\frac{(600.\ \text{cm}^{-1})}{(0.695\ \text{cm}^{-1}\ \text{K}^{-1})(298\ \text{K})}}}$$

$$= N(114\ \text{cm}^{-1})$$

Ensemble B will have the larger internal energy.

P32.8) The lowest four energy levels for atomic vanadium (V) have the following energies and degeneracies:

Level (n)	Energy (cm^{-1})	Degeneracy
0	0	4
1	137.38	6
2	323.46	8
3	552.96	10

What is the contribution to the average energy from electronic degrees of freedom for V when $T = 298$ K?

$$q = \sum_n g_n e^{-\beta \varepsilon_n} = 4 + 6e^{-\beta(137.38\ \text{cm}^{-1})} + 8e^{-\beta(323.46\ \text{cm}^{-1})} + 10e^{-\beta(552.96\ \text{cm}^{-1})}$$

$$\left(\frac{\partial q}{\partial \beta}\right)_V = -\left((824\ \text{cm}^{-1})e^{-\beta(137.38\ \text{cm}^{-1})} + (2588\ \text{cm}^{-1})e^{-\beta(323.46\ \text{cm}^{-1})} + (5530\ \text{cm}^{-1})e^{-\beta(552.96\ \text{cm}^{-1})}\right)$$

$$U = \frac{-N}{q}\left(\frac{\partial q}{\partial \beta}\right)_V = N\frac{(824\ \text{cm}^{-1})e^{-\beta(137.38\ \text{cm}^{-1})} + (2588\ \text{cm}^{-1})e^{-\beta(323.46\ \text{cm}^{-1})} + (5530\ \text{cm}^{-1})e^{-\beta(552.96\ \text{cm}^{-1})}}{4 + 6e^{-\beta(137.38\ \text{cm}^{-1})} + 8e^{-\beta(323.46\ \text{cm}^{-1})} + 10e^{-\beta(552.96\ \text{cm}^{-1})}}$$

$$= N(143\ \text{cm}^{-1})$$

converting to J:

$$U = N(143\ \text{cm}^{-1})hc = nN_A(143\ \text{cm}^{-1})(6.626 \times 10^{-34}\ \text{J s})(3.00 \times 10^{10}\ \text{cm s}^{-1}) = n(1.71\ \text{kJ mol}^{-1})$$

$$U_m = 1.71\ \text{kJ mol}^{-1}$$

P32.10) Consider an ensemble of units in which the first excited electronic state at energy ε_1 is m_1-fold degenerate, and the energy of the ground state is m_0-fold degenerate with energy ε_0.

a) Demonstrate that if $\varepsilon_0 = 0$, the expression for the electronic partition function is

$$q_E = m_0\left(1 + \frac{m_1}{m_0}e^{-\varepsilon_1/kT}\right)$$

b) Determine the expression for the internal energy U of an ensemble of N such units. What is the limiting value of U as the temperature approaches zero and infinity?

a)

$$q = m_0 e^{-\beta\varepsilon_0} + m_1 e^{-\beta\varepsilon_1} = m_0 + m_1 e^{-\beta\varepsilon_1}$$

$$= m_0 + m_0\left(\frac{m_1}{m_0}\right)e^{-\beta\varepsilon_1}$$

$$= m_0\left(1 + \left(\frac{m_1}{m_0}\right)e^{-\beta\varepsilon_1}\right)$$

$$= m_0\left(1 + \left(\frac{m_1}{m_0}\right)e^{-\frac{\varepsilon_1}{kT}}\right)$$

b)

$$U = \frac{-N}{q}\left(\frac{\partial q}{\partial \beta}\right)_V = \frac{-N}{q}\left(\frac{\partial}{\partial \beta}\left(m_0\left(1 + \left(\frac{m_1}{m_0}\right)e^{-\beta\varepsilon_1}\right)\right)\right)_V$$

$$= \frac{-N}{q}\left(-m_1\varepsilon_1 e^{-\beta\varepsilon_1}\right)$$

$$= \frac{Nm_1\varepsilon_1 e^{-\beta\varepsilon_1}}{m_0\left(1 + \left(\frac{m_1}{m_0}\right)e^{-\beta\varepsilon_1}\right)} = \frac{Nm_1\varepsilon_1 e^{-\frac{\varepsilon_1}{kT}}}{m_0\left(1 + \left(\frac{m_1}{m_0}\right)e^{-\frac{\varepsilon_1}{kT}}\right)}$$

Looking at the limiting behavior with temperature:

$$\lim_{T\to 0} U == \lim_{T\to 0}\frac{Nm_1\varepsilon_1 e^{-\frac{\varepsilon_1}{kT}}}{m_0\left(1 + \left(\frac{m_1}{m_0}\right)e^{-\frac{\varepsilon_1}{kT}}\right)} = \lim_{T\to 0}\frac{Nm_1\varepsilon_1}{m_0\left(e^{\frac{\varepsilon_1}{kT}} + \left(\frac{m_1}{m_0}\right)\right)} = 0$$

$$\lim_{T\to\infty} U == \lim_{T\to 0}\frac{Nm_1\varepsilon_1 e^{-\frac{\varepsilon_1}{kT}}}{m_0\left(1 + \left(\frac{m_1}{m_0}\right)e^{-\frac{\varepsilon_1}{kT}}\right)} = \frac{Nm_1\varepsilon_1}{m_0 + m_1}$$

P32.14) Determine the vibrational contribution to C_V for a mole of HCl ($\tilde{v} = 2886$ cm^{-1}) over a temperature range from 500 to 5000 K in 500-K intervals and plot your result. At what temperature do you expect to reach the high-temperature limit for the vibrational contribution to C_V?

The problem requires evaluation of the following expression versus temperature:

$$C_V = \frac{N}{kT^2}(hc\tilde{\nu})^2 \frac{e^{\frac{hc\tilde{\nu}}{kT}}}{\left(e^{\frac{hc\tilde{\nu}}{kT}} - 1\right)^2}$$

Using Excel or a similar program, the following plot of the molar heat capacity versus temperature can be constructed.

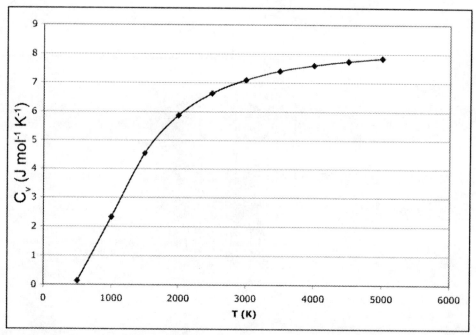

The high-temperature limit value for the molar heat capacity is (1 mol) $\times$ R = 8.314 J K^{-1}. Comparison of this value to the graph illustrates that the high-temperature limit will be value for temperatures well above 6000. Specifically, the high-temperature limit is applicable when $T > 10\Theta_V$ or ~40,000 K for HCl.

P32.15) Determine the vibrational contribution to C_V for HCN where $\tilde{\nu}_1 = 2041\,\mathrm{cm}^{-1}$, $\tilde{\nu}_2 = 712\,\mathrm{cm}^{-1}$ (doubly degenerate), and $\tilde{\nu}_3 = 3369\,\mathrm{cm}^{-1}$ at T = 298, 500., and 1000. K.

The total vibrational heat capacity will be equal to the sum of heat capacity contributions from each vibrational degree of freedom. Keeping track of the degeneracy of the 712 cm^{-1} mode, the total heat capacity can be written as:

$$C_{V,total} = C_{V,\tilde{\nu}_1} + 2C_{V,\tilde{\nu}_2} + C_{V,\tilde{\nu}_3}$$

where the heat capacity for a specific mode is determined using:

$$C_V = \frac{N}{kT^2}(hc\tilde{\nu})^2 \frac{e^{\frac{hc\tilde{\nu}}{kT}}}{\left(e^{\frac{hc\tilde{\nu}}{kT}} - 1\right)^2}$$

Evaluating this expression for the 2041 cm^{-1} mode at 500. K yields:

$$C_V = \frac{N}{kT^2}(hc\tilde{v})^2 \frac{e^{\frac{hc\tilde{v}}{kT}}}{\left(e^{\frac{hc\tilde{v}}{kT}} - 1\right)^2}$$

$$= \frac{N}{(1.38\times10^{-23}\text{ J K}^{-1})(500.\text{ K})^2}\left((6.626\times10^{-34}\text{ J s})(3.00\times10^{10}\text{ cm s}^{-1})(2041\text{ cm}^{-1})\right)^2$$

$$\times \frac{e^{\frac{(6.626\times10^{-34}\text{ J s})(3.00\times10^{10}\text{ cm s}^{-1})(2041\text{ cm}^{-1})}{(1.38\times10^{-23}\text{ J K}^{-1})(500.\text{ K})}}}{\left(e^{\frac{(6.626\times10^{-34}\text{ J s})(3.00\times10^{10}\text{ cm s}^{-1})(2041\text{ cm}^{-1})}{(1.38\times10^{-23}\text{ J K}^{-1})(500.\text{ K})}} - 1\right)^2}$$

$$= N\left(4.77\times10^{-22}\text{ J K}^{-1}\right)\left(2.81\times10^{-3}\right)$$

$$= nN_A\left(2.29\times10^{-24}\text{ J K}^{-1}\right)$$

$$= nN_A\left(2.29\times10^{-24}\text{ J K}^{-1}\right)$$

$$= n\left(0.811\text{ J mol}^{-1}\text{ K}^{-1}\right)$$

Similar calculations for the other vibrational degrees of freedom and temperatures of interest yields the following table of molar constant volume heat capacities (units of J mol^{-1} K^{-1})

	298 K	500 K	1000 K
2041 cm^{-1}	0.042	0.811	4.24
712 cm^{-1}	3.37	5.93	7.62
3369 cm^{-1}	0.000	0.048	1.56
Total	6.78	12.7	21.0

P32.19) The speed of sound is given by the relationship

$$c_{sound} = \left(\frac{\dfrac{C_P}{C_V}RT}{M}\right)^{1/2}$$

where C_p is the constant pressure heat capacity (equal to $C_V + R$), R is the ideal gas constant, T is temperature, and M is molar mass.
a) What is the expression for the speed of sound for an ideal monatomic gas?
b) What is the expression for the speed of sound of an ideal diatomic gas?
c) What is the speed of sound in air at 298 K, assuming that air is mostly made up of nitrogen ($B = 2.00$ cm^{-1} and $\tilde{v} = 2359$ cm^{-1})?

a) For a monatomic gas, only translational degrees of freedom contribute to the C_V:

$$C_V = \frac{3}{2}Nk = \frac{3}{2}nR$$

$$\frac{C_P}{C_V} = \frac{\frac{5}{2}nR}{\frac{3}{2}nR} = \frac{5}{3}$$

$$c_{sound} = \left(\frac{\frac{C_P}{C_V}RT}{M}\right)^{1/2} = \left(\frac{\frac{5}{3}RT}{M}\right)^{1/2}$$

b) In addition to translations, rotational and vibrational degrees of freedom will also contribute to C_V:

$$C_V = C_{V,trans} + C_{V,rot} + C_{V,vib} = \frac{3}{2}nR + nR + nR\beta^2\left(hc\tilde{v}\right)^2\frac{e^{\beta hc\tilde{v}}}{\left(e^{\beta hc\tilde{v}}-1\right)^2}$$

$$= nR\left(\frac{5}{2} + \beta^2\left(hc\tilde{v}\right)^2\frac{e^{\beta hc\tilde{v}}}{\left(e^{\beta hc\tilde{v}}-1\right)^2}\right)$$

$$\frac{C_P}{C_V} = \frac{\left(\frac{7}{2} + \beta^2\left(hc\tilde{v}\right)^2\frac{e^{\beta hc\tilde{v}}}{\left(e^{\beta hc\tilde{v}}-1\right)^2}\right)}{\left(\frac{5}{2} + \beta^2\left(hc\tilde{v}\right)^2\frac{e^{\beta hc\tilde{v}}}{\left(e^{\beta hc\tilde{v}}-1\right)^2}\right)}$$

$$c_{sound} = \left(\frac{\frac{C_P}{C_V}RT}{M}\right)^{1/2} = \left(\frac{\left(\frac{\left(\frac{7}{2} + \beta^2\left(hc\tilde{v}\right)^2\frac{e^{\beta hc\tilde{v}}}{\left(e^{\beta hc\tilde{v}}-1\right)^2}\right)}{\left(\frac{5}{2} + \beta^2\left(hc\tilde{v}\right)^2\frac{e^{\beta hc\tilde{v}}}{\left(e^{\beta hc\tilde{v}}-1\right)^2}\right)}\right)RT}{M}\right)^{1/2}$$

c) First, evaluation of the vibrational contribution to C_V demonstrates that this contribution is small relative to the contribution from translational and rotational degrees of freedom, and can be neglected to good approximation:

$$\frac{C_{V,vib}}{nR} = \beta^2 \left(hc\tilde{v}\right)^2 \frac{e^{\beta hc\tilde{v}}}{\left(e^{\beta hc\tilde{v}} - 1\right)^2}$$

$$= \left(\frac{\left(6.626 \times 10^{-34} \text{ J s}\right)\left(3.00 \times 10^{10} \text{ cm s}^{-1}\right)\left(2359 \text{ cm}^{-1}\right)}{\left(1.38 \times 10^{-23} \text{ J K}^{-1}\right)\left(298 \text{ K}\right)}\right)^2$$

$$\times \frac{e^{\frac{\left(6.626 \times 10^{-34} \text{ J s}\right)\left(3.00 \times 10^{10} \text{ cm s}^{-1}\right)\left(2359 \text{ cm}^{-1}\right)}{\left(1.38 \times 10^{-23} \text{ J K}^{-1}\right)\left(298 \text{ K}\right)}}}{\left(e^{\frac{\left(6.626 \times 10^{-34} \text{ J s}\right)\left(3.00 \times 10^{10} \text{ cm s}^{-1}\right)\left(2359 \text{ cm}^{-1}\right)}{\left(1.38 \times 10^{-23} \text{ J K}^{-1}\right)\left(298 \text{ K}\right)}} - 1\right)^2}$$

$$= 1.47 \times 10^{-3} << \frac{5}{2}$$

Therefore, keeping only the translational and rotational contributions to C_V yields the following value for the speed of sound in N_2:

$$c_{sound} = \left(\frac{\frac{C_p}{C_V} RT}{M}\right)^{1/2} \cong \left(\frac{\frac{7}{5} RT}{M}\right)^{1/2} = \left(\frac{\frac{7}{5}\left(8.314 \text{ J mol}^{-1} \text{ K}^{-1}\right)\left(298 \text{ K}\right)}{\left(0.028 \text{ kg mol}^{-1}\right)}\right)^{1/2}$$

$$= 352 \text{ m s}^{-1}$$

P32.22) Determine the molar entropy for 1 mol of gaseous Ar at 200., 300., and 500. K and $V = 1000.$ cm^3 assuming that Ar can be treated as an ideal gas. How does the result of this calculation change if the gas is Kr instead of Ar?

Determining the molar entropy for gaseous Ar at 200. K yields:

$$S = \frac{U}{T} + k \ln Q = \frac{3}{2} Nk + k \ln \left(\frac{q^N}{N!} \right) = \frac{3}{2} Nk + NK \ln q - k \ln (N!)$$

$$= \frac{3}{2} Nk + Nk \ln q - k (N \ln N - N)$$

$$= \frac{5}{2} nR + nR \ln q - nR \ln(nN_A)$$

$$S_m = \frac{5}{2} R + R \ln \left(\frac{V}{\Lambda^3} \right) - R \ln \left(6.022 \times 10^{23} \right)$$

$$= 20.79 \text{ J mol}^{-1} \text{ K}^{-1} + R \ln \left(\frac{V}{\left(\frac{h^2}{2\pi mkT} \right)^{3/2}} \right) - 456 \text{ J mol}^{-1} \text{ K}^{-1}$$

$$= 20.79 \text{ J mol}^{-1} \text{ K}^{-1}$$

$$+ R \ln \left(\frac{1.00 \times 10^{-3} \text{ m}^3}{N_A \left(\frac{\left(6.626 \times 10^{-34} \text{ J s} \right)^2}{2\pi \left(\frac{0.040 \text{ kg mol}^{-1}}{N_A} \right) \left(1.38 \times 10^{-23} \text{ J K}^{-1} \right) (200. \text{ K})} \right)^{3/2}} \right) - 456 \text{ J mol}^{-1} \text{ K}^{-1}$$

$$= 123 \text{ J mol}^{-1} \text{ K}^{-1}$$

Repeating the calculation for $T = 300$ and 500 K, and the molar entropy is found to be 128 J mol^{-1} K^{-1} and 135 J mol^{-1} K^{-1}, respectively. Kr is heavier than Ar; therefore, the thermal wavelength will be shorter, and the translational partition function will correspondingly be larger. Since the molar entropy is linear related to $\ln(q)$, we would expected the molar entropy for Kr to be greater than that of Ar. This expectation can be confirmed by repeating the above calculation for Kr, or by simply looking at the difference in entropy between Kr and Ar:

$$S_{Kr} - S_{Ar} = \left(\frac{U}{T} + k \ln Q \right)_{Kr} - \left(\frac{U}{T} + k \ln Q \right)_{Ar}$$

$$= k \ln \left(\frac{Q_{Kr}}{Q_{Ar}} \right) = Nk \ln \left(\frac{q_{Kr}}{q_{Ar}} \right) = Nk \ln \left(\frac{\Lambda_{Ar}^3}{\Lambda_{Kr}^3} \right) = Nk \ln \left(\frac{m_{Kr}}{m_{Ar}} \right)^{3/2}$$

$$S_{m,Kr} - S_{m,Ar} = \frac{3}{2} R \ln \left(\frac{m_{Kr}}{m_{Ar}} \right) = \frac{3}{2} R \ln (2.00)$$

$$S_{m,Kr} = 8.64 \text{ J mol}^{-1} \text{ K}^{-1} + S_{m,Ar}$$

P32.24) Determine the standard molar entropy of N₂O, a linear triatomic molecule at 298 K. For this molecule, $B = 0.419$ cm^{-1} and $\tilde{v}_1 = 1285$ cm^{-1}, $\tilde{v}_2 = 589$ cm^{-1} (doubly degenerate), and $\tilde{v}_3 = 2224$ cm^{-1}.

At 298 K both the translational and rotational degrees of freedom will be in the high-temperature limit, but the vibrational contributions must be calculated. Using the standard volume of 24.5 L the entropy is determined as follows:

$$U_m^\circ = U_{T,m}^\circ + U_{R,m}^\circ + U_{V,m}^\circ + U_{E,m}^\circ$$

$$= \frac{3}{2}RT + RT + N_A hc \left[\left(\frac{\tilde{v}_1}{e^{\beta hc \tilde{v}_1} - 1} \right) + 2 \left(\frac{\tilde{v}_2}{e^{\beta hc \tilde{v}_2} - 1} \right) + \left(\frac{\tilde{v}_3}{e^{\beta hc \tilde{v}_3} - 1} \right) \right]$$

$$= \frac{5}{2}R(298 \text{ K}) + 906 \text{ J mol}^{-1} = 7.10 \text{ kJ mol}^{-1}$$

$$S_m^\circ = \frac{U_m^\circ}{T} + k \ln Q = \frac{U_m^\circ}{T} + k \ln \left(\frac{q_{total}^N}{N!} \right) = \frac{U_m^\circ}{T} + Nk \ln(q_{total}) - k \ln(N!)$$

$$= \frac{7.10 \text{ kJ mol}^{-1}}{298 \text{ K}} + R \ln(q_T q_R q_V q_E) - R \ln((1 \text{ mol}) N_A) + R$$

$$= -424 \text{ J mol}^{-1} \text{ K}^{-1} + R \ln(q_T q_R q_V q_E)$$

$$q_T = \left(\frac{V}{\Lambda^3} \right) = \frac{0.0245 \text{ m}^3}{3.55 \times 10^{-33} \text{ m}^3} = 6.91 \times 10^{30}$$

$$q_R = \left(\frac{kT}{\sigma B} \right) = \frac{(0.695 \text{ cm}^{-1} \text{ K}^{-1})(298.15 \text{ K})}{0.42 \text{ cm}^{-1}} = 493$$

$$q_V = \left(\frac{1}{1 - e^{-\beta \tilde{v}_1}} \right) \left(\frac{1}{1 - e^{-\beta \tilde{v}_2}} \right)^2 \left(\frac{1}{1 - e^{-\beta \tilde{v}_3}} \right)$$

$$= \left(\frac{1}{1 - e^{-\frac{1285 \text{ cm}^{-1}}{(0.695 \text{ cm}^{-1})(298 \text{ K})}}} \right) \left(\frac{1}{1 - e^{-\frac{589 \text{ cm}^{-1}}{(0.695 \text{ cm}^{-1})(298 \text{ K})}}} \right)^2 \left(\frac{1}{1 - e^{-\frac{2224 \text{ cm}^{-1}}{(0.695 \text{ cm}^{-1})(298 \text{ K})}}} \right)$$

$$= 1.00$$

$$q_E = 1.00$$

$$S_m^\circ = -424 \text{ J mol}^{-1} \text{ K}^{-1} + R \ln(q_T q_R q_V q_E) = -424 \text{ J mol}^{-1} \text{ K}^{-1} + 643 \text{ J mol}^{-1} \text{ K}^{-1}$$

$$S_m^\circ = 219 \text{ J mol}^{-1} \text{ K}^{-1}$$

P32.30) The standard molar entropy of CO is 197.7 J mol^{-1} K^{-1}. How much of this value is due to rotational and vibrational motion of CO?

The contribution of rotational and vibrational energetic degrees of freedom is equal to the difference between the standard molar entropy and the contribution from translational degrees of freedom. The translational contribution to the molar entropy can be calculated using the Sackur–Tetrode equation as follows:

$$S = nR \ln\left[\frac{RTe^{5/2}}{\Lambda^3 N_A P}\right]$$

$$S_m = R \ln\left[\frac{RTe^{5/2}}{\Lambda^3 N_A P}\right]$$

$$\Lambda^3 = \left(\frac{h^2}{2\pi mkT}\right)^{3/2} = \left(\frac{\left(6.626\times10^{-34}\text{ J s}\right)^2}{2\pi\left(4.65\times10^{-26}\text{ kg}\right)\left(1.38\times10^{-23}\text{ J K}^{-1}\right)\left(298.15\text{ K}\right)}\right)^{3/2} = 6.99\times10^{-33}\text{ m}^3$$

$$S_m = R \ln\left[\frac{RTe^{5/2}}{\Lambda^3 N_A P}\right] = nR \ln\left[\frac{\left(8.21\times10^{-5}\text{ m}^3\text{ atm mol}^{-1}\text{ K}^{-1}\right)\left(298.15\text{ K}\right)e^{5/2}}{\left(6.99\times10^{-33}\text{ m}^3\right)N_A\left(1\text{ atm}\right)}\right] = 148.6\text{ J mol}^{-1}\text{ K}^{-1}$$

$$S_{rot,vib} = S_{total} - S_{trans} = 197.7\text{ J mol}^{-1}\text{ K}^{-1} - 148.6\text{ J mol}^{-1}\text{ K}^{-1} = 49.1\text{ J mol}^{-1}\text{ K}^{-1}$$

P32.32) The molecule NO has a ground electronic level that is doubly degenerate, and a first excited level at 121.1 cm^{-1} that is also twofold degenerate. Determine the contribution of electronic degrees of freedom to the standard molar entropy of NO. Compare your result to $R\ln(4)$. What is the significance of this comparison?

$$q_E = g_0 + g_1 e^{-\beta\varepsilon_1} = 2 + 2e^{-\beta hc\left(121.1\text{ cm}^{-1}\right)} = 3.11$$

$$U_E = \frac{-N}{q_E}\left(\frac{\partial q_E}{\partial\beta}\right)_V = \frac{2Nhc\left(121.1\text{ cm}^{-1}\right)e^{-\beta hc\left(121.1\text{ cm}^{-1}\right)}}{2 + 2e^{-\beta hc\left(121.1\text{ cm}^{-1}\right)}}$$

$$= \frac{Nhc\left(121.1\text{ cm}^{-1}\right)e^{-\beta hc\left(121.1\text{ cm}^{-1}\right)}}{1 + e^{-\beta hc\left(121.1\text{ cm}^{-1}\right)}}$$

$$U_{E,m} = \frac{N_A hc\left(121.1\text{ cm}^{-1}\right)e^{-\beta hc\left(121.1\text{ cm}^{-1}\right)}}{1 + e^{-\beta hc\left(121.1\text{ cm}^{-1}\right)}} = 518\text{ J mol}^{-1}$$

$$S_{E,m} = \frac{U_{E,m}}{T} + R\ln\left(q_E\right) = \frac{518\text{ J mol}^{-1}}{298.15\text{ K}} + R\ln\left(3.11\right)$$

$$= 1.73\text{ J mol}^{-1}\text{ K}^{-1} + 9.43\text{ J mol}^{-1}\text{ K}^{-1}$$

$$= 11.2\text{ J mol}^{-1}\text{ K}^{-1}$$

$$R\ln(4) = 11.5\text{ J mol}^{-1}\text{ K}^{-1}$$

At sufficiently high temperatures, $q_E = 4$, and the contribution of the two lowest electronic energy levels to the molar entropy would equal $R\ln(4)$. The similarity between the calculated and limiting values demonstrates that this limiting behavior is being approached at this temperature.

P32.37) Calculate the standard Helmholtz energy for molar ensembles of Ne and Kr at 298 K. First, performing the calculation for Ne ($M = 0.020$ kg mol^{-1}):

$$A = -kT \ln Q = -kT \ln\left(\frac{q^N}{N!}\right)$$

$$= -NkT \ln q + kT \ln(N!)$$

$$= -NkT \ln q + kT(N \ln N - N)$$

$$= -NkT(\ln q - \ln N + 1)$$

$$q = q_T = \frac{V}{\Lambda^3} = \frac{0.0245 \text{ m}^3}{\left(\dfrac{h^2}{2\pi mkT}\right)^{3/2}} = \frac{0.0245 \text{ m}^3}{1.08 \times 10^{-32} \text{ m}^3} = 2.27 \times 10^{30}$$

$$N = n \times N_A = 6.022 \times 10^{23}$$

$$A = -nRT(\ln q - \ln N + 1)$$

$$A_m^\circ = -(8.314 \text{ J mol}^{-1} \text{ K}^{-1})(298.15 \text{ K})\left(\ln(2.27 \times 10^{30}) - \ln(6.022 \times 10^{23}) + 1\right)$$

$$A_m^\circ = -40.0 \text{ kJ mol}^{-1}$$

This calculation can be repeated for Kr ($M = 0.083$ kg mol^{-1}), or the difference between the Helmholtz energy of Kr and Ne can be determined:

$$A_{Kr} - A_{Ar} = -kT(\ln Q_{Kr} - \ln Q_{Ar})$$

$$= -kT \ln\left(\frac{Q_{Kr}}{Q_{Ar}}\right)$$

$$= -NkT \ln\left(\frac{q_{Kr}}{q_{Ar}}\right)$$

$$= -NkT \ln\left(\left(\frac{\Lambda_{Ar}}{\Lambda_{Kr}}\right)^3\right)$$

$$= -NkT \ln\left(\left(\frac{m_{Kr}}{m_{Ar}}\right)^{3/2}\right)$$

$$= -\frac{3}{2} NkT \ln\left(\frac{m_{Kr}}{m_{Ar}}\right)$$

$$= -\frac{3}{2} nRT \ln\left(\frac{0.083}{0.020}\right)$$

$$A_{Kr,m}^\circ = -\frac{3}{2} RT \ln(4.15) + A_{Ar,m}^\circ = -45.3 \text{ kJ mol}^{-1}$$

P32.41) Determine the equilibrium constant for the dissociation of sodium at 298 K:
$$Na_2(g) \rightleftharpoons 2Na(g).$$

For Na_2, $B = 0.155$ cm^{-1}, $\tilde{\nu} = 159$ cm^{-1}, the dissociation energy is 70.4 kJ/mol, and the ground-state electronic degeneracy for Na is 2.

$$K = \frac{\left(\dfrac{q}{N_A}\right)^2_{Na}}{\left(\dfrac{q}{N_A}\right)_{Na_2}} e^{-\beta \varepsilon_D}$$

$$q_{Na} = q_T q_E = \left(\frac{V}{\Lambda^3}\right)(2) = \left(\frac{0.0245 \text{ m}^3}{9.38 \times 10^{-33} \text{ m}^3}\right)(2) = 5.22 \times 10^{30}$$

$$q_{Na_2} = q_T q_R q_V q_E = \left(\frac{V}{\Lambda^3}\right)\left(\frac{kT}{\sigma B}\right)\left(\frac{1}{1 - e^{-\beta hc\tilde{\nu}}}\right)(1)$$

$$= \left(\frac{0.0245 \text{ m}^3}{3.32 \times 10^{-33} \text{ m}^3}\right)\left(\frac{(0.695 \text{ cm}^{-1} \text{ K}^{-1})(298.15 \text{ K})}{(2)(0.155 \text{ cm}^{-1})}\right)$$

$$\times \left(\frac{1}{1 - e^{-\frac{(6.626 \times 10^{-34} \text{ J s})(3.00 \times 10^{10} \text{ cm s}^{-1})(159 \text{ cm}^{-1})}{(1.38 \times 10^{-23} \text{ J K}^{-1})(298.15 \text{ K})}}}\right)(1)$$

$$= 9.22 \times 10^{33}$$

$$K = \frac{\left(\dfrac{5.22 \times 10^{30}}{N_A}\right)^2_{Na}}{\left(\dfrac{9.22 \times 10^{33}}{N_A}\right)_{Na_2}} e^{-\frac{70,400 \text{ J mol}^{-1}/N_A}{(1.38 \times 10^{-23} \text{ J K}^{-1})(298.15 \text{ K})}}$$

$$= 2.25 \times 10^{-9}$$

Computational Problems

Before solving the computational problems, it is recommended that students work through Tutorials 1–3 under the Help menus in Spartan Student Edition to gain familiarity with the program.

Computational Problem 32.1: Perform a Hartree–Fock 6-31G* calculation on acetonitrile and determine the contribution of translations to the molar enthalpy and entropy under standard thermodynamic conditions. How does the calculation compare to the values determined by hand computation using the expressions for enthalpy and entropy presented in this chapter.

Procedure

Step 1: Create a new file and build acetonitrile (H_3CCN).

Step 2: Go to "**Setup > Calculations.**" Set the calculation type to equilibrium geometry and the method to **Hartree–Fock 6-31G***.

Step 3: In the Calculations dialog window, make sure "**Equilibrium Geometry**" is selected under the Calculate pull-down menu, check the box next to "IR."

Step 4: Click the "**Submit**" button in the Calculations dialog window. Once you provide a file name the calculation will begin.

Step 5: When the calculation is finished, go to "**Display > Output**" to view the output file for the calculation. Scroll down until you locate the field labeled "Standard Thermodynamic quantities at 298.15 K and 1.00 atm." In this field you will find the standard molar enthalpy and entropy. Record these values.

Computational Problem 32.2: Perform a Hartree–Fock 3-21G calculation on cyclohexane and determine the frequency of the lowest-frequency mode consistent with the reaction coordinate corresponding to the "boat-to-boat" conformational change. Using the calculated frequency of this vibrational degree of freedom, determine how much this mode contributes the vibrational heat capacity at 298 K.

Procedure

Step 1: Create a new file and build cyclohexane (this is most efficiently done by using the pull-down menu under "**Groups**" in the organic section of the building toolbox).

Step 2: Go to "**Setup > Calculations.**" Set the calculation type to equilibrium geometry and the method to **Hartree–Fock 3-21G**.

Step 3: In the Calculations dialog window, make sure "Equilibrium Geometry" is selected under the Calculate pull-down menu, check the box next to "IR."

Step 4: Click the "**Submit**" button in the Calculations dialog window. Once you provide a file name the calculation will begin.

Step 5: When the calculation is finished, go to "**Display > Spectra**" to view the calculated vibrational modes.

Step 6: In the Spectra dialog box, click "**Draw IR Spectra.**"

Step 7: In the Spectra dialog box, check the box next to the first calculated vibrational mode (should be around 248 cm^{-1}). After checking the box, the mode will be animated.

Step 8: Continue to select modes until you identify the mode associated with the "boat-to-boat" conformational change. If you are having difficulty visualizing the modes, you can increase the amplitude in the Spectra dialog box using the "Amp" window.

Step 9: Using the frequency for the boat-to-boat mode identified in Step 8, calculate the contribution to C_V for this mode.

Computational Problem 32.3: In the treatment of C_V presented in this chapter the contribution of electronic degrees of freedom was ignored. To further explore this issue, perform a Hartree–Fock 6-31G* calculation on anthracene and determine the HOMO-LUMO energy gap, which provides a measure of the gap between the ground and first excited electronic states. Using this energy gap, determine the temperature where 2% of the anthracene molecules populate the excited electronic state, and using this temperature determine the contribution of electronic degrees of freedom to C_V. Reflecting on this result, why can we generally ignore electronic degrees of freedom when calculating standard thermodynamic properties?

Procedure

Step 1: Create a new file and build anthracene (this is most efficiently done by using the pull-down menu under "Groups" in the organic section of the building toolbox).

Step 2: Go to "Setup > Calculations." Set the calculation type to equilibrium geometry and the method to **Hartree–Fock 6-31G***.

Step 3: In the Calculations dialog window, make sure "**Equilibrium Geometry**" is selected under the Calculate pull-down menu.

Step 4: Click the "Submit" button in the Calculations dialog window. Once you provide a file name the calculation will begin.

Step 5: When the calculation is finished, go to "**Display > Output**" and scroll down until you locate the energies of the HOMO and LUMO. Record the difference in energy.

Step 6: Using the difference in energy determined in Step 5, calculate the temperature at which 2% of the molecules occupy the LUMO, and use this temperature and energy gap to determine the contribution of the electronic energy levels to C_V.

Computational Problem 32.4: In 2001 astronomers discovered the presence of vinyl alcohol (C_2H_3OH) in an interstellar cloud near the center of the Milky Way Galaxy. This complex is a critical piece of the puzzle regarding the origin of complex organic molecules in space. The temperature of an interstellar cloud is ~20. K. Determine the vibrational frequencies of this compound using a Hartree–Fock 3-21G basis set, and determine if any of the vibrational modes of vinyl alcohol contribute more than $0.1R$ to C_V at this temperature.

Procedure

Step 1: Create a new file and build vinyl alcohol (Hint: a Lewis dot structure may help in defining the types of bonds in this molecule). In addition, make sure you have the appropriate bond angles for the alcohol group to minimize steric repulsion.

Step 2: Go to "**Setup > Calculations**." Set the calculation type to equilibrium geometry and the method to **Hartree–Fock 3-21G**.

Step 3: In the Calculations dialog window, make sure "**Equilibrium Geometry**" is selected under the Calculate pull-down menu, check the box next to "IR."

Step 4: Click the "**Submit**" button in the Calculations dialog window. Once you provide a file name the calculation will begin.

Step 5: When the calculation is finished, go to "**Display > Spectra**" and view the vibrational frequencies in the Spectra dialog box.

Step 6: Record the vibrational frequencies for vinyl alcohol, and determine which modes will contribute 0.1R or greater to C_V at 20. K.

Computational Problem 32.5: Nitrous acid (HONO) is of interest in atmospheric chemistry as a reservoir compound contributing to the NO_x cycle. Perform a Hartree-Fock 6-31G* calculation on the cis conformer of this compound and determine the standard molar entropy of this compound. What are the contributions of vibrational and rotational degrees of freedom to the entropy? Using the techniques presented in this chapter and treating HONO as an ideal gas, determine the translational contribution to the standard molar entropy and compare to the computational result.

Procedure

Step 1: Create a new file and build HONO (Hint: a Lewis dot structure will help in identifying the bonding in this molecule). Make sure you are working on the lowest-energy configuration of the molecule where the terminal hydrogen and oxygen are trans about the central N-O bond.
Step 2: Go to "**Setup > Calculations.**" Set the calculation type to equilibrium geometry and the method to **Hartree–Fock 6-31G***.
Step 3: In the Calculations dialog window, make sure "**Equilibrium Geometry**" is selected under the Calculate pull-down menu, check the box next to "**IR.**"
Step 4: Click the "**Submit**" button in the Calculations dialog window. Once you provide a file name the calculation will begin.
Step 5: When the calculation is finished, go to "Display > Output" to view the output file for the calculation. Scroll down until you locate the field labeled "Standard Thermodynamic quantities at 298.15 K and 1.00 atm." In this field you will find the contributions to the standard molar entropy from the various degrees of freedom. Record these values.

The calculated values for the contribution to the standard molar entropy are:

$$S^\circ_{trans} = 156.7635 \text{ J mol}^{-1} \text{ K}^{-1}$$
$$S^\circ_{rot} = 85.7990 \text{ J mol}^{-1} \text{ K}^{-1}$$
$$S^\circ_{vib} = 3.0936 \text{ J mol}^{-1} \text{ K}^{-1}$$

Calculating the translational contribution to the entropy:

$$S^\circ_{trans} = R \ln\left[\frac{RTe^{5/2}}{\Lambda^3 N_A P}\right]$$

$$\Lambda^3 = \left(\frac{h^2}{2\pi mkT}\right)^{3/2} = \left(\frac{\left(6.626\times10^{-34} \text{ J s}\right)^2}{2\pi\left(7.80\times10^{-26}\text{kg}\right)\left(1.38\times10^{-23} \text{ J K}^{-1}\right)\left(298 \text{ K}\right)}\right)^{3/2} = 3.22\times10^{-33} \text{ m}^3$$

$$S^\circ_{trans} = R \ln\left[\frac{RTe^{5/2}}{\Lambda^3 N_A P}\right] = R \ln\left[\frac{\left(8.21\times10^{-5} \text{ m}^3 \text{ atm mol}^{-1} \text{ K}^{-1}\right)\left(298 \text{ K}\right)e^{5/2}}{\left(3.22\times10^{-33} \text{ m}^3\right)N_A \left(1 \text{ atm}\right)}\right] = 156.7 \text{ J mol}^{-1} \text{ K}^{-1}$$

Computational Problem 32.6: For dibromine oxide (BrOBr), perform a Hartree–Fock 6-31G* calculation and determine the translational, rotational, and vibrational contribution to the standard molar entropy. Using the techniques presented in this chapter, calculate these values and compare the analytical and computational results.

Procedure

Step 1: Create a new file and build BrOBr.
Step 2: Go to "**Setup > Calculations.**" Set the calculation type to equilibrium geometry and the method to **Hartree–Fock 6-31G***.
Step 3: In the Calculations dialog window, make sure "**Equilibrium Geometry**" is selected under the Calculate pull-down menu, check the box next to "**IR.**"
Step 4: Click the "**Submit**" button in the Calculations dialog window. Once you provide a file name the calculation will begin.

Step 5: When the calculation is finished, go to "**Display > Output**" to view the output file for the calculation. Scroll down until you locate the field labeled "Standard Thermodynamic quantities at 298.15 K and 1.00 atm." In this field you will find the contributions to the standard molar entropy from the various degrees of freedom. Record these values.

Chapter 33: Kinetic Theory of Gases

P33.1) Consider a collection of gas particles confined to translate in two dimensions (for example, a gas molecule on a surface). Derive the Maxwell speed distribution for such a gas.

Beginning with the Maxwell-Boltzmann velocity distribution in one-dimension

$$f(v_j) = \left(\frac{m}{2\pi \, kT}\right)^{1/2} e^{-\frac{m}{2\,kT}v_j^2}$$

and the definition of speed in 2 dimensions

$$v = \left(v_x^2 + v_y^2\right)^{1/2}$$

The speed distribution in 2 dimensions is given by:

$$F dv = f(v_x) f(v_y) dv_x dv_y$$

where dv_j is the differential of velocity in the jth direction. Thus

$$F dv = \left(\frac{m}{2\pi \, kT}\right)^{1/2} e^{-\frac{m}{2\,kT}v_x^2} \left(\frac{m}{2\pi \, kT}\right)^{1/2} e^{-\frac{m}{2\,kT}v_y^2} dv_x dv_y$$

$$= \left(\frac{m}{2\pi \, kT}\right) e^{-\frac{m}{2\,kT}\left(v_x^2 + v_y^2\right)} dv_x dv_y$$

$$F dv = \left(\frac{m}{2\pi \, kT}\right) e^{-\frac{m}{2\,kT}v^2} dv_x dv_y$$

The differential is defined as:

$$dv_x dv_y = 2\pi v dv$$

Substituting this into the expression for *Fdv*:

$$F dv = \left(\frac{m}{2\pi \, kT}\right) e^{-\frac{m}{2\,kT}v^2} (2\pi v) dv = \frac{m}{kT} e^{-\frac{m}{2\,kT}v^2} v dv$$

P33.5) Compare the average speed and average kinetic energy of O_2 with that of CCl_4 at 298 K.
$M_{O_2} = 0.0320$ kg mol^{-1} $M_{CCl_4} = 0.154$ kg mol^{-1}

At the same temperature, the speed for two particles of different mass is related by the square root of the mass ratios. For this case of O_2 and CCl_4:

$$v_{CCl_4} = \left(\frac{M_{O_2}}{M_{CCl_4}}\right)^{1/2} v_{O_2}$$

$$v_{CCl_4} = \left(\frac{0.0320 \text{ kg mol}^{-1}}{0.154 \text{ kg mol}^{-1}}\right)^{1/2} v_{O_2} = (0.456) v_{O_2}$$

The average speed for O_2 at 298 K is:

$$v_{ave} = \sqrt{\frac{8RT}{\pi M}} = \sqrt{\frac{8\left(8.314 \text{ J mol}^{-1} \text{ K}^{-1}\right)\left(298 \text{ K}\right)}{\pi \cdot 0.0320 \text{ kg mol}^{-1}}} = 444 \text{ m s}^{-1}$$

Using this result, the average speed for CCl_4 at this same temperature is:

$$v_{ave,CCl_4} - 0.456 \times v_{ave,O_2}$$

$$v_{ave,CCl_4} = (0.456)\left(444 \text{ m s}^{-1}\right)$$

$$v_{ave,CCl_4} = 203 \text{ m s}^{-1}$$

The average kinetic energy is mass independent, thus for a given temperature, all gases have the same kinetic energy. The average kinetic energy per gas particle is therefore:

$$\langle KE \rangle = \frac{3}{2} kT = \frac{3}{2}\left(1.38 \times 10^{-23} \text{ J K}^{-1}\right)\left(298 \text{ K}\right)$$

$$\langle KE \rangle = 6.17 \times 10^{-21} \text{ J}$$

P33.11) The probability that a particle will have a velocity in the x direction in the range of $-v_{x0}$ and v_{x0}

is given by

$$f(-v_{x_0} \le v_x \le v_{x_0}) = \left(\frac{m}{2\pi kT}\right)^{1/2} \int_{-v_{x_0}}^{v_{x_0}} e^{-mv_x^2/2kT} dv_x$$

$$= \left(\frac{2m}{\pi kT}\right)^{1/2} \int_0^{v_{x_0}} e^{-mv_x^2/2kT} dv_x$$

The preceding integral can be rewritten using the following substitution: $\xi^2 = mv_x^2/2kT$, resulting in

$f(-v_{x0} \le v_x \le v_{x0}) = 2/\sqrt{\pi} \left(\int_0^{\xi_0} e^{-\xi^2} d\xi\right)$, which can be evaluated using the error function defined as

$\text{erf}(z) = 2/\sqrt{\pi} \left(\int_0^z e^{-x^2} dx\right)$. The complementary error function is defined as $\text{erfc}(z) = 1 - \text{erf}(z)$. Finally, a plot

of both $\text{erf}(z)$ and $\text{erfc}(z)$ as a function of z is shown here:

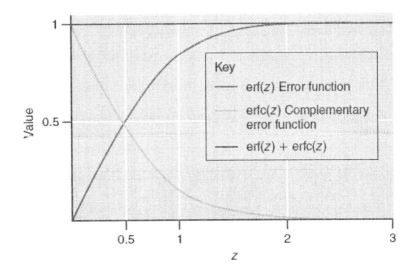

Using this graph of erf(z), determine the probability that $|v_x| \leq (2kT/m)^{1/2}$. What is the probability that

$|v_x| > (2kT/m)^{1/2}$?

The probability that a particle will have a velocity in the x direction in the range of $-v_{x_0}$ and v_{x_0} is given by

$$f(-v_{x_0} \leq v_x \leq v_{x_0}) = \left(\frac{m}{2\pi kT}\right)^{1/2} \int_{-v_{x_0}}^{v_{x_0}} e^{\frac{-mv_x^2}{2kT}} dv_x = \left(\frac{2m}{\pi kT}\right)^{1/2} \int_0^{v_{x_0}} e^{\frac{-mv_x^2}{2kT}} dv_x$$

The preceding integral can be rewritten using the following substitution: $\xi^2 = \frac{mv_x^2}{2kT}$, resulting

in $f(-v_{x_o} \leq v_x \leq v_{x_o}) = \frac{2}{\sqrt{\pi}} \int_0^{\xi_o} e^{-\xi^2} d\xi$, which can be evaluated using the error function defined

as $\mathrm{erf}(z) = \frac{2}{\sqrt{\pi}} \int_0^z e^{-x^2} dx$. The complementary error function is defined as $\mathrm{erfc}(z) = 1 - \mathrm{erf}(z)$. Finally, a plot of both erf(z) and erfc(z) as a function of z is shown in the text (tabulated values are available in the Math Supplement, Appendix A): Using this graph of erf(z), determine the probability that $|v_x| \leq (2kT/m)^{1/2}$. What is the probability that $|v_x| > (2kT/m)^{1/2}$?

$$\text{If } |v_x| \leq \left(\frac{2\,kT}{m}\right)^{1/2}, \text{ then}$$

$$\xi^2 \leq \frac{m}{2\,kT}\left(\frac{2\,kT}{m}\right) = 1$$

and

$$\xi_0 \leq \sqrt{1} \leq 1$$

The probability that the particle has $|v_x| \le \left(\dfrac{2kT}{m}\right)^{1/2}$ is given by:

$$f_{\le} = \frac{2}{\sqrt{\pi}} \int_0^1 e^{-\xi^2} d\xi$$

$$f_{\le} = \mathrm{erf}\,(1)$$

$$f_{\le} - 0.8427$$

The probability that the particle would have a velocity $|v_x| > \left(\dfrac{2kT}{m}\right)^{1/2}$ is found from the previous part, since the total probability must be 1:

$$f_> = 1 - 0.8427$$

$$f_> = 0.1573$$

P33.12) The speed of sound is given by $v_{sound} = \sqrt{\dfrac{\gamma kT}{m}} = \sqrt{\dfrac{\gamma RT}{M}}$, where $\gamma = C_P/C_V$.

a) What is the speed of sound in Ne, Kr, and Ar at 1000. K?
b) At what temperature will the speed of sound in Kr equal the speed of sound in Ar at 1000 K?

a) $M_{Ne} = 0.0202$ kg mol^{-1} $M_{Kr} = 0.0840$ kg mol^{-1} $M_{Ar} = 0.0400$ kg mol^{-1}

The heat capacities for the three gases are the same:

$C_V = 12.5$ J K^{-1}mol^{-1} and $C_p = 20.8$ J K^{-1}mol^{-1}

Thus, $\gamma = 1.6666$.
a) Ne:

$$V_{sound} = \sqrt{\frac{1.67\left(8.314 \text{ J mol}^{-1}\text{K}\right)(1000.\text{ K})}{0.0202 \text{ kg mol}^{-1}}} = 829 \text{ m s}^{-1}$$

Kr:

$$V_{sound} = \sqrt{\frac{1.67\left(8.314 \text{ J mol}^{-1}\text{K}^{-1}\right)(1000.\text{ K})}{0.0840 \text{ kg mol}^{-1}}} = 407 \text{ m s}^{-1}$$

Ar:

$$V_{sound} = \sqrt{\frac{1.67\left(8.314 \text{ J mol}^{-1}\text{K}^{-1}\right)(1000.\text{ K})}{0.0400 \text{ kg mol}}} = 589 \text{ m s}^{-1}$$

b) Setting the speed of sound equal for Kr and Ar:

$$\sqrt{\frac{\gamma RT_{Kr}}{M_{Kr}}} = \sqrt{\frac{\gamma RT_{Ar}}{M_{Ar}}}$$

$$T_{Kr} = \frac{M_{Kr}}{M_{Ar}} \cdot T_{Ar}$$

Therefore:

$$T_{Kr} = \left(\frac{0.0840 \text{ kg mol}^{-1}}{0.0400 \text{ kg mol}^{-1}}\right) 1000. \text{ K}$$

$$T_{Kr} = (2.10)1000. \text{ K}$$

$$T_{Kr} = 2100 \text{ K}$$

P33.19) Starting with the Maxwell speed distribution, demonstrate that the probability distribution for translational energy for $\varepsilon_{Tr} \gg kT$ is given by:

$$f(\varepsilon_{Tr}) d\varepsilon_{Tr} = 2\pi \left(\frac{1}{\pi kT}\right)^{3/2} e^{-\varepsilon_{Tr}/kT} \varepsilon_{Tr}^{1/2} d\varepsilon_{Tr}$$

The translational energy of a particle can be related to the velocity of the particle by the expressions

$$\varepsilon_{Tr} = \frac{1}{2} mv^2$$

$$v = \sqrt{\frac{2\varepsilon_{T}}{m}}$$

$$dv = \sqrt{\frac{2}{m}} \cdot \frac{1}{2} \sqrt{\frac{1}{\varepsilon_{Tr}}} \, d\varepsilon_{Tr} = \frac{1}{2} \sqrt{\frac{2}{m\varepsilon_{Tr}}} \, d\varepsilon_{Tr}$$

Substituting this result into the Maxwell speed distribution:

$$f(\varepsilon_{Tr}) d\varepsilon_{Tr} = 4\pi \left(\frac{m}{2\pi kT}\right)^{3/2} \left(\frac{2\varepsilon_{Tr}}{m}\right) e^{-\frac{m}{2kT}\left(\frac{2\varepsilon_{Tr}}{m}\right)} \left(\frac{1}{2}\sqrt{\frac{2}{m\varepsilon_{Tr}}} \, d\varepsilon_{Tr}\right)$$

$$= 2\pi \left(\frac{1}{\pi kT}\right)^{3/2} e^{-\frac{\varepsilon_{Tv}}{kT}} \varepsilon_{T}^{1/2} \, d\varepsilon_{Tv}$$

P33.23) Imagine a cubic container with sides 1 cm in length that contains 1 atm of Ar at 298 K. How many gas–wall collisions are there per second?

The collisional rate is given by:

$$\frac{dN_c}{dt} = \frac{PAv_{ave}}{4kT} = \frac{PA N_A v_{avg\backslash c}}{4RT}$$

With $M_{Ar} = 0.0400$ kg mol^{-1} and $T = 298$ K the average speed is:

Substituting into the expression for the collisional rate:

$$\frac{dN_c}{dt} = \frac{\left(1 \text{ atm} \cdot \frac{101.325 \times 10^3 \text{ N m}^{-2}}{1 \text{ atm}}\right)\left(1 \text{ cm}^2\left(\frac{1 \text{ m}}{100 \text{ cm}}\right)^2\right)\left(6.022 \times 10^{23} \text{ mol}^{-1}\right)\left(398 \text{ m s}^{-1}\right)}{4\left(8.314 \text{ J mol}^{-1}\text{K}^{-1}\right)\left(298 \text{ K}\right)}$$

$$\frac{dN_c}{dt} = 2.45 \times 10^{23} \text{ coll. per sec. per wall}$$

Taking into account the six walls that comprise the container, the total collisional rate is:

$$\frac{dN_c}{dt} = 6 \text{ walls} \cdot 2.45 \times 10^{23} \text{ coll. s}^{-1} \text{ wall}^{-1}$$

$$\frac{dN_c}{dt} = 1.47 \times 10^{24} \text{ coll. s}^{-1}$$

P33.26)

a) How many molecules strike a 1-cm^2 surface during 1 min if the surface is exposed to O_2 at 1 atm and 298 K?

b) Ultrahigh vacuum studies typically employ pressures on the order of 10^{-10} Torr. How many collisions will occur at this pressure at 298 K?

For O_2, $M = 0.0320$ kg mol^{-1} and $Z_c = \dfrac{PN_A}{\left(2\pi MRT\right)^{1/2}}$

a) @ 1 atm: $Z_c = \dfrac{\left(1.00 \text{ atm}\right)\left(\dfrac{101.325 \times 10^3 \text{ Pa}}{1 \text{ atm}}\right)\left(6.022 \times 10^{23} \text{ mol}^{-1}\right)}{\left(2\pi\left(0.032 \text{ kg mol}^{-1}\right)\left(8.314 \text{ J mol}^{-1}\text{K}^{-1}\right)\left(298 \text{ K}\right)\right)^{1/2}}$

$$Z_c = 2.73 \times 10^{27} \text{ m}^{-2}\text{s}^{-1}$$

$$\frac{dN_c}{dt} = Z_c \times A = \left(2.73 \times 10^{27} \text{ m}^{-2}\text{s}^{-1}\right)\left(1.00 \text{ cm}^2\right)\left(\frac{1 \text{ m}}{100 \text{ cm}}\right)^2$$

$$\frac{dN_c}{dt} = 2.73 \times 10^{23} \text{ coll. s}^{-1}$$

b) @ 10^{-10} torr: $Z_c = \dfrac{\left(10^{-10} \text{ torr }\right)\left(\dfrac{133.32 \text{ Pa}}{1 \text{ torr}}\right)\left(6.022\times10^{23} \text{ mol}^{-1}\right)}{\left(2\pi\left(0.032 \text{ kg mol}^{-1}\right)\left(8.314 \text{ J mol}^{-1}\text{K}^{-1}\right)\left(298 \text{ K}\right)\right)^{1/2}}$

$$Z_c = 3.60\times10^{14} \text{ m}^{-2}\text{s}^{-1}$$

$$\frac{dN_c}{dt} = Z_c \times A = \left(3.60\times10^{14} \text{ m}^{-2}\text{s}^{-1}\right)\left(1.00 \text{ cm}^2\right)\left(\frac{1 \text{ m}}{100 \text{ cm}}\right)^2$$

$$\frac{dN_c}{dt} = 3.60\times10^{14} \text{ coll s}^{-1}$$

P33.28) Many of the concepts developed in this chapter can be applied to understanding the atmosphere. Because atmospheric air is comprised primarily of N_2 (roughly 78% by volume), approximate the atmosphere as consisting only of N_2 in answering the following questions:

a) What is the single-particle collisional frequency at sea level, with $T = 298$ K and $P = 1$ atm? The corresponding single-particle collisional frequency is reported as 10^{10} s^{-1} in the *CRC Handbook of Chemistry and Physics* (62nd ed., p. F-171).
b) At the tropopause (11 km in altitude), the collisional frequency decreases to 3.16×10^9 s^{-1}, primarily due to a reduction in temperature and barometric pressure (i.e., fewer particles). The temperature at the tropopause is ~220 K. What is the pressure of N_2 at this altitude?
c) At the tropopause, what is the mean free path for N_2?

The collisional cross section of N_2 is $\sigma = 4.3 \times 10^{-19}$ m^2, and M = 0.0280 kg mol^{-1}.

a) $z_{11} = \sqrt{2}\sigma\dfrac{PN_A}{RT}\left(\dfrac{8RT}{\pi M}\right)^{1/2}$

$= \sqrt{2}\left(4.30\times10^{-19} \text{ m}^2\right)\dfrac{\left(1.00 \text{ atm}\right)\left(6.022\times10^{23} \text{ mol}^{-1}\right)}{\left(8.21\times10^{-2} \text{ L atm mol}^{-1} \text{ K}^{-1}\right)\left(298 \text{ K}\right)}\left(\dfrac{8\left(8.314 \text{ J mol}^{-1} \text{ K}^{-1}\right)\left(298 \text{ K}\right)}{\pi\left(0.028 \text{ kg mol}^{-1}\right)}\right)^{1/2}$

$= \sqrt{2}\left(2.46\times10^{22} \text{ L}^{-1}\right)\left(4.3\times10^{-19} \text{ m}^2\right)\left(475 \text{ m s}^{-1}\right)\left(\dfrac{1000 \text{ L}}{\text{m}^3}\right)$

$z_{11} = 7.1\times10^9 \text{ s}^{-1}$

b) $3.16\times10^9 \text{ s}^{-1} = \dfrac{\sqrt{2}\left(P\right)\left(6.022\times10^{23} \text{ mol}^{-1}\right)}{\left(8.21\times10^{-2} \text{ L atm mol}^{-1} \text{ K}^{-1}\right)\left(220. \text{ K}\right)}\left(4.30\times10^{-19} \text{ m}^2\right)\left(\dfrac{8\left(8.314 \text{ J mol}^{-1} \text{ K}^{-1}\right)\left(220. \text{ K}\right)}{\pi\left(0.0280 \text{ kg mol}^{-1}\right)}\right)^{1/2}$

$P = \dfrac{\left(3.16\times10^9 \text{ s}^{-1}\right)}{\sqrt{2}\left(3.34\times10^{22} \text{ L}^{-1} \text{ atm}^{-1}\right)\left(4.3\times10^{-19} \text{ m}^2\right)\left(408 \text{ m s}^{-1}\right)}\left(\dfrac{1 \text{ m}^3}{1000 \text{ L}}\right)$

$$P = 0.38 \text{ atm}$$

c) $\lambda = \left(\dfrac{RT}{PN_A}\right)\dfrac{1}{\sqrt{2}\sigma} = \left(\dfrac{\left(8.21\times10^{-2}\ \text{L atm mol}^{-1}\ \text{s}^{-1}\right)\left(220.\ \text{K}\right)}{\left(0.38\ \text{atm}\right)\left(6.022\times10^{23}\ \text{mol}^{-1}\right)}\right)\dfrac{1}{\sqrt{2}\left(4.30\times10^{-19}\ \text{m}^2\right)}$

$\lambda = \left(\dfrac{18.1\ \text{L atm mol}^{-1}}{1.40\times10^{5}\ \text{atm mol}^{-1}\ \text{m}^2}\right)\left(\dfrac{1\ \text{m}^3}{1000\ \text{L}}\right)$

$\lambda - 1.3\times10^{-7}\ \text{m}$

P33.32) Determine the mean free path for Ar at 298 K at the following pressures:
a) 0.500 atm
b) 0.00500 atm
c) 5.00×10^{-6} atm

For Ar, $\sigma = 3.6\times10^{-19}\ \text{m}^2$ and M = 0.040 kg mol^{-1}

a) $\lambda = \left(\dfrac{RT}{PN_A}\right)\dfrac{1}{\sqrt{2}\sigma} = \left(\dfrac{\left(8.21\times10^{-2}\ \text{L atm mol}^{-1}\ \text{K}^{-1}\right)\left(298\ \text{K}\right)}{\left(0.500\ \text{atm}\right)\left(6.022\times10^{23}\ \text{mol}^{-1}\right)}\right)\dfrac{1}{\sqrt{2}\ \left(3.6\times10^{-19}\ \text{m}^2\right)}\left(\dfrac{1\ \text{m}^3}{1000\ \text{L}}\right)$

$\lambda = 1.6\times10^{-7}\ \text{m}$

b) Since the mean free path is inversely proportional to pressure, the result from part (a) can be used to determine the mean free path at pressures specified in parts (b) and (c) as follows:

$\lambda_{0.005} = \lambda_{0.5}\left(\dfrac{0.500\ \text{atm}}{0.00500\ \text{atm}}\right) = 1.6\times10^{-7}\ \text{m}\ (100)$

$\lambda_{0.005} = 1.6\times10^{-5}\ \text{m}$

c) $\lambda_{5\times10^{-6}} = \lambda_{0.5}\left(\dfrac{0.500\ \text{atm}}{5.00\times10^{-6}\ \text{atm}}\right) = 1.60\times10^{-7}\ \text{m}\ \left(10^5\right)$

$\lambda_{5\times10^{-6}} = 1.6\times10^{-2}\ \text{m}$

P33.35) A comparison of v_{ave}, v_{mp}, and v_{rms} for the Maxwell speed distribution reveals that these three quantities are not equal. Is the same true for the one-dimensional velocity distributions?

$v_{avg} = \langle v \rangle = \displaystyle\int_{-\infty}^{\infty} v_x \left(\dfrac{M}{2\pi RT}\right)^{1/2} e^{-\frac{M}{2RT}v_x^2}\ dv_x$

$= \left(\dfrac{M}{2\pi RT}\right)^{1/2} \displaystyle\int_{-\infty}^{\infty} v_x e^{-\frac{M}{2RT}v_x^2}\ dv_x$

$v_{avg} = 0$

$v_{mp} \Rightarrow 0 = \dfrac{\partial}{\partial v_x}\left[\left(\dfrac{M}{2\pi RT}\right)^{1/2} e^{-\frac{M}{2RT}v_x^2}\right]$

$0 = \left(\dfrac{M}{2\pi RT}\right)^{1/2}\left(\dfrac{M}{2\pi RT}v_x\right)e^{-\frac{M}{2RT}v_x^2}$

The above equality will be true when $v_x = 0$; therefore, $v_{mp} = 0$.

$$v_{rms} = \left\langle v^2 \right\rangle^{1/2} = \left[\int_{-\infty}^{\infty} v_x^2 \left(\frac{M}{2\pi RT} \right)^{1/2} e^{-\frac{M}{2RT} v_x^2} \, dv_x \right]^{1/2}$$

$$= \left[\left(\frac{\beta}{\pi} \right)^{1/2} \int_{-\infty}^{\infty} v_x^2 e^{-\beta v_x^2} \, dv_x \right]^{1/2} \quad \text{for } \beta = \frac{M}{2RT}$$

$$= \left[\left(\frac{\beta}{\pi} \right)^{1/2} \left(2 \int_0^{\infty} v_x^2 e^{-\beta v_x^2} \, dv_x \right) \right]^{1/2}$$

$$= \left[\left(\frac{\beta}{\pi} \right)^{1/2} \cdot \frac{1}{2} \left(\frac{\pi}{\beta^3} \right)^{1/2} \right]^{1/2}$$

$$= \left[\frac{1}{2} \beta^{-1} \right]^{1/2}$$

$$v_{rms} = \sqrt{\frac{RT}{M}}$$

Chapter 34: Transport Phenomena

P34.3)

a) The diffusion coefficient for Xe at 273 K and 1 atm is 0.5×10^{-5} m^2 s^{-1}. What is the collisional cross section of Xe?

b) The diffusion coefficient of N$_2$ is threefold greater than that of Xe under the same pressure and temperature conditions. What is the collisional cross section of N$_2$?

a)

$$D_{Xe} = 0.5 \times 10^{-5} \, \text{m}^2 \text{s}^{-1} \quad @ \; 273 \text{ K and 1 atm}$$

$$\sigma = \frac{1}{3} \sqrt{\frac{8kT}{\pi M}} \left(\frac{RT}{PN_A} \right) \frac{1}{\sqrt{2}D}$$

$$= \frac{1}{3\sqrt{2}} \sqrt{\frac{8(8.314 \, \text{J mol}^{-1} \, \text{K}^{-1})(273 \, \text{K})}{\pi(0.131 \, \text{kg mol}^{-1})}} \left(\frac{(8.21 \times 10^{-2} \, \text{L atm mol}^{-1} \, \text{K}) \left(\frac{1 \, \text{m}^3}{1000 \, \text{L}} \right)(273 \, \text{K})}{(1 \, \text{atm})(6.022 \times 10^{23} \, \text{mol}^{-1})(0.5 \times 10^{-5} \, \text{m}^2 \, \text{s}^{-1})} \right)$$

$$\sigma = 0.368 \, \text{nm}^2 \approx 0.4 \, \text{nm}^2$$

b) The ratio of collisional cross sections is given by:

$$\frac{\sigma_{N_2}}{\sigma_{Xe}} = \frac{D_{Xe}}{D_{N_2}} \sqrt{\frac{M_{Xe}}{M_{N_2}}}$$

$$\sigma_{N_2} = \sigma_{Xe} \left(\frac{D_{Xe}}{D_{N_2}} \right) \sqrt{\frac{M_{Xe}}{M_{N_2}}}$$

$$= (0.368 \, \text{nm}^2) \left(\frac{1}{3} \right) \sqrt{\frac{0.131 \, \text{kg mol}^{-1}}{0.028 \, \text{kg mol}^{-1}}}$$

$$\sigma_{N_2} = 0.265 \, \text{nm}^2 \approx 0.3 \, \text{nm}^2$$

P34.7)

A thermopane window consists of two sheets of glass separated by a volume filled with air (which we will model as N$_2$ where $\kappa = 0.0240$ J K^{-1} m^{-1} s^{-1}). For a thermopane window that is 1 m^2 in area with a separation between glass sheets of 3 cm, what is the loss of energy when:

a) the exterior of the window is at a temperature of 10°C and the interior of the window is at a temperature of 22°C?

b) the exterior of the window is at a temperature of –20°C and the interior of the window is at a temperature of 22°C?

c) the same temperature differential as in part (b) is used but the window is filled with Ar ($\kappa = 0.0163$ J K^{-1} m^{-1} s^{-1}) rather than N$_2$?

The energy flux is given by

$$J = -\kappa \left(\frac{dT}{dx} \right) = -\kappa \left(\frac{\Delta T}{\Delta x} \right)$$

The loss in energy is equal to the flux times the area (A) through which the energy loss occurs:

$$\Delta E = -\kappa \left(\frac{\Delta T}{\Delta x}\right) \cdot A$$

a) $\Delta E = -(0.0240 \text{ J K}^{-1} \text{ m}^{-1} \text{ s}^{-1})\left(\frac{12 \text{ K}}{3 \text{ cm}}\right)\left(\frac{100 \text{ cm}}{1 \text{ m}}\right)(1 \text{ m}^2)$

$= -9.60 \text{ J s}^{-1}$

b) $\Delta T = 42 \text{ K}$

$\Delta E = -(0.0240 \text{ J K}^{-1} \text{ m}^{-1} \text{ s}^{-1})\left(\frac{42 \text{ K}}{3 \text{ cm}}\right)\left(\frac{100 \text{ cm}}{1 \text{ m}}\right)(1 \text{ m}^2)$

$= -33.6 \text{ J s}^{-1}$

c) $\Delta T = 42 \text{ K} \qquad \kappa = 0.0163 \text{ J K}^{-1}\text{m}^{-1}\text{s}^{-1}$

$\Delta E = -(0.0163 \text{ J K}^{-1}\text{m}^{-1}\text{s}^{-1})\left(\frac{42 \text{ K}}{3 \text{ cm}}\right)\left(\frac{100 \text{ cm}}{1 \text{ m}}\right)(1 \text{ m}^2)$

$= -22.8 \text{ J s}^{-1}$

P34.10) Determine the thermal conductivity of the following species at 273 K and 1 atm:
a) Ar ($\sigma = 0.36 \text{ nm}^2$)
b) Cl_2 ($\sigma = 0.93 \text{ nm}^2$)
c) SO_2 ($\sigma = 0.58 \text{ nm}^2$, geometry: bent)
You will need to determine $C_{V,m}$ for the species listed. You can assume that the translational and rotational degrees of freedom are in the high-temperature limit, and that the vibrational contribution to $C_{V,m}$ can be ignored at this temperature.

$$\kappa = \frac{1}{3}\frac{C_{vm}}{N_A}\cdot\left(\frac{8RT}{\pi M}\right)^{1/2}\frac{1}{\sqrt{2}\sigma}$$

a) $C_{V,m}^{,Av} = \frac{3}{2}R$

$\kappa = \frac{1}{3}\left(\frac{3}{2}\frac{R}{N_A}\right)\cdot\left(\frac{8RT}{\pi M}\right)^{1/2}\frac{1}{\sqrt{2}\sigma}$

$= \frac{8.314 \text{ J mol}^{-1}\text{K}^{-1}}{2(6.022\times10^{23} \text{ mol}^{-1})}\cdot\left(\frac{8(8.314 \text{ J mol}^{-1} \text{ K}^{-1})(273 \text{ K})}{\pi(0.040 \text{ kg mol}^{-1})}\right)^{1/2}\frac{1}{\sqrt{2}(0.36 \text{ nm}^2)}\cdot\left(\frac{10^9 \text{ nm}}{1 \text{ m}}\right)^2$

$- 0.00516 \text{ J K}^{-1} \text{ m}^{-1} \text{ s}^{-1}$

b) $C_{V,m}^{Cl_2} = C_V^T + C_V^R = \frac{3}{2}R + R = \frac{5}{2}R$

$$\kappa = \frac{1}{3}\left(\frac{5}{2}\frac{R}{N_A}\right)\left(\frac{8RT}{\pi M}\right)^{1/2}\frac{1}{\sqrt{2}\sigma}$$

$$-\frac{5\left(8.314\ \text{J mol}^{-1}\ \text{K}^{-1}\right)}{6\left(6.022\times10^{23}\ \text{mol}^{-1}\right)}\left(\frac{8\left(8.314\ \text{J mol}^{-1}\ \text{K}^{-1}\right)(273\ \text{K})}{\pi\left(0.071\ \text{kg mol}^{-1}\right)}\right)^{1/2}\frac{1}{\sqrt{2}\left(0.93\ \text{nm}^2\right)}\left(\frac{10^9\ \text{nm}}{1\ \text{m}}\right)^2$$

$$= 0.00249\ \text{J K}^{-1}\ \text{m}^{-1}\ \text{s}^{-1}$$

c) $C_{V,m}^{SO_2} = C_V^T + C_V^R = \frac{3}{2}R + \frac{3}{2}R = 3R$

$$\kappa = \frac{1}{3}\left(\frac{3R}{N_A}\right)\left(\frac{8RT}{\pi M}\right)^{1/2}\frac{1}{\sqrt{2}\sigma}$$

$$= \frac{\left(8.314\ \text{J mol}^{-1}\ \text{K}^{-1}\right)}{3\left(6.022\times10^{23}\ \text{mol}^{-1}\right)}\left(\frac{8\left(8.314\ \text{J mol}^{-1}\ \text{K}^{-1}\right)(273\ \text{K})}{\pi\left(0.064\ \text{kg mol}^{-1}\right)}\right)^{1/2}\frac{1}{\sqrt{2}\left(0.58\ \text{nm}^2\right)}\left(\frac{10^9\ \text{nm}}{1\ \text{m}}\right)^2$$

$$= 0.0050\ \text{J K}^{-1}\ \text{m}^{-1}\ \text{s}^{-1}$$

P34.11) The thermal conductivity of Kr is 0.0087 J K^{-1} m^{-1} s^{-1} at 273 K and 1 atm. Estimate the collisional cross section of Kr.

Treating Kr as an ideal monatomic gas, $C_{V,m} = \frac{3}{2}R$, and the thermal conductivity is:

$$\kappa = \frac{1}{3}C_{V,m}v_{ave}\frac{1}{\sqrt{2}\sigma} = \frac{1}{3}\left(\frac{3}{2}\frac{R}{N_A}\right)\left(\frac{8RT}{\pi M}\right)^{1/2}\frac{1}{\sqrt{2}\sigma}$$

Rearranging to isolate the collisional cross section:

$$\kappa = \frac{1}{3}C_{V,m}v_{ave}\frac{1}{\sqrt{2}\sigma} = \frac{1}{3}\left(\frac{3}{2}\frac{R}{N_A}\right)\left(\frac{8RT}{\pi M}\right)^{1/2}\frac{1}{\sqrt{2}\sigma}$$

$$\sigma = \frac{1}{3}\left(\frac{3}{2}\frac{R}{N_A}\right)\left(\frac{8RT}{\pi M}\right)^{1/2}\frac{1}{\sqrt{2}\kappa}$$

$$= \frac{1}{2}\left(\frac{R}{N_A}\right)\left(\frac{8\left(8.314\ \text{J mol}^{-1}\ \text{K}^{-1}\right)(273\ \text{K})}{\pi\left(0.0838\ \text{kg mol}^{-1}\right)}\right)^{1/2}\frac{1}{\sqrt{2}\left(0.0087\ \text{J K}^{-1}\ \text{m}^{-1}\ \text{s}^{-1}\right)} = 1.47\times10^{-19}\ \text{m}^2$$

P34.15) The thermal conductivities of acetylene (C_2H_2) and N_2 at 273 K and 1 atm are 0.01866 and 0.0240 J m^{-1} s^{-1} K^{-1}, respectively. Based on these data, what is the ratio of the collisional cross section of acetylene relative to N_2?

$$\frac{\kappa_{C_2H_2}}{\kappa_{N_2}} = \frac{\dfrac{C_{V,m}^{C_2H_2}}{3N_A}\left(\dfrac{8RT}{\pi M_{C_2H_2}}\right)^{1/2}\dfrac{1}{\sqrt{2}\sigma_{C_2H_2}}}{\dfrac{C_{V,m}^{N_2}}{3N_A}\left(\dfrac{8RT}{\pi M_{N_2}}\right)^{1/2}\dfrac{1}{\sqrt{2}\sigma_{N_2}}}$$

$$= \frac{C_{V,m}^{C_2H_2}}{C_{V,m}^{N_2}}\left(\frac{M_{N_2}}{M_{C_2H_2}}\right)^{1/2}\frac{\sigma_{N_2}}{\sigma_{C_2H_2}}$$

Rearranging to isolate the ratio of collisional cross sections:

$$\frac{\sigma_{C_2H_2}}{\sigma_{N_2}} = \frac{C_{V,m}^{C_2H_2}}{C_{V,m}^{N_2}}\left(\frac{M_{N_2}}{M_{C_2H_2}}\right)^{1/2}\frac{\kappa_{N_2}}{\kappa_{C_2H_2}}$$

Both C_2H_2 and N_2 are linear molecules, and will therefore have the same heat capacity value so that the collision cross section ratio depends only on the mass and thermal conductivity ratios:

$$\frac{\sigma_{C_2H_2}}{\sigma_{N_2}} = \left(\frac{0.0280 \text{ kg mol}^{-1}}{0.0260 \text{ kg mol}^{-1}}\right)^{1/2}\left(\frac{0.0240 \text{ J m}^{-1}\text{ s}^{-1}\text{ K}^{-1}}{0.01866 \text{ J m}^{-1}\text{ s}^{-1}\text{ K}^{-1}}\right)$$

$$\frac{\sigma_{C_2H_2}}{\sigma_{N_2}} = 1.33$$

P34.19) The viscosity of H_2 at 273 K at 1 atm is 84.0 μP. Determine the viscosities of D_2 and HD.

The expression for viscosity is

$$\eta = \frac{1}{3}\left(\frac{8RT}{\pi M}\right)^{1/2}\frac{1}{\sqrt{2}\sigma}\frac{M}{N_A}$$

Taking the ratio of viscosities for two species (denoted as 1 and 2) yields

$$\frac{\eta_2}{\eta_1} = \sqrt{\frac{M_2}{M_1}}\left(\frac{\sigma_1}{\sigma_2}\right)$$

Assuming that the collisional cross sections for the species are the same, the ratio of velocities reduces to:

$$\frac{\eta_2}{\eta_1} = \sqrt{\frac{M_2}{M_1}}$$

Substituting the molecular weights into the above expression yields the following viscosities for D_2 and HD:

$$\eta_{D_2} = \eta_{H_2} \sqrt{\frac{M_{D_2}}{M_{H_2}}}$$

$$= (84\ \mu P) \sqrt{\frac{4.04\ \text{g mol}^{-1}}{2.02\ \text{g mol}^{-1}}}$$

$$\eta_{D_2} = 118\ \mu P$$

$$\eta_{HD} = 84\ \mu P \sqrt{\frac{3.03\ \text{g mol}^{-1}}{2.02\ \text{g mol}^{-1}}}$$

$$= 103\ \mu P$$

P34.21) How long will it take to pass 200. mL of H_2 at 273 K through a 10-cm-long capillary tube of 0.250 mm if the gas input and output pressures are 1.05 and 1.00 atm, respectively?

The flow rate is given as

$$\frac{\Delta V}{\Delta t} = \frac{\pi r^4}{8\eta} \left(\frac{P_2 - P_1}{x_2 - x_1} \right)$$

Substituting into this expression and solving for Δt yields:

$$\frac{(0.2\ \text{L})}{\Delta t} = \frac{\pi (2.5 \times 10^{-4}\ \text{m})^4}{8(84 \times 10^{-6}\text{P}) \left(\dfrac{0.1\ \text{kg m}^{-1}\text{s}^{-1}}{1\ \text{P}} \right)} \left(\frac{1.05\ \text{atm} - 1.00\ \text{atm}}{0.1\ \text{m}} \right)$$

$$\frac{(0.2\ \text{L})}{\Delta t} = 9.13 \times 10^{-11}\ \text{m}^4\ \text{kg}^{-1}\ \text{s} \cdot \frac{101,325\ \text{N m}^{-2}}{1\ \text{atm}} \cdot$$

$$\frac{(0.2\ \text{L})}{\Delta t} = 9.13 \times 10^{-6}\ \text{m}^3\ \text{s}^{-1}$$

$$\frac{(0.2\ \text{L}) \left(\dfrac{1\ \text{m}^3}{1000\ \text{L}} \right)}{9.13 \times 10^{-5}\ \text{m}^3\ \text{s}^{-1}} = 21.9\ \text{s} = \Delta t$$

P34.25) Poiseuille's law can be used to describe the flow of blood through blood vessels. Using Poiseuille's law, determine the pressure drop accompanying the flow of blood through 5 cm of the aorta ($r = 1$ cm). The rate of blood flow through the body is 0.08 L s^{-1} and the viscosity of blood is approximately 4 cP at 310 K.

$$\frac{\Delta V}{\Delta t} = \frac{\pi r^4}{8\eta}\left(\frac{P_2 - P_1}{x_2 - x_1}\right) = \frac{\pi r^4}{8\eta}\left(\frac{\Delta P}{\Delta x}\right)$$

$$\Delta P = \left(\frac{\Delta V}{\Delta t}\right)\frac{8\eta\Delta x}{\pi r^4} = \left(8.00\times10^{-5}\ \text{m}^3\ \text{s}^{-1}\right)\frac{8\left(0.00400\ \text{kg m}^{-1}\ \text{s}^{-1}\right)\left(0.050\ \text{m}\right)}{\pi\left(0.0100\ \text{m}\right)^4} = 4.07\ \text{Pa}$$

P34.27) Myoglobin is a protein that participates in oxygen transport. For myoglobin in water at 20°C, $\overline{s} = 2.04 \times 10^{-13}$ s, $D = 1.13 \times 10^{-10}$ m² s⁻¹, and $\overline{V} = 0.740$ cm³ g⁻¹. The density of water is 0.998 g cm³ and the viscosity is 1.002 cP at this temperature.

a) Using the information provided, estimate the size of myoglobin.
b) What is the molecular weight of myoglobin?

a) Using the Stokes-Einstein equation, the radius of myoglobin is:

$$r = \frac{kT}{6\pi\eta D}$$

$$= \frac{\left(1.38\times10^{-23}\ \text{J K}^{-1}\right)\left(293\ \text{K}\right)}{6\pi\left(0.01002\ \text{P}\right)\left(\dfrac{0.1\ \text{kg m}^{-1}\text{s}^{-1}}{1\ \text{P}}\right)\left(1.13\times10^{-10}\ \text{m}^2\ \text{s}^{-1}\right)}$$

$$= 1.89\times10^{-9}\ \text{m}$$

$$= 1.89\ \text{nm}$$

b) The molecular weight of myoglobin can be found as follows:

$$M = \frac{RT\overline{s}}{D\left(1-\overline{V}\rho\right)}$$

$$= \frac{\left(8.314\ \text{J mol}^{-1}\ \text{K}^{-1}\right)\left(293\ \text{K}\right)\left(2.04\times10^{-13}\ \text{s}\right)}{\left(1.13\times10^{-10}\ \text{m}^2\ \text{s}^{-1}\right)\left(1-\left(0.740\ \text{cm g}^{-1}\right)\left(0.998\ \text{g cm}^{-3}\right)\right)}$$

$$= 16.8\ \text{kg mol}^{-1}$$

P34.30) Boundary centrifugation is performed at an angular velocity of 40,000. rpm to determine the sedimentation coefficient of cytochrome c (M = 13,400 g mol⁻¹) in water at 20°C (ρ = 0.998 g cm⁻³, η = 1.002 cP). The following data are obtained on the position of the boundary layer as a function of time:

Time (h)	x_b (cm)
0	4.00
2.5	4.11
5.2	4.23
12.3	4.57
19.1	4.91

a) What is the sedimentation coefficient for cytochrome c under these conditions?
b) The specific volume of cytochrome c is 0.728 cm^3 g^{-1}. Estimate the size of cytochrome c.

a) Using the data from the table, a plot of $ln\left(\dfrac{x_b}{x_{b,t=0}}\right)$ versus t can be constructed, the slope of which is

equal to $\omega^2 \overline{s}$:

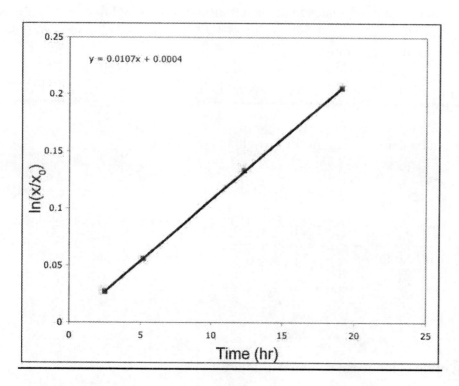

The slope from the best fit to the line is 0.0107 hr^{-1}. Using this slope, the sedimentation coefficient is determined as follows:

$$\omega^2\overline{s} = 0.0107 \text{ hr}^{-1}$$

$$\overline{s} = \frac{0.0107 \text{ hr}^{-1}}{\omega^2}\left(\frac{1 \text{ hr}}{3600 \text{ s}}\right) = \frac{2.97\times10^{-6} \text{ s}^{-1}}{\omega^2}$$

$$= \frac{2.97\times10^{-6} \text{ s}^{-1}}{\left(4.00\times10^4 \text{ rev min}^{-1}\right)^2\left(\dfrac{2\pi \text{ rad}}{1 \text{ rev}}\right)^2\left(\dfrac{1 \text{ min}}{60 \text{ sec}}\right)^2}$$

$$= 1.70\times10^{-13} \text{ s}$$

b) First, the frictional force is calculated as follows:

$$f = \frac{m\left(1-\overline{V}\rho\right)}{\overline{s}}$$

$$= \frac{\left(\dfrac{13.4 \text{ kg mol}^{-1}}{6.022\times10^{23} \text{ mol}^{-1}}\right)\left(1-\left(0.728 \text{ cm}^3\text{g}^{-1}\right)\left(0.998 \text{ g cm}^{-3}\right)\right)}{1.70\times10^{-13} \text{ s}}$$

$$f = 3.58\times10^{-11} \text{ kg s}^{-1}$$

With the frictional force, the particle radius can be determined:

$$6\pi\eta r = f$$

$$r = \frac{f}{6\pi\eta} = \frac{3.58\times10^{-11} \text{ kg s}^{-1}}{6\pi\left(0.01005 \text{ P}\right)\left(\dfrac{0.1 \text{ kg m}^{-1}\text{ s}^{-1}}{\text{P}}\right)}$$

$$= 1.89\times10^{-9} \text{ m}$$

$$= 1.89 \text{ nm}$$

P34.35) The molar conductivity of sodium acetate, CH_3COONa, is measured as a function of concentration in water at 298 K, and the following data are obtained:

Concentration (M)	Λ_m (S m^2 mol^{-1})
0.0005	0.00892
0.001	0.00885
0.005	0.00857
0.01	0.00838
0.02	0.00812
0.05	0.00769
0.1	0.00728

Is sodium acetate a weak or strong electrolyte? Determine Λ_m^0 using appropriate methodology depending on your answer.

If the electrolyte is strong, a plot of Λ_m versus $\sqrt{\dfrac{c}{c_0}}$ (assuming $c_0 = 1$ M) should yield a straight line.

The corresponding plot is shown below:

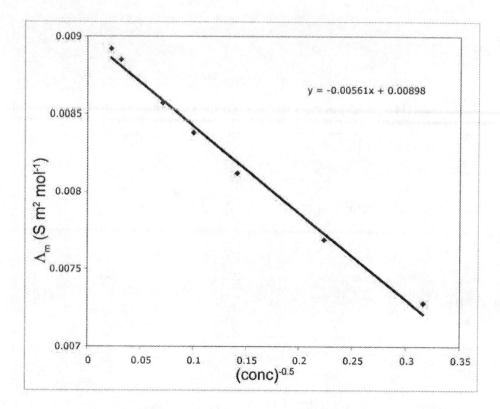

The linearity of the plot demonstrates that sodium acetate is a strong electrolyte. Best fit by a straight line yields the following:

$$\Lambda_m = -0.00561 \cdot \sqrt{\frac{c}{c_0}} + 0.00898 \quad \left(\text{S m}^2 \text{ mol}^{-1}\right)$$

Therefore, for sodium acetate $\Lambda_m^0 = 0.00898$ S m^2 mol^{-1}.

P34.39) In the determination of molar conductivities, it is convenient to define the cell constant, K, as $K = \dfrac{l}{A}$, where l is the separation between the electrodes in the conductivity cell, and A is the area of the electrodes.

a) A standard solution of KCl (conductivity or $\kappa = 1.06296 \times 10^{-6}$ S m^{-1} at 298 K) is employed to standardize the cell, and a resistance of 4.2156 Ω is measured. What is the cell constant?
b) The same cell is filled with a solution of HCl and a resistance of 1.0326 Ω is measured. What is the conductivity of the HCl solution?

a) The conductivity is defined as

$$\kappa = \frac{\ell}{R \cdot A} = \frac{K}{R}$$

where R is the resistance and K is the cell constant. Using this relationship, the cell constant is determined as follows:

$$\kappa = 1.06296 \times 10^{-6} \text{ s m}^{-1} \qquad R = 4.2156 \Omega$$

$$K = \kappa R = \left(1.06296 \times 10^{-6} \text{ S m}^{-1}\right)\left(4.2156 \ \Omega\right)$$

$$K = 4.48 \times 10^{-6} \text{ S m}^{-1} \ \Omega$$

b)

$$\kappa = \frac{K}{R} = \frac{4.48 \times 10^{-6} \text{ S m}^{-1} \ \Omega}{1.0326 \ \Omega}$$

$$= 4.34 \times 10^{-6} \text{ S m}^{-1}$$

Chapter 35: Elementary Chemical Kinetics

P35.2) Consider the first-order decomposition of cyclobutane at 438°C at constant volume:

$$C_4H_8(g) \longrightarrow 2C_2H_4(g)$$

a) Express the rate of the reaction in terms of the change in total pressure as a function of time.

b) The rate constant for the reaction is $2.48 \times 10^{-4} \, s^{-1}$. What is the half-life?

c) After initiation of the reaction, how long will it take for the initial pressure of C_4H_8 to drop to 90.0% of its initial value?

a) The rate is given as

$$R = -\frac{1}{RT} \frac{d\,P_{C_4H_8}}{dt} = \frac{1}{2RT} \frac{d\,P_{C_2H_4}}{dt}$$

The pressure at time t is given as $P_t = P_{t=0} - P_{C_4H_8} + P_{C_2H_4}$. Where $P_{t=0}$ is the initial pressure, and $P_{C_4H_8}$ and $P_{C_2H_4}$ represent the pressures of the individual gases as a specific time. By stoichiometry $P_{C_4H_8} = \frac{1}{2} P_{C_2H_4}$ so that:

$$P_t = P_{t=0} - \frac{1}{2} P_{C_2H_4} + P_{C_2H_4} = P_{t=0} + \frac{1}{2} P_{C_2H_4}$$

Since $P_{C_2H_4}$ is dependent on time and the initial pressure is time independent so that:

$$\frac{dP_t}{dt} = \frac{1}{2} \frac{d\,P_{C_2H_4}}{dt}$$

And

$$R = \frac{1}{RT} \frac{dP_t}{dt}$$

b) The reaction is first order with respect to C_4H_8; therefore:

$$t_{1/2} = \frac{\ln 2}{k}$$

so

$$t_{1/2} = \frac{0.693}{2.48 \times 10^{-4} \, s^{-1}}$$
$$= 2.79 \times 10^3 \, s$$

c) Using the integrated rate law

$$-kt = \ln\left(\frac{[A]}{[A]_0}\right)$$

Where

$$\frac{[A]}{[A]_0} = 0.900$$

one finds

$$t = -\frac{\ln(0.900)}{2.48 \times 10^{-4} \text{ s}^{-1}}$$

$$= \frac{0.105}{2.48 \times 10^{-4} \text{ s}^{-1}}$$

$$= 425 \text{ s}$$

P35.4) Consider the following reaction involving bromophenol blue (BPB) and OH: $BPB(aq) + OH^-(aq) \longrightarrow BPBOH^-(aq)$.

The concentration of BPB can be monitored by following the absorption of this species and using the Beer–Lambert law. In this law, absorption, A, and concentration are linearly related.

a) Express the reaction rate in terms of the change in absorbance as a function of time.

b) Let A_o be the absorbance due to BPB at the beginning of the reaction. Assuming that the reaction is first order with respect to both reactants, how is the absorbance of BPB expected to change with time?

c) Given your answer to part (b), what plot would you construct to determine the rate constant for the reaction?

a) Beer's Law states that

$$A = \varepsilon b[\text{BPB}]$$

where ε is the molar absorptivity of BPB and b is the cell path length. The rate of reaction is

$$Rate = -\frac{d[\text{BPB}]}{dt}$$

Thus

$$Rate = -\frac{d[\text{BPB}]}{dt}$$

$$= -\frac{d\left(\dfrac{A}{\varepsilon b}\right)}{dt}$$

$$Rate = -\frac{1}{\varepsilon b}\frac{dA}{dt}$$

b) Let A_0 be the initial absorbance of BPB and $A(t)$ be the absorbance at time, t. Since the reaction is a second order reaction of type II, the integrate rate equation has the form

$$\frac{1}{[OH^-]_0 - [BPB]_0} \ln\left(\frac{[OH^-]/[OH^-]_0}{[BPB]/[BPB]_0}\right) = kt$$

Substituting into the above expression for [BPB] yields:

$$\frac{1}{[OH^-]_0 - \frac{A_0}{b\varepsilon}} \ln\left(\frac{[OH^-]/[OH^-]_0}{A/A_0}\right) = kt$$

$$\ln\left(\frac{A}{A_0}\right) = \ln\left(\frac{[OH^-]}{[OH^-]_0}\right) - \left([OH^-] - \frac{A_0}{b\varepsilon}\right)kt$$

c) A plot of $\ln\left(\frac{A}{A_0}\right)$ vs. t is predicted to be linear.

P35.8) The disaccharide lactose can be decomposed into its constituent sugars galactose and glucose. This decomposition can be accomplished through acid-based hydrolysis, or by the enzyme lactase. Lactose intolerance in humans is due to the lack of lactase production by cells in the small intestine. However, the stomach is an acidic environment; therefore, one might expect lactose hydrolysis to still be an efficient process. The following data were obtained on the rate of lactose decomposition as a function of acid and lactose concentration. Using this information, determine the rate law expression for the acid-based hydrolysis of lactose.

Initial Rate (M s^{-1})	[lactose]$_0$ (M^{-1})	[H$^+$] (M^{-1})
0.00116	0.01	0.001
0.00232	0.02	0.001
0.00464	0.01	0.004

The rate doubles when the lactose concentration is doubled, so the reaction is first order in lactose. The rate quadruples when the H$^+$ concentration is multiplied by four, so the reaction is first order in H$^+$:

$$Rate = k \, [lactose] \, [H^+]$$

P35.9) (Challenging) The first-order thermal decomposition of chlorocyclohexane is as follows:
$C_6H_{11}Cl(g) \longrightarrow C_6H_{10}(g) + HCl(g)$. For a constant volume system the following total pressure was measured as a function of time:

Time (s)	P (torr)	Time (min)	P (torr)
3.00	237.2	24.0	332.1
6.00	255.3	27.0	341.1
9.00	271.3	30.0	349.3
12.0	285.8	33.0	356.9
15.0	299.0	36.0	363.7
18.0	311.2	39.0	369.9
21.0	322.2	42.0	375.5

a) Derive the following relationship for a first-order reaction:

$P(t_2) - P(t_1) = \left(P(t_\infty) - P(t_0)\right)e^{-kt_1}\left(1 - e^{-k(t_2 - t_1)}\right)$. In this relation, $P(t_1)$ and $P(t_2)$ are the pressures at two specific times; $P(t_0)$ is the initial pressure when the reaction is initiated, $P(t_\infty)$ is the pressure at the completion of the reaction, and k is the rate constant for the reaction. To derive this relationship:

i. Given the first-order dependence of the reaction, write down the expression for the pressure of chlorocyclohexane at a specific time t_1.
ii. Write the expression for the pressure at another time t_2, which is equal to $t_1 + \Delta$ where delta is a fixed quantity of time.
iii. Write down expressions for $P(t_\infty) - P(t_1)$ and $P(t_\infty) - P(t_2)$.
iv. Subtract the two expressions from part (iii).

b) Using the natural log of the relationship from part (a) and the data provided in the table given earlier in this problem, determine the rate constant for the decomposition of chlorocyclohexane. (*Hint:* Transform the data in the table by defining $t_2 - t_1$ to be a constant value, for example, 9 s.)

a)   Since the reaction is first order, we can write:

$$P(t_\infty) - P(t_1) = \left(P(t_\infty) - P(t_0)\right)e^{-kt_1}$$
$$P(t_\infty) - P(t_2) = \left(P(t_\infty) - P(t_0)\right)e^{-kt_2} = \left(P(t_\infty) - P(t_0)\right)e^{-k(t_1+\Delta)}$$

subtracting the two previous equations:

$$P(t_2) - P(t_1) = \left(P(t_\infty) - P(t_0)\right)e^{-kt_1}\left(1 - e^{-k\Delta}\right)$$

where Δ is the difference in time between t_2 and t_1.

b)   Taking the natural log of the previous expression yields:

$$\ln\left(P(t_2) - P(t_1)\right) = \ln\left(P(t_\infty) - P(t_0)\right) - kt_1 + \ln\left(1 - e^{-k\Delta}\right)$$
$$= \ln\left[\left(P(t_\infty) - P(t_0)\right)\left(1 - e^{-k\Delta}\right)\right] - kt_1$$

251

Therefore, a plot of the difference in pressure at fixed difference in time versus t_1 should yield a straight line with slope equal to $-k$. Using a difference in time of 9 s yields the following table of the difference in pressures versus time:

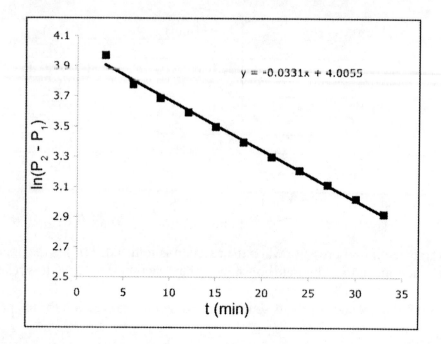

The slope of the best fit line is -0.0311 min^{-1}; therefore, the rate constant is 0.0311 min^{-1}, or 5.18×10^{-4} s^{-1}.

P35.13) The half-life of ^{238}U is 4.5×10^9 years. How many disintegrations occur in 1 minute for a 10.0-mg sample of this element?

The rate constant can be found from the half-life:

$$t_{1/2} = \frac{\ln 2}{k} \quad or \quad k = \frac{\ln 2}{t_{1/2}}$$

Thus

$$k = \frac{0.693}{4.5 \times 10^9 \text{ yr}} \left(\frac{1 \text{ yr}}{364.25 \text{ day}} \right) \left(\frac{1 \text{ day}}{24 \text{ hr}} \right) \left(\frac{1 \text{ hr}}{60 \text{ min}} \right)$$

$$k = 2.94 \times 10^{-16} \text{ min}^{-1}$$

Converting 10.0 mg into the number of ^{238}U atoms:

$$N_{238_U} = \left(1 \times 10^{-2} \text{ g} \right) \left(238 \text{ g mol}^{-1} \right) \left(6.022 \times 10^{23} \text{ mol}^{-1} \right) = 1.43 \times 10^{24}$$

Employing the rate law, the number of disintegrations in 1 minute is:

$$N = N_o e^{-kt}$$
$$= \left(1.43 \times 10^{24}\right) e^{-2.94 \times 10^{-16} \text{ min}^{-1} \cdot 1 \text{ min}}$$
$$= 1.43 \times 10^{24}$$

The rate constant is so small that a negligible number of disintegrations occur in one minute.

P35.18) Show that the ratio of the half-life to the three-quarter life, $t_{1/2}/t_{3/4}$, for a reaction that is *n*th order ($n > 1$) in reactant *A* can be written as a function of *n* alone (that is, there is no concentration dependence in the ratio).

In general, for order *n*,

$$kt = \frac{1}{n-1}\left\{\frac{1}{[A]^{n-1}} - \frac{1}{[A]_0^{n-1}}\right\}$$

$$t_{1/2} \Rightarrow [A] = \frac{1}{2}[A]_0$$

$$kt_{1/2} = \frac{1}{n-1}\left\{\frac{2^{n-1}}{[A]_0^{n-1}} - \frac{1}{[A]_0^{n-1}}\right\}$$

and

$$t_{1/2} = \frac{2^{n-1}-1}{k(n-1)[A]_0^{n-1}}$$

$$t_{3/4} \Rightarrow [A] = \frac{1}{4}[A]_0$$

$$kt_{3/4} = \frac{1}{n-1}\left\{\frac{4^{n-1}}{[A]_0^{n-1}} - \frac{1}{[A]_0^{n-1}}\right\}$$

and

$$t_{3/4} = \frac{4^{n-1}-1}{k(n-1)[A]_0^{n-1}}$$

Thus,

$$\frac{t_{1/2}}{t_{3/4}} = \frac{\dfrac{2^{n-1}-1}{k(n-1)[A]_0^{n-1}}}{\dfrac{4^{n-1}-1}{k(n-1)[A]_0^{n-1}}}$$

$$\frac{t_{1/2}}{t_{3/4}} = \frac{2^{n-1}-1}{4^{n-1}-1}$$

P35.20) For the sequential reaction A $\xrightarrow{k_A}$ B $\xrightarrow{k_B}$ C , the rate constants are $k_A = 5 \times 10^6$ s^{-1} and $k_B = 3 \times 10^6$ s^{-1}. Determine the time at which [B] is at a maximum.

For a sequential reaction

$$[B] = \frac{k_A}{k_B - k_A}\left(e^{-k_A t} - e^{-k_B t}\right)[A_\circ]$$

The maximum occurs when

$$\frac{d[B]}{dt} = 0 = \frac{k_A}{k_B - k_A}[A]_0 \frac{d}{dt}\left(e^{-k_A t} - e^{-k_B t}\right)$$

$$0 = \frac{k_A}{k_B - k_A}[A]_0\left(-k_A e^{-k_A t} + k_B e^{-k_B t}\right)$$

The above equality will be true when the term in parenthesis equals zero, therefore:

$$k_A e^{-k_A t} = k_B e^{-k_B t}$$

$$\ln k_A - k_A t = \ln k_B - k_B t$$

$$\ln k_A - \ln k_B = (k_A - k_B)t$$

$$\frac{1}{k_A - k_B}\ln\left(\frac{k_A}{k_B}\right) = t$$

Substituting the values of k_A and k_B into the previous expression yields:

$$t = \frac{1}{\left(5\times10^6 \text{ s}^{-1}\right) - \left(3\times10^6 \text{ s}^{-1}\right)}\ln\left(\frac{5\times10^6 \text{ s}^{-1}}{3\times10^6 \text{ s}^{-1}}\right)$$

$$= \left(5\times10^{-7} \text{ s}\right)(0.511)$$

$$t = 2.6\times10^{-7} \text{ s}$$

P35.23) For a type II second-order reaction, the reaction is 60% complete in 60 seconds when $[A]_0 = 0.1$ M and $[B]_0 = 0.5$ M.
a) What is the rate constant for this reaction?
b) Will the time for the reaction to reach 60% completion change if the initial reactant concentrations are decreased by a factor of two?

a) The integrated rate-law expression for a second-order reaction of type II is:

$$kt = \frac{1}{[B]_0 - [A]_0}\ln\left(\frac{[B]/[B]_0}{[A]/[A]_0}\right)$$

at $t = 60$ s, $[A] = 0.04$ M with 1:1 stoichiometry so that $[B] = 0.44$ M. Substituting these values into the above expression and using $t = 60$ s yields:

$$k = \frac{1}{60 \text{ s } (0.5 \text{ M} - 0.1 \text{ M})} \ln\left(\frac{0.44 \text{ M}/0.5 \text{ M}}{0.04 \text{ M}/0.1 \text{ M}}\right)$$

$$= (0.0417 \text{ M}^{-1} \text{ s}^{-1})(0.788)$$

$$= 0.0329 \text{ M}^{-1} \text{ s}^{-1} \approx 0.03 \text{ M}^{-1} \text{ s}^{-1}$$

b) The time will double, assuming the k value is the same. Numerically checking this expectation:

$$t = \frac{1}{k([B]_0 - [A]_0)} \ln\left(\frac{[B]/[B]_0}{[A]/[A]_0}\right)$$

$$= \frac{1}{(0.0329 \text{ M}^{-1} \text{ s}^{-1})(0.25 \text{ M} - 0.05 \text{ M})} \ln\left(\frac{0.22 \text{ M}/0.25 \text{ M}}{0.02 \text{ M}/0.05 \text{ M}}\right)$$

$$= (152 \text{ s})(0.788)$$

$$= 120 \text{ s}$$

P35.26) In the stratosphere, the rate constant for the conversion of ozone to molecular oxygen by atomic chlorine is $Cl + O_3 \rightarrow ClO + O_2$ $[(k = 1.7 \times 10^{10} \text{ M}^{-1} \text{ s}^{-1}) e^{-260 K/T}]$.

a) What is the rate of this reaction at 20 km where $[Cl] = 5 \times 10^{-17}$ M, $[O_3] = 8 \times 10^{-9}$ M, and $T = 220$. K?

b) The actual concentrations at 45 km are $[Cl] = 3 \times 10^{-15}$ M and $[O_3] = 8 \times 10^{-11}$ M. What is the rate of the reaction at this altitude where $T = 270$. K?

c) (Optional) Given the concentrations in part (a), what would you expect the concentrations at 45 km to be assuming that the gravity represents the operative force defining the potential energy?

a) Based on the units of k, the reaction is second order overall so that the rate law expression is:

$Rate = k[Cl][O_3]$

For $[Cl] = 5 \times 10^{-17}$ M, $[O_3] = 8 \times 10^{-9}$ M, and $T = 220$. K

$$k = 1.7 \times 10^{10} \text{ M}^{-1} \text{ s}^{-1} e^{-260. \text{ K}/220. \text{ K}}$$

$$k = 5.21 \times 10^9 \text{ M}^{-1} \text{ s}^{-1}$$

$$Rate = 5.21 \times 10^9 \text{ M}^{-1} \text{ s}^{-1} (5 \times 10^{-17} \text{ M})(8 \times 10^{-9} \text{ M})$$

$$= 2.08 \times 10^{-15} \text{ M s}^{-1}$$

b) $[Cl] = 3 \times 10^{-15}$ M $\quad$ $[O_3] = 8 \times 10^{-11}$ M $\quad$ $T = 270.$ K

$$k = 1.7 \times 10^{10} \text{ M}^{-1} \text{ s}^{-1} e^{-260. \text{ K}/270. \text{ K}}$$
$$k = 6.49 \times 10^{9} \text{ M}^{-1} \text{ s}^{-1}$$
$$Rate = 6.49 \times 10^{9} \text{ M}^{-1} \text{ s}^{-1} \left(3 \times 10^{-15} \text{ M}\right)\left(8 \times 10^{-11} \text{ M}\right)$$
$$= 1.56 \times 10^{-15} \text{ M s}^{-1}$$

b) The ratio of pressures at two altitudes is given by:

$$\frac{[P]_1}{[P]_2} = e^{-\frac{mg(h_1 - h_2)}{kT}}$$

Using this expression to determine the difference in concentration for Cl at 45 versus 20 km yields:

$$\frac{[Cl]_{45}}{[Cl]_{20}} = e^{-\frac{(0.035 \text{ kg mol}^{-1})\left(\frac{1}{N_A}\right)(9.80 \text{ m s}^{-2})(2.5 \times 10^4 \text{ m})}{(1.38 \times 10^{-23} \text{ J K}^{-1})(270 \text{ K})}} = 0.0219$$

$$[Cl]_{45} = \left(5 \times 10^{-17} \text{ M}\right)(0.0219) = 1.10 \times 10^{-18} \text{ M}$$

Performing the same calculation for O_3:

$$\frac{[O_3]_{45}}{[O_3]_{20}} = e^{-\frac{(0.048 \text{ kg mol}^{-1})\left(\frac{1}{N_A}\right)(9.80 \text{ m s}^{-2})(2.5 \times 10^4 \text{ m})}{(1.38 \times 10^{-23} \text{ J K}^{-1})(270 \text{ K})}} = 0.0053$$

$$[O_3]_{45} = \left(8 \times 10^{-9} \text{ M}\right)(0.0053) = 4.24 \times 10^{-11} \text{ M}$$

Finally, the rate is:

$$Rate = 6.49 \times 10^{9} \text{ M}^{-1} \text{ s}^{-1} \left(1.10 \times 10^{-18} \text{ M}\right)\left(4.24 \times 10^{-11} \text{ M}\right)$$
$$= 3.03 \times 10^{-19} \text{ M s}^{-1}$$

Notice that since this simple model for the concentration dependence versus altitude significantly underestimates the concentration of Cl, the rate of ozone depletion by reaction with Cl is also significantly underestimated.

P35.28) A standard "rule of thumb" for thermally activated reactions is that the reaction rate doubles for every 10 K increase in temperature. Is this statement true independent of the activation energy (assuming that the activation energy is positive and independent of temperature)?

The analysis is best performed numerically

$$\frac{k_2}{k_1} = \frac{e^{-E_a/RT_2}}{e^{-E_a/RT_1}} = e^{-\frac{E_a}{R}\left(\frac{1}{T_2} - \frac{1}{T_1}\right)}$$

T_1 (K)	T_2 (K)	E_a (J mol^{-1})	$\dfrac{k_2}{k_1}$
100	110	50,000	236
300	310	50,000	1.91
1000	1010	50,000	1.06
300	310	500,000	6.43
300	310	5,000	1.06

Inspection of the table demonstrates that the rule is only valid for temperatures around room temperature, and moderate E_a values ($\sim$50,000 J mol^{-1}).

P35.32) At 552.3 K, the rate constant for the thermal decomposition of SO_2Cl_2 is 1.02×10^{-6} s^{-1}. If the activation energy is 210. kJ mol^{-1}, calculate the Arrhenius preexponential factor and determine the rate constant at 600. K.

$$k_{552.3\ K} = 1.02 \times 10^{-6}\ s^{-1} \qquad E_a = 210\ kJ\ mol^{-1}$$

$$1.02 \times 10^{-6}\ s^{-1} = Ae^{-2.10\times10^5\ J\ mol^{-1}/(8.314\ J\ mol^{-1}\ K^{-1})(552.3\ K)}$$

$$A = \left(1.02 \times 10^{-6}\ s^{-1}\right)e^{2.10\times10^5\ J\ mol^{-1}/(8.314\ J\ mol^{-1}\ K^{-1})(552.3\ K)}$$

$$A = 7.38 \times 10^{13}\ s^{-1}$$

$$k_{600\ K} = \left(7.38 \times 10^{13}\ s^{-1}\right)e^{-2.10\times10^5\ J\ mol^{-1}/(8.314\ J\ mol^{-1}K^{-1})(600.\ K)}$$

$$= \left(7.38 \times 10^{13}\ s^{-1}\right)\left(5.23 \times 10^{-19}\right)$$

$$= 3.86 \times 10^{-5}\ s^{-1}$$

P35.34) Consider the reaction $A + B \underset{k'}{\overset{k}{\rightleftarrows}} P$. A temperature-jump experiment is performed where the

relaxation time constant is measured to be 310 μs, resulting in an equilibrium where $K_{eq} = 0.7$ with $[P]_{eq}$ = 0.2 M. What are k and k'? (Watch the units!)

$$\tau = 310 \times 10^{-6}\ s \qquad K_{eq} = 0.7 \qquad [P]_{eq} = 0.2\ M$$

Assuming the following rate law

$$\frac{d[A]}{dt} = -k^+[A][B] + k^{+'}[P] = 0,$$

The post-jump equilibrium concentrations with respect to the initial concentrations and concentration shift are:

$$[A] - \xi = [A]_{eq}$$
$$[B] - \xi = [B]_{eq}$$
$$[P] + \xi = [P]_{eq}$$

Therefore, the differential rate expression for the concentration shift, ξ, is:

$$\frac{d\xi}{dt} = -k^+ \left([A]_{eq} + \xi\right)\left([B]_{eq} + \xi\right) + k^{+'}\left([P]_{eq} - \xi\right)$$

$$\frac{d\xi}{dt} = -\xi\left(k^+[A]_{eq} + k^+[B]_{eq} + k^+\xi + k^{+'}\right)$$

$$= -\xi\left(k^+\left([A]_{eq} + k^+[B]_{eq}\right) + k^{+'}\right) + O\left(\xi^2\right)$$

Ignoring terms on the order ξ^2, the relaxation time is:

$$\tau = \left[k^+\left([A]_{eq} + k^+[B]_{eq}\right) + k^{+'}\right]^{-1}$$

Next, the equilibrium constant is given by:

$$K = \frac{k^+}{k^{+'}} = \frac{[P]_{eq}}{[A]_{eq}[B]_{eq}} = 0.7$$

If we assume that $[A]_o = [B]_o$ then $[A]_{eg} = [B]_{eg}$ and

$$0.7 = \frac{0.2 \text{ M}}{x^2} \Rightarrow x = 0.535 \text{ M} = [A]_{eq} = [B]_{eq}$$

And using the expression for K, we know that $k^+ = 0.7\, k^{+'}$. Use these last two results in the expression for the relaxation time yields:

$$310 \times 10^{-6} \text{ s} = \frac{1}{k^+\left(0.535 + 0.535\right) + k^{+'}}$$

$$310 \times 10^{-6} \text{ s} = \frac{1}{\left(0.7 k^{+'}\right)\left(1.070\right) + k^{+'}}$$

$$310 \times 10^{-6} \text{ s} = \frac{1}{k^{+'}\left(1.749\right)}$$

$$k^{+'} = 1845 \text{ s}^{-1}$$

$$k^+ = 0.7 k^{+'} = 1291 \text{ M}^{-1} \text{ s}^{-1}$$

The units of the rate constants are consistent with the forward reaction being second order, and the reverse reaction being first order.

P35.42) Consider the "unimolecular" isomerization of methylcyanide, a reaction that will be discussed in detail in the subsequent chapter:

$$CH_3NC(g) \longrightarrow CH_3CN(g)$$

The Arrhenius parameters for this reaction are $A = 2.5 \times 10^{16} \text{ s}^{-1}$ and $E_a = 272 \text{ kJ mol}^{-1}$. Determine the Eyring parameters $\Delta H^{\ddagger}$ and $\Delta S^{\ddagger}$ for this reaction with $T = 300.$ K.

For a unimolecular gas phase reaction

$$E_a = \Delta H^{\ddagger} + RT$$

$$\Delta H^{\ddagger} = 272 \times 10^3 \text{ J mol}^{-1} - \left(8.314 \text{ J mol}^{-1} \text{ K}^{-1}\right)\left(300 \text{ K}\right)$$

$$= 272 \times 10^3 \text{ J mol}^{-1} - 2.49_4 \times 10^3 \text{ J mol}^{-1}$$

$$= 269.5 \times 10^3 \text{ J mol}^{-1}$$

$$A = \frac{e k_B T}{h} e^{\Delta S^{\ddagger}/R}$$

$$\Delta S^{\ddagger} = R \ln\left(\frac{Ah}{e k_B T}\right)$$

$$= 8.314 \text{ J mol}^{-1}\text{K}^{-1} \ln\left(\frac{\left(2.5 \times 10^{16} \text{ s}^{-1}\right)\left(6.626 \times 10^{-34} \text{ J s}\right)}{e\left(1.38 \times 10^{-23} \text{ J K}^{-1}\right)\left(300. \text{ K}\right)}\right)$$

$$= \left(8.314 \text{ J mol}^{-1} \text{ K}^{-1}\right) \ln\left(1472\right)$$

$$= 60.6 \text{ J mol}^{-1}\text{K}^{-1}$$

P35.44) Chlorine monoxide (ClO) demonstrates three bimolecular self-reactions:

$$Rxn_1: \quad ClO \cdot (g) + ClO \cdot (g) \xrightarrow{k_1} Cl_2(g) + O_2(g)$$

$$Rxn_2: \quad ClO \cdot (g) + ClO \cdot (g) \xrightarrow{k_2} Cl \cdot (g) + ClOO \cdot (g)$$

$$Rxn_3: \quad ClO \cdot (g) + ClO \cdot (g) \xrightarrow{k_3} Cl \cdot (g) + OClO \cdot (g)$$

The following table provides the Arrhenius parameters for this reaction:

	A (M^{-1} s^{-1})	E_a (kJ/mol)
Rxn_1	6.08×10^8	13.2
Rxn_2	1.79×10^{10}	20.4
Rxn_3	2.11×10^8	11.4

a) For which reaction is $\Delta H^{\ddagger}$ greatest and by how much relative to the next closest reaction?

b) For which reaction is $\Delta S^{\ddagger}$ the smallest and by how much relative to the next closest reaction?

a) Since each $\Delta H^{\ddagger}$ depends linearly on the T and E_a, at the same T, the largest E_a corresponds to the largest $\Delta H^{\ddagger}$. Therefore, Rxn_2 will have the largest $\Delta H^{\ddagger}$ by 7.2 kJ mol^{-1} relative to Rxn_1.

b) $\Delta S^{\ddagger}$ depends linearly on $\ln A$; therefore, the smallest $\Delta S^{\ddagger}$ corresponds to the reaction with the smallest A, or Rxn_3. The next smallest $\Delta S^{\ddagger}$ is Rxn_1, and the difference is:

$$\ln(6.08 \times 10^8) - \ln(2.11 \times 10^8) = \ln\left(\frac{6.08}{2.11}\right) = \ln(2.88) = 1.06$$

Computational Problems

Before solving the computational problems, it is recommended that students work through Tutorials 1–3 under the Help menus in Spartan Student Edition to gain familiarity with the program.

Computational Problem 35.1: Chlorofluorocarbons are a potential source of atomic chlorine in the stratosphere. In this problem the energy needed to dissociate the C—Cl bond in CF_3Cl will be determined.

a. Perform a Hartree–Fock 3-21G calculation on the freon CF_3Cl and determine the minimum energy of this compound.

b. Select the C—Cl bond and calculate the ground-state potential energy surface along this coordinate by determining the energy of compound for the following C—Cl bond lengths:

r_{CCl} (Å)	E (Hartree)	r_{CCl} (Å)	E (Hartree)
1.40		2.40	
1.50		2.60	
1.60		2.80	
1.70		3.00	
1.80		4.00	
2.00		5.00	
2.20		6.00	

c. Using an energy barrier for dissociation calculated as the difference between the minimum of the potential energy surface to the energy at 6.00 Å, determine the barrier to dissociation.

d. Assuming an Arrhenius preexponential factor of 10^{12} s^{-1}, what is the expected rate constant for dissociation based on this calculation at 220. K? Is thermal dissociation of the C—Cl bond occurring to an appreciable extent in the stratosphere?

Procedure

Step 1: Create a new file and build CF_3Cl.
Step 2: Go to "**Setup > Calculations.**" Set the calculation type to equilibrium geometry and the method to **Hartree–Fock 3-21G**.
Step 3: In the Calculations dialog window, make sure "**Equilibrium Geometry**" is selected under the Calculate pull-down menu.

Step 4: Click the "**Submit**" button in the Calculations dialog window. Once you provide a file name the calculation will begin.

Step 5: When the calculation is finished, go to "**Display > Output**" to view the output file for the calculation. Scroll down until you locate the energy for the last step of the minimization process and record that energy.

Step 6: In the main window, click the "**<?>**" box on the toolbar to select the bond-length selection tool. Click on the C-Cl bond. The length of the bond will appear in the lower right hand corner of the main window. Record the value of the equilibrium geometry.

Step 7: In the bond length window enter 1.40 Å. The C-Cl bond length will shorten corresponding to this new distance.

Step 8: Go to "**Setup > Calculations**" and select "**Energy**" from the pull down menu for "**Calculate**" and then click "**Submit.**"

Step 9: When the calculation is finished, go to "**Display > Output**" to view the output file for the calculation. Scroll down until you locate the calculated energy for this new geometry and record that value.

Step 10: Repeat steps 6 through 9 for the bond lengths specified in the problem.

Computational Problem 35.2: Consider the dissociation of the C—F bond in $CFCl_3$. Using a standard bond dissociation energy of 485 kJ mol^{-1}, what would be the effect on the predicted rate constant for dissociation if zero-point energy along the C—F stretch coordinate were ignored? Performing a Hartree–Fock 6-31G* calculation, determine the frequency of the vibrational mode dominated by C—F stretch character. Calculate the dissociation rate using the Arrhenius expression without consideration of zero-point energy by adding the zero-point energy to the standard dissociation energy. Assume $A = 10^{10}$ s^{-1} and $T = 298$ K. Perform the corresponding calculation using the standard dissociation energy only. Does zero-point energy make a significant difference in the rate constant for this dissociation?

Procedure

Step 1: Create a new file and build $CFCl_3$.

Step 2: Go to "**Setup > Calculations.**" Set the calculation type to equilibrium geometry and the method to **Hartree–Fock 6-31G***.

Step 3: In the Calculations dialog window, make sure "**Equilibrium Geometry**" is selected under the Calculate pull-down menu, check the box next to "**IR.**"

Step 4: Click the "**Submit**" button in the Calculations dialog window. Once you provide a file name the calculation will begin.

Step 5: When the calculation is finished, go to "**Display > Spectra**" and view the vibrational frequencies in the Spectra dialog box.

Step 6: In the Spectra dialog box click "**Draw IR Spectrum.**" As you select a frequency, the corresponding vibrational mode will be animated in the main window. View the calculated modes until you identify the mode of predominately C-F stretch character and record the corresponding frequency.

Chapter 36: Complex Reaction Mechanisms

P36.1) A proposed mechanism for the formation of N_2O_5 from NO_2 and O_3 is

$$NO_2 + O_3 \xrightarrow{k_1} NO_3 + O_2$$
$$NO_3 + NO_2 + M \xrightarrow{k_2} N_2O_5 + M$$

Determine the rate law expression for the production of N_2O_5 given this mechanism.

$$\frac{d[N_2O_5]}{dt} = k_2[NO_2][NO_3]$$

$$\frac{d[NO_3]}{dt} = k_1[NO_2][O_3] - k_2[NO_2][NO_3]$$

Applying the steady state approximation to the intermediate NO_3 and substituting back into the differential rate expression for N_2O_5 yields:

$$\frac{d[NO_3]}{dt} = 0 = k_1[NO_2][O_3] - k_2[NO_2][NO_3]$$

$$k_2[NO_2][NO_3] = k_1[NO_2][O_3]$$

$$[NO_3] = \frac{k_1}{k_2}[O_3]$$

$$\frac{d[N_2O_5]}{dt} = k_2[NO_2][NO_3]$$

$$= k_2[NO_2]\left(\frac{k_1}{k_2}[O_3]\right)$$

$$= k_1[NO_2][O_3]$$

The mechanism predicts that the reaction is first order in NO_2 and O_3, second order overall.

P36.4) The hydrogen-bromine reaction corresponds to the production of HBr from H_2 and Br_2 as follows: $H_2 + Br_2 \longrightarrow 2HBr$. This reaction is famous for its complex rate law, determined by Bodenstein and Lind in 1906:

$$\frac{d[HBr]}{dt} = \frac{k[H_2][Br_2]^{1/2}}{1 + \frac{m[HBr]}{[Br_2]}}$$

where k and m are constants. It took 13 years for the correct mechanism of this reaction to be proposed, and this feat was accomplished simultaneously by Christiansen, Herzfeld, and Polyani. The mechanism is as follows:

$$Br_2 \underset{k_{-1}}{\overset{k_1}{\rightleftharpoons}} 2Br\cdot$$

$$Br\cdot + H_2 \xrightarrow{k_2} HBr + H\cdot$$

$$H\cdot + Br_2 \xrightarrow{k_3} HBr + Br\cdot$$

$$HBr + H\cdot \xrightarrow{k_4} H_2 + Br\cdot$$

Construct the rate law expression for the hydrogen-bromine reaction by performing the following steps:
a) Write down the differential rate expression for [HBr].
b) Write down the differential rate expressions for [Br] and [H].
c) Because Br and H are reaction intermediates, apply the steady-state approximation to the result of part (b).
d) Add the two equations from part (c) to determine [Br] in terms of [Br$_2$].
e) Substitute the expression for [Br] back into the equation for [H] derived in part (c) and solve for [H].
f) Substitute the expressions for [Br] and [H] determined in part (e) into the differential rate expression for [HBr] to derive the rate law expression for the reaction.

a) $$\frac{d[HBr]}{dt} = k_2[Br\cdot][H_2] + k_3[H\cdot][Br_2] - k_4[HBr][H\cdot]$$

b) $$\frac{d[Br\cdot]}{dt} = 2k_1[Br_2] - 2k_{-1}[Br\cdot]^2 - k_2[Br\cdot][H_2] + k_3[H\cdot][Br_2] + k_4[HBr][H\cdot]$$

$$\frac{d[H\cdot]}{dt} = k_2[Br\cdot][H_2] - k_3[H\cdot][Br_2] - k_4[HBr][H\cdot]$$

c) Applying the steady state approximation to [Br·] and [H·] and adding yields

$$0 = 2k_1[Br_2] - 2k_{-1}[Br\cdot]^2 - k_2[Br\cdot][H_2] + k_3[H\cdot][Br_2] + k_4[HBr][H\cdot]$$

$$0 = k_2[Br\cdot][H_2] - k_3[H\cdot][Br_2] - k_4[HBr][H\cdot]$$

d)

$$0 = 2k_1[Br_2] - 2k_{-1}[Br\cdot]^2$$

$$[Br\cdot] = \sqrt{\frac{k_1}{k_{-1}}}[Br_2]^{1/2}$$

e)

$$k_2[\text{Br}\cdot][\text{H}_2] = k_3[\text{H}\cdot][\text{Br}_2] + k_4[\text{HBr}][\text{H}\cdot]$$

$$\frac{k_2[\text{Br}\cdot][\text{H}_2]}{k_3[\text{Br}_2] + k_4[\text{HBr}]} = [\text{H}\cdot]$$

$$\frac{k_2\sqrt{\dfrac{k_1}{k_{-1}}}[\text{Br}_2]^{1/2}[\text{H}_2]}{k_3[\text{Br}_2] + k_4[\text{HBr}]} = [\text{H}\cdot]$$

f) Now, substitution into (a) yields

$$\frac{d[\text{HBr}]}{dt} = k_2[\text{Br}][\text{H}_2] + k_3[\text{H}\cdot][\text{Br}_2] - k_4[\text{HBr}][\text{H}\cdot]$$

$$= k_2\sqrt{\frac{k_1}{k_{-1}}}[\text{Br}_2]^{1/2}[\text{H}_2] + \frac{k_3 k_2\sqrt{\dfrac{k_1}{k_{-1}}}[\text{Br}_2]^{3/2}[\text{H}_2]}{k_3[\text{Br}_2] + k_4[\text{HBr}]} - \frac{k_4 k_2\sqrt{\dfrac{k_1}{k_{-1}}}[\text{Br}_2]^{1/2}[\text{H}_2][\text{HBr}]}{k_3[\text{Br}_2] + k_4[\text{HBr}]}$$

$$= k_2\sqrt{\frac{k_1}{k_{-1}}}[\text{Br}_2]^{1/2}[\text{H}_2]\left(\frac{k_3[\text{Br}_2] + k_4[\text{HBr}]}{k_3[\text{Br}_2] + k_4[\text{HBr}]}\right) + \frac{k_3 k_2\sqrt{\dfrac{k_1}{k_{-1}}}[\text{Br}_2]^{3/2}[\text{H}_2]}{k_3[\text{Br}_2] + k_4[\text{HBr}]}$$

$$\quad - \frac{k_4 k_2\sqrt{\dfrac{k_1}{k_{-1}}}[\text{Br}_2]^{1/2}[\text{H}_2][\text{HBr}]}{k_3[\text{Br}_2] + k_4[\text{HBr}]}$$

$$= \frac{2k_3 k_2\sqrt{\dfrac{k_1}{k_{-1}}}[\text{Br}_2]^{3/2}[\text{H}_2]}{k_3[\text{Br}_2] + k_4[\text{HBr}]}$$

$$= \frac{2k_2\sqrt{\dfrac{k_1}{k_{-1}}}[\text{Br}_2]^{1/2}[\text{H}_2]}{1 + \dfrac{k_4[\text{HBr}]}{k_3[\text{Br}_2]}}$$

P36.6) For the reaction $\text{I}^-(aq) + \text{OCl}^-(aq) \rightleftharpoons \text{OI}^-(aq) + \text{Cl}^-(aq)$ occurring in aqueous solution, the following mechanism has been proposed:

$$\text{OCl}^- + \text{H}_2\text{O} \underset{k_{-1}}{\overset{k_1}{\rightleftharpoons}} \text{HOCl} + \text{OH}^-$$

$$\text{I}^- + \text{HOCl} \xrightarrow{k_2} \text{HOI} + \text{Cl}^-$$

$$\text{HOI} + \text{OH}^- \xrightarrow{k_3} \text{H}_2\text{O} + \text{OI}^-$$

a) Derive the rate law expression for this reaction based on this mechanism. (*Hint:* [OH$^-$] should appear in the rate law.)

264

b) The initial rate of reaction was studied as a function of concentration by Chia and Connick [*J. Phys. Chem.* 63 (1959), 1518], and the following data were obtained:

$[I^-]_0$ (M)	$[OCl^-]_0$ (M)	$[OH^-]_0$ (M)	Initial Rate (M s^{-1})
2.0×10^{-3}	1.5×10^{-3}	1.00	1.8×10^{-4}
4.0×10^{-3}	1.5×10^{-3}	1.00	3.6×10^{-4}
2.0×10^{-3}	3.0×10^{-3}	2.00	1.8×10^{-4}
4.0×10^{-3}	3.0×10^{-3}	1.00	7.2×10^{-4}

Is the predicted rate law expression derived from the mechanism consistent with these data?

a) $$\frac{d[OI^-]}{dt} = k_3[HOI][OH^-]$$

The intermediate species have rate expressions (with steady state approximation)

$$\frac{d[HOCl]}{dt} = 0 = k_1[OCl^-][H_2O] - k_{-1}[HOCl][OH^-] - k_2[I^-][HOCl]$$

$$\frac{d[HOI]}{dt} = k_2[I^-][HOCl] - k_3[HOI][OH^-] = 0$$

Solving the last expression for [HOI] and substituting into the differential rate expression for [OI$^-$] yields:

$$\frac{d[OI^-]}{dt} = k_2[I^-][HOCl]$$

Next, [HOCl] must be expressed in terms of reactants and [OH$^-$]. Rearranging the steady-state approximation applied to the differential rate expression for [HOCl] results in the following:

$$[HOCl] = \frac{k_1[OCl^-][H_2O]}{k_{-1}[OH^-] + k_2[I^-]}$$

This result is substituted into the differential rate expression for [OI$^-$] to yield:

$$\frac{d[OI^-]}{dt} = \frac{k_1 k_2[OCl^-][H_2O][I^-]}{k_{-1}[OH^-] + k_2[I^-]}$$

Inspection of the concentrations employed in the table suggests that $k_2[I^-] \ll k_1[OI^-]$ resulting in:

$$\frac{d[OI^-]}{dt} = \frac{k_1 k_2[I^-][OCl^-][H_2O]}{k_{-1}[OH^-]}$$

A markedly similar expression is obtained using the pre-equilibrium approximation to determine [HOCl]

b) Consider set 1 & 2: $[I^-]_0$ is doubled, doubling the rate.

set 1 & 3: Doubling $[OCl^-]$ and $[OH^-]$ results in no net change in rate.
set 1 & 4: Doubling $[I^-]$ and $[OCl^-]$ quadruples the rate.

These results confirm the rate expression.

P36.8) Consider the following mechanism, which results in the formation of product P:

$$A \underset{k_{-1}}{\overset{k_1}{\rightleftarrows}} B \underset{k_{-2}}{\overset{k_2}{\rightleftarrows}} C$$

$$B \xrightarrow{k_3} P$$

If only the species A is present at $t = 0$, what is the expression for the concentration of P as a function of time? You can apply the pre-equilibrium approximation in deriving your answer.

Using the pre-equilibrium approximation, we can express [B] and [C] in terms of [A] as follows:

$$[B] = \frac{k_1}{k_{-1}}[A] = K_1[A]$$

$$[C] = \frac{k_2}{k_{-2}}[B] = K_2[B] = K_1 K_2[A]$$

$$\frac{d[P]}{dt} = k_3[B] = k_3 K_1[A]$$

Next, from mass conservation:

$$[A]_0 = [A] + [B] + [C] + [P]$$

$$\frac{d[A]_0}{dt} = 0 = \frac{d[A]}{dt} + \frac{d[B]}{dt} + \frac{d[C]}{dt} + \frac{d[P]}{dt}$$

$$\frac{d[P]}{dt} = -\left(\frac{d[A]}{dt} + \frac{d[B]}{dt} + \frac{d[C]}{dt} \right)$$

$$\frac{d[P]}{dt} = -\left(\frac{d[A]}{dt} + \frac{d(K_1[A])}{dt} + \frac{d(K_1 K_2[A])}{dt} \right)$$

$$= -(1 + K_1 + K_1 K_2) \frac{d[A]}{dt}$$

Setting the two differential rate expressions for [P] and integrating yields:

$$-\left(1+K_1+K_1K_2\right)\frac{d[A]}{dt}=k_3K_1[A]$$

$$\int_{[A]_0}^{[A]}\frac{d[A]}{[A]}=\int_0^t\frac{-k_3K_1}{\left(1+K_1+K_1K_2\right)}dt$$

$$[A]=[A]_0\,e^{\frac{-k_3K_1t}{\left(1+K_1+K_1K_2\right)}}$$

If a pre-equilibrium is established rapidly before any product formation:

$$[A]_0=[A]_0^{eq}+[B]_0^{eq}+[C]_0^{eq}$$

$$=\left(1+K_1+K_1K_2\right)[A]_0^{eq}$$

$$[P]=[A]_0-\left([A]+[B]+[C]\right)$$

$$=[A]_0-\left([A]+K_1[A]+K_1K_2[A]\right)$$

$$=[A]_0-\left(1+K_1+K_1K_2\right)[A]$$

$$=[A]_0-\left(1+K_1+K_1K_2\right)\frac{[A]_0}{\left(1+K_1+K_1K_2\right)}e^{\frac{-k_3K_1t}{\left(1+K_1+K_1K_2\right)}}$$

$$=[A]_0\left(1-e^{\frac{-k_3K_1t}{\left(1+K_1+K_1K_2\right)}}\right)$$

P36.12) The enzyme fumarase catalyzes the hydrolysis of fumarate:

$$\text{Fumarate}+H_2O\longrightarrow\text{L-malate}$$

The turnover number for this enzyme is $2.5\times10^3\ s^{-1}$, and the Michaelis constant is 4.2×10^{-6} M. What is the rate of fumarate conversion if the initial enzyme concentration is 1×10^{-6} M and the fumarate concentration is 2×10^{-4} M?

Recognizing that k_2 is the turnover number, the values provided in the problem can be used directly in the rate expression to determine the initial rate of reaction:

$$k_2=2.5\times10^3\ s^{-1},\ K_m=4.2\times10^{-6}\ \text{M},\ [E]_0=1\times10^{-6}\ \text{M},\ [S]_0=2\times10^{-4}\ \text{M}$$

$$rate_o=\frac{k_2[S]_0[E]_0}{[S]_0+K_m}$$

$$=\frac{\left(2.5\times10^3\ s^{-1}\right)\left(2\times10^{-4}\ \text{M}\right)\left(1\times10^{-6}\ \text{M}\right)}{\left(2\times10^{-4}\ \text{M}\right)+\left(4.2\times10^{-6}\ \text{M}\right)}$$

$$=2.45\times10^{-3}\ \text{M s}^{-1}$$

P36.17) The enzyme glycogen synthase kinase (GSK-3β) plays a central role in Alzheimer's disease. The onset of Alzheimer's disease is accompanied by the production of highly phosphorylated forms of a protein referred to as "τ." GSK-3β contributes to the hyperphosphorylation of τ such that inhibiting the activity of this enzyme represents a pathway for the development of an Alzheimer's drug. A compound known as Ro 31-8220 is a competitive inhibitor of GSK-3β. The following data were obtained for the rate of GSK-3β activity in the presence and absence of Ro 31-8220 [A. Martinez *et al.*, *J. Medicinal Chemistry* 45 (2002), 1292]:

[S] (μM)	$Rate_0$ (μM s^{-1}), [I] = 0	$Rate_0$ (μM s^{-1}) [I] = 200 μM
66.7	4.17×10^{-8}	3.33×10^{-8}
40.0	3.97×10^{-8}	2.98×10^{-8}
20.0	3.62×10^{-8}	2.38×10^{-8}
13.3	3.27×10^{-8}	1.81×10^{-8}
10.0	2.98×10^{-8}	1.39×10^{-8}
6.67	2.31×10^{-8}	1.04×10^{-8}

Determine K_m and $rate_{max}$ for GSK-3β and, using the data with the inhibitor, determine K_m^* and K_i.

Analyzing the data without inhibitor using a Lineweaver–Burk plot yields:

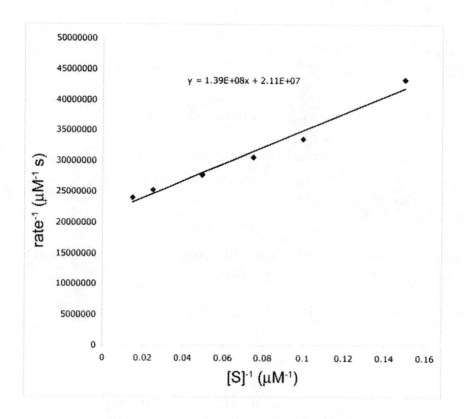

$$y = 1.39E{+}08x + 2.11E{+}07$$

The best-fit straight line to the data yields the following equation:

$$\frac{1}{rate_0} = 1.39 \times 10^8 \text{ s} \frac{1}{[\text{S}]_0} + 2.11 \times 10^6 \, \mu\text{M}^{-1} \text{ s}$$

The maximum rate is equal to the inverse of the y-intercept:

$$rate_{max} = \frac{1}{y-\text{int}} = \frac{1}{2.11 \times 10^7 \, \mu\text{M}^{-1}\text{s}}$$

$$rate_{max} = 4.74 \times 10^{-8} \, \mu\text{M s}^{-1}$$

With the maximum rate and slope of the best-fit line, the Michaelis constant obtained as follows:

$$K_m = (\text{slope}) \times (rate_{max}) = \left(1.39 \times 10^8 \text{ s}\right) \times \left(4.74 \times 10^{-8} \, \mu\text{M s}^{-1}\right)$$

$$K_m = 6.49 \, \mu\text{M}$$

Using the inhibited data, the Lineweaver–Burk plot is:

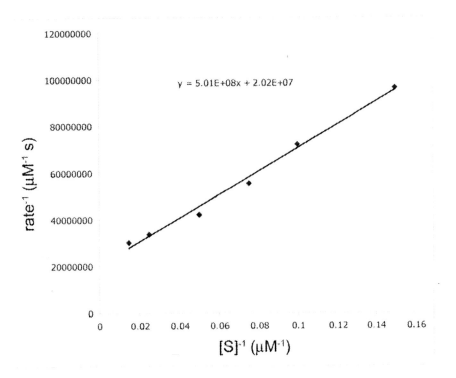

Best fit by a straight line to the data yields the following equation:

$$\frac{1}{rate_0} = 5.01 \times 10^8 \text{ s} \frac{1}{[\text{S}]_0} + 2.01 \times 10^7 \, \mu\text{M}^{-1}\text{s}$$

The maximum rate with inhibitor is equal to the inverse of the y-intercept:

$$rate_{max} = \frac{1}{2.01 \times 10^7 \ \mu M^{-1} \ s}$$

$$= 4.98 \times 10^{-8} \ \mu M^{-1} \ s$$

The apparent Michaelis constant is given by:

$$K_m^* = (slope) \times (rate_{max})$$

$$= (5.01 \times 10^8 \ s) \times (4.98 \times 10^{-8} \ \mu M^{-1} \ s)$$

$$K_m^* = 24.9 \ \mu M$$

Finally, the K_I value is given by:

$$K_I = \frac{[I]}{\dfrac{K_m^*}{K_m} - 1} = \frac{200 \mu M}{\left(\dfrac{24.9 \ \mu M}{6.49 \ \mu M}\right) - 1}$$

$$= 70.4 \ \mu M$$

P36.19) Reciprocal plots provide a relatively straightforward way to determine if an enzyme demonstrates Michaelis–Menten kinetics and to determine the corresponding kinetic parameters. However, the slope determined from these plots can require significant extrapolation to regions corresponding to low substrate concentrations. An alternative to the reciprocal plot is the Eadie–Hofstee plot where the reaction rate is plotted versus the rate divided by the substrate concentration and the data are fit to a straight line.

a. Beginning with the general expression for the reaction rate given by the Michaelis–Menten mechanism:

$$R_0 = \frac{R_{max}[S]_0}{[S]_0 + K_m}$$

rearrange this equation to construct the following expression which is the basis for the Eadie–Hofstee plot:

$$R_0 = R_{max} - K_m \left(\frac{R_0}{[S]_0}\right)$$

b. Using an Eadie–Hofstee plot, determine R_{max} and K_m for hydrolysis of sugar by the enzyme invertase using the following data:

[Sucrose]$_0$ (M)	Rate (M s^{-1})
0.029	0.182
0.059	0.266
0.088	0.310
0.117	0.330
0.175	0.362
0.234	0.361

a)

$$R_0 = \frac{R_{max}[S]_0}{[S]_0 + K_m}$$

$$R_0\left([S]_0 + K_m\right) = R_{max}[S]_0$$

$$R_0[S]_0 = R_{max}[S]_0 - R_0 K_m$$

$$R_0 = R_{max} - K_m\left(\frac{R_0}{[S]_0}\right)$$

a) Using the data provided, a plot of $rate_0$ versus $rate_0/[S]_0$ is constructed:

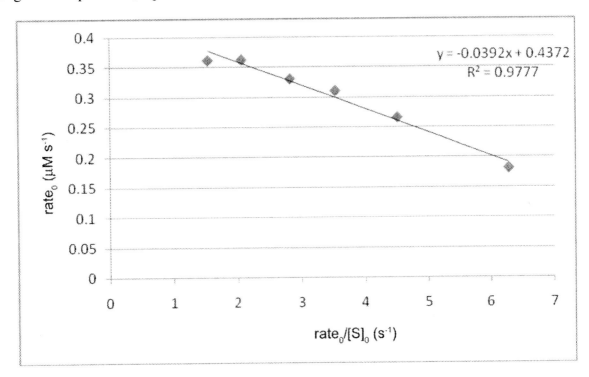

Best fit to a straight line yields a y-intercept of 0.437 M s^{-1}, which is equal to $Rate_{max}$. The slope of the line is equal to -0.0392 M, which is equal to $-K_m$ such that $K_m = 0.0392$ M.

P36.20) Determine the predicted rate law expression for the following radical-chain reaction:

$$A_2 \xrightarrow{k_1} 2A \cdot$$

$$A \cdot \xrightarrow{k_2} B \cdot + C$$

$$A \cdot + B \cdot \xrightarrow{k_3} P$$

$$A \cdot + P \xrightarrow{k_4} B \cdot$$

The differential rate of P formation is

$$\frac{d[P]}{dt} = k_3[A \cdot][B \cdot] - k_4[A \cdot][P]$$

The rate expression for A· and B· are

$$\frac{d[A \cdot]}{dt} = 2k_1[A_2] - k_2[A \cdot] - k_3[A \cdot][B \cdot] - k_4[A \cdot][P]$$

$$\frac{d[B \cdot]}{dt} = k_2[A \cdot] - k_3[A \cdot][B \cdot] + k_4[A \cdot][P]$$

Applying the steady state approximation for [B·]

$$k_3[A \cdot][B \cdot] - k_4[A \cdot][P] = k_2[A \cdot]$$

$$[B \cdot] = \frac{k_2 + k_4[P]}{k_3}$$

Substituting this result into the differential rate expression for P yields:

$$\frac{d[P]}{dt} = k_3[A \cdot]\left(\frac{k_2 + k_4[P]}{k_3}\right) - k_4[A \cdot][P]$$

$$= k_2[A \cdot]$$

Now, adding the steady-state expressions for $[A \cdot]$ and $[B \cdot]$ yields:

$$0 = 2k_2[A_2] - 2k_3[A \cdot][B \cdot]$$

$$[A \cdot] = \frac{k_2[A_2]}{k_3[B \cdot]} = \frac{k_2[A_2]}{k_2 + k_4[P]}$$

Substituting this expression into the differential rate expression for P yields the final result:

$$\frac{d[P]}{dt} = \frac{k_2 k_1[A_2]}{k_2 + k_4[P]}$$

P36.24) The adsorption of ethyl chloride on a sample of charcoal at 0°C measured at several different pressures is as follows:

$P_{C_2H_5Cl}$ (Torr)	V_{ads} (mL)
20	3.0
50	3.8
100	4.3
200	4.7
300	4.8

Using the Langmuir isotherm, determine the fractional coverage at each pressure and V_M.

The Langmuir equation can be written as

$$\frac{P}{V} = \frac{P}{V_m} + \frac{1}{KV_m}$$

And the fractional coverage, θ, is simply the ratio of adsorbed volume to the volume of maximum adsorption (V_m):

$$\theta = \frac{V}{V_m}$$

Therefore, V_m is required to determine the fractional coverage versus pressure. The plot of $\frac{P}{V}$ vs. P should yield a straight line with slope equal to the inverse of V_m. This plot is as follows:

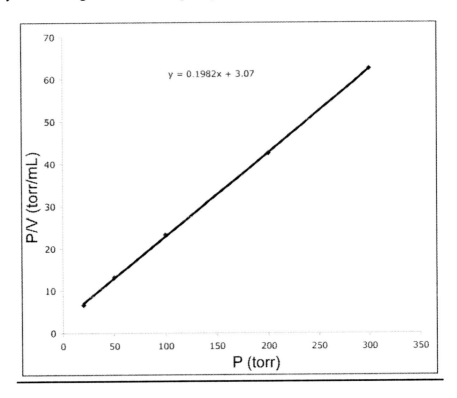

y = 0.1982x + 3.07

The equation for the best-fit line is:

$$\frac{P}{V} = 0.198 \ \text{mL}^{-1}(P) + 3.07 \ \text{torr mL}^{-1}$$

Thus, the V_m value is

$$V_m = \frac{1}{slope}$$

$$V_m = 5.04 \ \text{mL}$$

With V_m, θ can be determined resulting in the following:

P (atm)	θ
20.	0.595
50.	0.754
100.	0.853
200.	0.932
300.	0.952

P36.27) Many surface reactions require the adsorption of two or more different gases. For the case of two gases, assuming that the adsorption of a gas simply limits the number of surface sites available for adsorption, and derive expressions for the fractional coverage of each gas.

If we assume that the only effect of the gases is to limit the number of sites available for adsorption, then the Langmuir model yields the two following relationships:

$$k_{d1}\theta_1 = k_{a1}P_1\left(1-\theta_1-\theta_2\right)$$
$$k_{d2}\theta_2 = k_{a2}P_2\left(1-\theta_1-\theta_2\right)$$

Where k_a, k_d, and θ are the rate of adsorption, desorption, and fractional coverage, respectively. In addition, the subscript denotes either gas 1 or 2 in the two gas mixture. The two equations can be solved to yield θ_1 and θ_2 as follows:

$$\theta_1 = \frac{K_1P_1}{1+K_1P_1+K_2P_2}$$

$$\theta_2 = \frac{K_2P_2}{1+K_1P_1+K_2P_2}$$

P36.30) If $\tau_f = 1 \times 10^{-10}$ s and $k_{ic} = 5 \times 10^8$ s^{-1}, what is Φ_f? Assume that the rate constants for intersystem crossing and quenching are sufficiently small that these processes can be neglected.

Φ_f is related to k_f and τ_f by the expression

$$\Phi_f = k_f \tau_f$$

and k_f is related to τ_f by

$$\frac{1}{\tau_f} = k_f + k_{ic} \text{ (assuming } k_{isc}^3, k_q \text{ are small)}$$

$$\tau_f = 1 \times 10^{-10} \text{ s} \quad \text{and} \quad k_{ic} = 5 \times 10^8 \text{ s}^{-1}$$

Thus

$$\frac{1}{1 \times 10^{-10} \text{ s}} = k_f + 5 \times 10^8 \text{ s}^{-1}$$

and

$$k_f = 1 \times 10^{10} \text{ s}^{-1} - 5 \times 10^8 \text{ s}^{-1}$$
$$k_f = 9.5 \times 10^9 \text{ s}^{-1}$$

And finally,

$$\Phi_f = k_f \tau_f$$
$$= 9.5 \times 10^9 \text{ s}^{-1} \cdot 1 \times 10^{-10}$$
$$\Phi_f = 0.95$$

P36.32) If 10% of the energy of a 100-W incandescent bulb is in the form of visible light having an average wavelength of 600 nm, how many quanta of light are emitted per second from the light bulb?

Assuming that 10% of energy from the bulb is in the form of visible light, the power of the light is given by:

$$P_\ell = 0.100(P)$$
$$= 0.100(100. \text{ W})$$
$$= 10.0 \text{ J s}^{-1}$$

The energy per photon at 600. nm is:

$$E = \frac{hc}{\lambda} = \frac{(6.626 \times 10^{-34} \text{ J s})(3.00 \times 10^8 \text{ m s}^{-1})}{600. \times 10^{-9} \text{ m}} = 3.31 \times 10^{-19} \text{ J}$$

Therefore, the number of photons per second is given by:

$$n = \frac{10.0 \text{ J s}^{-1}}{3.13 \times 10^{-19} \text{ J photon}^{-1}} = 3.02 \times 10^{19} \text{ photon s}^{-1}$$

P36.35) A central issue in the design of aircraft is improving the lift of aircraft wings. To assist in the design of more efficient wings, wind-tunnel tests are performed in which the pressures at various parts of the wing are measured generally using only a few localized pressure sensors. Recently, pressure-sensitive paints have been developed to provide a more detailed view of wing pressure. In these paints, a luminescent molecule is dispersed into an oxygen-permeable paint and the aircraft wing is painted. The wing is placed into an airfoil, and luminescence from the paint is measured. The variation in O_2 pressure is measured by monitoring the luminescence intensity, with lower intensity demonstrating areas of higher O_2 pressure due to quenching.

a) The use of platinum octaethylporphyrin (PtOEP) as an oxygen sensor in pressure-sensitive paints was described by Gouterman and coworkers [*Review of Scientific Instruments* 61 (1990), 3340]. In this work, the following relationship between luminescence intensity and pressure was derived: $\dfrac{I_0}{I} = A + B\left(\dfrac{P}{P_0}\right)$, where I_0 is the fluorescence intensity at ambient pressure P_0, and I is the fluorescence intensity at an arbitrary pressure P. Determine coefficients A and B in the preceding expression using the Stern–Volmer equation: $k_{total} = \dfrac{1}{\tau_l} = k_l + k_q[Q]$. In this equation τ_l is the luminescence lifetime, k_r is the luminescent rate constant, and k_q is the quenching rate constant. In addition, the luminescent intensity ratio is equal to the ratio of luminescence quantum yields at ambient pressure, Φ_0, and an arbitrary pressure, Φ:

$$\frac{\Phi_0}{\Phi} = \frac{I_0}{I}.$$

b) Using the following calibration data of the intensity ratio versus pressure observed for PtOEP, determine A and B:

I_0/I	P/P_0	I_0/I	P/P_0
1.0	1.0	0.65	0.46
0.9	0.86	0.61	0.40
0.87	0.80	0.55	0.34
0.83	0.75	0.50	0.28
0.77	0.65	0.46	0.20
0.70	0.53	0.35	0.10

c) At an ambient pressure of 1 atm, $I_0 = 50{,}000$ (arbitrary units) and 40,000 at the front and back of the wing. The wind tunnel is turned on to a speed of Mach 0.36 and the measured luminescence intensity is 65,000 and 45,000 at the respective locations. What is the pressure differential between the front and back of the wing?

a) Starting with the version of the Stern–Volmer Eq. provided in the problem:

$$k_{tot} = k_l + k_q[Q]$$

The luminescence quantum yield can be expressed in terms of k_l and k_{total} as:

$$\Phi = \frac{k_l}{k_{total}}$$

Therefore:

$$\frac{\Phi_0}{\Phi} = \frac{k_{total}}{k_{total_0}}$$

$$= \frac{k_l + k_q P}{k_l + k_q P_0}$$

$$= \frac{k_l}{k_l + k_q P_0} + \frac{k_q P_0}{k_l + k_q P_0}\left(\frac{P}{P_0}\right)$$

$$= A + B\left(\frac{P}{P_0}\right)$$

b) The plot of (I_0/I) versus (P/P_0) is as follows:

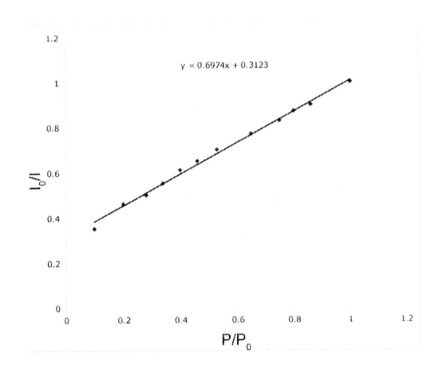

Best fit by a straight line to the data yields the following equation:

$$\frac{I_0}{I} = 0.697\left(\frac{P}{P_0}\right) + 0.312$$

Thus, $A = 0.312$ and $B = 0.697$.

c)

$$\left(\left(\frac{I_0}{I}\right)_{front} - \left(\frac{I_0}{I}\right)_{back} =\right) = 0.697 \left(\frac{P_{front} - P_{back}}{P_0}\right)$$

$$\left(\frac{50,000}{65,000} - \frac{40,000}{45,000}\right) = 0.697 \left(\frac{P_{front} - P_{back}}{1 \text{ atm}}\right)$$

$$-0.120 = 0.697 \left(\frac{P_{front} - P_{back}}{1 \text{ atm}}\right)$$

$$-0.172 \text{ atm} = P_{front} - P_{back}$$

P36.37) The pyrene/coumarin FRET pair ($r_0 = 39$ Å) is used to study the fluctuations in enzyme structure during the course of a reaction. Computational studies suggest that the pair will be separated by 35 Å in one conformation, and 46 Å in a second configuration. What is the expected difference in FRET efficiency between these two conformational states?

$$Eff = \frac{r_0^6}{\left(r_0^6 + r^6\right)}$$

Using the above expression to calculate the difference in FRET efficiency:

$$Eff\left(35\,\text{Å}\right) - Eff\left(46\,\text{Å}\right) = \frac{\left(39\,\text{Å}\right)^6}{\left(\left(39\,\text{Å}\right)^6 + \left(35\,\text{Å}\right)^6\right)} - \frac{\left(39\,\text{Å}\right)^6}{\left(\left(39\,\text{Å}\right)^6 + \left(46\,\text{Å}\right)^6\right)}$$

$$= 0.386 \approx 0.39$$